ADVANCES IN SPACE RESEARCH

The Official Journal of the Committee on Space Research (COSPAR)
A Scientific Committee of the International Council of Scientific Unions (ICSU)

VOLUME 22, NUMBER 7

X-RAY TIMING AND COSMIC GAMMA RAY BURSTS

Symposium E1.1
Sponsors

COMMITTEE ON SPACE RESEARCH (COSPAR)

Program Committee

S. S. Holt, U.S.A.
J. H. Swank, U.S.A.
E. van der Heuvel, The Netherlands
G. Hasinger, Germany
L. Piro, Italy
A. Parmar, The Netherlands
P. Mandrou, France
M. Gilfanov, Russia
M. Watson, U.K.
F. Nagase, Japan
H. Bradt, U.S.A.
N. White, U.S.A.

Symposium E1.2
Sponsors

COMMITTEE ON SPACE RESEARCH (COSPAR)
EUROPEAN SPACE AGENCY (ESA)
INTERNATIONAL ASTRONOMICAL UNION (IAU)

Program Committee

C. Kouveliotou, U.S.A.
K. Hurley, U.S.A.
C. Meegan, U.S.A.
C. Winkley, The Netherlands
M. Rees, U.K.
E. Feninore, U.S.A.
R. Sunyaev, Russia

X-RAY TIMING AND COSMIC GAMMA RAY BURSTS

Proceedings of the E1.1 and E1.2 Symposia of COSPAR Scientific Commission E which were held during the Thirty-first COSPAR Scientific Assembly, Birmingham, U.K., 14–21 July 1996

Edited by

J. H. SWANK

Code 662, NASA, Goddard Space Flight Center, Greenbelt, MD 20771, U.S.A.

C. KOUVELIOTOU

USRA, ES-84, NASA/MSFC, Huntsville, AL 35812, U.S.A.

and

K. HURLEY

UC Berkeley, Space Sciences Laboratory, Berkeley, CA 94720-7450, U.S.A.

Published for

THE COMMITTEE ON SPACE RESEARCH

PERGAMON

U.K.   Elsevier Science Ltd, The Boulevard, Langford Lane,
            Kidlington, Oxford OX5 1GB, U.K.

U.S.A.   Elsevier Science Inc., 660 White Plains Road,
            Tarrytown, New York 10591-5153, U.S.A.

JAPAN   Elsevier Science Japan, Tsunashima Building Annex,
            3-20-12 Yushima, Bunkyo-ku, Tokyo 113, Japan

© 1998 COSPAR

First edition 1998

ISBN 0-08-043471-1

*In order to make this volume available as economically and as
rapidly as possible the author's typescript has been reproduced in
its original form. This method unfortunately has its typographical
limitations but it is hoped that they in no way distract the reader.*

NOTICE TO READERS

*If your library is not already a subscriber to this series, may we
recommend that you place a subscription order to receive
immediately upon publication all new issues. Should you find that
these issues no longer serve your needs your order can be cancelled
at any time without notice. All these conference proceedings issues
are also available separately to non-subscribers. Write to your
nearest Elsevier Science office for further details.*

*ContentsDirect, the free e-mail alerting service, delivers the table
of contents for this journal directly to your PC, prior to
publication. The quickest way to register for ContentsDirect is
via the Internet at: http://www.elsevier.nl/locate/ContentsDirect.
If you don't have access to the Internet you can register for this
service by sending an e-mail to cdsubs@elsevier.co.uk—specifying
the title of the publication you wish to register for.*

Transferred to digital print 2009
Printed and bound in Great Britain by CPI Antony Rowe, Chippenham and Eastbourne

CONTENTS

COSMIC GAMMA RAY BURSTS

Pergamon

Adv. Space Res. Vol. 22, No. 7, p. 919, 1998
Published by Elsevier Science Ltd on behalf of COSPAR
Printed in Great Britain
0273-1177/98 $19.00 + 0.00

PII: S0273-1177(98)00211-7

EDITORIAL COMMENT

This is the 42nd issue of *Advances in Space Research* devoted to the papers presented at the 31st COSPAR Assembly held in Birmingham, UK in July 1996 and the fourth issue to include late papers published in an Appendix.

The COSPAR Editorial Committee, in an effort to decrease publication delays, has recommended that individual editors establish a deadline for submission of manuscripts for each session being published. However, because of unanticipated situations such as mailing problems, it was also decided to provide for publication of accepted manuscripts received after the editor's deadline.

Papers from the E2.1 session on The Sun and its Atmosphere were published in *Advances in Space Research*, Vol. 20, No. 12. One additional manuscript was received after the initial package was submitted for publication. This present issue of *Advances in Space Research* is devoted to other papers from Scientific Commission E, Research in Astrophysics from Space, and is an appropriate publication to include this late paper. Readers are urged to refer to the *Advances in Space Research*, Vol. 20, No. 12 for the majority of E2.1 papers. Other related papers have also appeared in Vol. 19, No. 6 (The Heliosphere at Solar Minimum and Beyond), and Vol. 20, No. 1 (The Sun and its Role in the Heliosphere).

Late papers were received for several sessions. All of these papers have been published as appendices in issues of *Advances in Space Research* associated with the appropriate COSPAR Scientific Commission.

M. A. Shea
Editor-in-Chief

X-RAY TIMING

Proceedings of the E1.1 Symposium of COSPAR Scientific Commission E which was held during the Thirty-first COSPAR Scientific Assembly, Birmingham, U.K., 14–21 July 1996

Edited by

J. H. SWANK

Code 662, NASA, Goddard Space Flight Center, Greenbelt, MD 20771, U.S.A.

 Pergamon

Adv. Space Res. Vol. 22, No. 7, p. 923, 1998
Published by Elsevier Science Ltd on behalf of COSPAR
Printed in Great Britain
0273-1177/98 $19.00 + 0.00

PII: S0273-1177(98)00124-0

PREFACE

Within a few months of the launches of NASA's Rossi X-Ray Timing Explorer and the BeppoSAX mission of the Italian Space Agency and the Netherlands Agency for Aereospace Programs a meeting on X-Ray Timing contained new results from the new missions, and both new details and reviews from the on-going ASCA, ROSAT, Compton Gamma-Ray Observatory, Granat, and other missions. RXTE's observations of low-mass X-ray binaries at frequencies above the Nyquist frequencies of previous missions had shown by then that kilohertz oscillations were a general property of these systems and that the neutron star compact stars in the systems are probably rotating with periods of 1.5-3 milliseconds. The exploration of these phenomena, their explanation, and the implications for the physics of neutron stars and their evolution continue to be major RXTE objectives. The review that was the first given after RXTE's launch is updated in this volume. BeppoSAX at the time of the meeting was barely finishing check-out of the instrument performance after launch. It is interesting that four days after the meeting, the first Gamma-ray burst was recorded which was simultaneously watched in the Wide Field Camera down to X-ray energies near 1 keV at which the brightness persisted hundreds of seconds.

This volume contains papers on the variability of black hole candidates as seen by Ginga, ASCA, and RXTE, variability in the form of state changes, power spectra, and correlations between the behavior of different emission components. Counterposed is a review of the range of theoretical possibilities that are believed possible with different accretion rates and boundary conditions. Several classes of neutron stars manifest themselves through X-ray timing and are reviewed here. Those with strong magnetic fields include accretion-powered and rotation-powered pulsars. The fast rotating bursters must have weak fields. Contrary indications leave uncertain the nature of a class of younger, relatively slow, and not very luminous pulsars. Ginga results on cyclotron lines are reviewed, as are the observations and theoretical interpretations of quasi-periodic oscillations in the flux from pulsars, oscillations presumably related to the magnetosphere's interaction with a disk, but in most cases difficult to interpret. In the "Intermediate Polars" a white dwarf is exchanged for the neutron star. Reviews are included here of ROSAT's substantial increase in the known systems and the beat implications of multiple periods that reveal how the accretion occurs.

The X-ray sky includes many X-ray emitting regions, including jets, whose nature we attempt to determine from spatial resolution and spectra as well as variations. X-ray instruments offer a range of information contributing to understanding their properties, as discussed in one paper here. A final paper, reminding us that not all that varies is compact or large, uses an excitation mechanism usually neglected. The explanation of ROSAT's discovery of X-rays from comets has not been obvious and one explanation figuring in current arguments was presented in this conference on X-ray Timing.

Of the 40 oral and 22 poster papers presented, 22 appear in this volume. To these authors, who submitted their papers at or soon after the meeting, and to the potential readers who would have benefited by timely publication, the editor extends apology for the delay. The pace of operations of a mission, new results, meetings, and publications has been fast. This led to many of the papers on new results appearing in other publications and to the delay in editing this volume. However, the reviews here capture the knowledge gleaned from observations before RXTE and BeppoSAX and present the framework from which we must advance with new observations and theories. The organizors thank the program committee, the session chairs and especially the speakers for a stimulating meeting. The editor thanks referees, including A. Baykal, J. Cannizzo, W. Focke, D. Gruber, C. Markwardt, M. Stark, T. Strohmayer, and L. Titarchuk. Finally the editor especially thanks Sandra Shrader for essential help in the process and the Editor-in-Chief for patience and advice.

Pergamon

Adv. Space Res. Vol. 22, No. 7, pp. 925–934, 1998
© 1998 COSPAR. Published by Elsevier Science Ltd. All rights reserved
Printed in Great Britain
0273-1177/98 $19.00 + 0.00

PII: S0273–1177(98)00125–2

KILOHERTZ QUASI-PERIODIC OSCILLATIONS IN LOW-MASS X-RAY BINARIES

M. van der Klis

*Astronomical Institute "Anton Pannekoek" and Center for High-Energy Astrophysics,
University of Amsterdam, Kruislaan 403, 1098 SJ Amsterdam, The Netherlands*

ABSTRACT

I review the current observational status of the new kilohertz quasi-periodic oscillations in accreting low-magnetic field neutron stars, and critically discuss some of the proposed models.

©1998 COSPAR. Published by Elsevier Science Ltd.

INTRODUCTION

The main motivation for studying X-ray binaries is not that they exhibit a wide range of complex phenomenology, which they do, but that they contain neutron stars (and black holes), objects of fundamental physical interest, and allow to derive information about the equation of state of high-density matter and perform tests of general relativity in the strong-field regime. In this talk, I shall be discussing low-mass X-ray binaries (LMXBs) containing neutron stars exclusively, as it is in the understanding of the physics of these systems that great progress has recently become possible by the discovery, with NASA's Rossi X-ray Timing Explorer (RXTE), of a new phenomenon, kilohertz quasi-periodic oscillations (kHz QPO).

In these X-ray binary systems matter is transferred from a low-mass ($\lesssim 1 M_\odot$) star to a neutron star by way of an accretion disk. The X-rays originate from the hot ($\sim 10^7$ K) plasma comprising the inner few 10^1 kilometers of the flow. This is very close to the neutron star, which itself has a radius, R, of order 10 km, so that by studying the properties of this flow one expects to be able to derive information about the star.

The high temperatures in the inner flow are caused by the release of large amounts of graviational energy when the matter descends into the neutron star's very deep gravitational potential well ($GM/R \sim 0.2c^2$; here and below I assume $M = 1.4 M_\odot$ for the neutron star's mass). The characteristic velocities near the star are of order $(GM/R)^{1/2} \sim 0.5c$. Therefore the dynamical time scale, the time scale for motion of matter through the emitting region, is short; $\tau_{dyn} \equiv (r^3/GM)^{1/2} \sim 0.1$ ms for r=10 km, and ~ 2 ms for r=100 km.

Up to a year ago, no direct information existed about the properties of these flows at these time scales. In this paper I report on how, since February 1996, we are for the first time actually observing time variability from accretion flows onto neutron stars at the expected millisecond time scales. A new rapid-variability phenomenon has been discovered, namely quasi-periodic oscillations in the X-ray flux with amplitudes of up to several 10% of the total flux, quality factors $Q \equiv \Delta\nu/\nu$ (see below) of up to several 100, and frequencies of up to ~ 1200 Hz. I shall call this phenomenon "kHz QPO" (kilohertz quasi-periodic oscillations) throughout the rest of this paper.

A great deal of information is available about the properties of LMXBs and the physics of accretion onto a neutron star. The last pre-kHz-QPO overview of rapid X-ray variability in X-ray binaries can be found in the Lewin et al. book "X-Ray Binaries" (van der Klis 1995; look here if you wish to find out about atoll sources, Z sources and the latters' 16–60 Hz horizontal-branch oscillations and the 6–20 Hz normal-flaring branch oscillations). For understanding what follows, it is useful to remind the reader of the usual terminology with respect to the subclasses of LMXBs (Hasinger and van der Klis 1989): Z sources are near-Eddington accretors and probably have somewhat stronger (1–$5 \; 10^9$ G) magnetic fields, atoll sources are often X-ray burst sources, have luminosities between $10^{-3} L_{Edd}$ and a few $10^{-1} L_{Edd}$, and are thought to have somewhat weaker magnetic fields (10^8–10^9 G).

X-ray astronomers are presently scrambling to try and make sense of the phenomenology of kHz QPO, which turn out to be at the same time highly suggestive of interpretation and very restrictive of possible models, and theorists have already begun working out sophisticated models. None of this has reached an equilibrium state yet, and what I report in this paper will necessarily be of a "snapshot" nature. What is clear at this point is that for the first time we are seeing a rapid X-ray variability phenomenon that is directly linked with a neutron star's most distinguishing characteristic (only shared among macroscopic objects with stellar-mass black holes): its compactness. This is particularly evident if the phenomena are in some way related to orbital motion. After all, a Keplerian orbital frequency $\nu_K = P_{orb}^{-1} = (GM/4\pi^2 r_K^3)^{1/2}$ of 1200 Hz around a $1.4 M_\odot$ neutron star as seen from infinity corresponds to an orbital radius $r_K = (GM/4\pi^2 \nu_K^2)^{1/3}$ of 15 km, directly constraining the equation of state of the bulk nuclear-density matter, and only just outside the general-relativistic marginally stable orbit. Whatever the model, for the first time we have to seriously worry about general-relativistic effects in describing the observable dynamics of the physical system.

EARLY DISCOVERIES

The first two sources in which kHz QPO were detected, during an exciting few days in February 1996, were 4U 1728–34 and Sco X-1. Tod Strohmayer of Goddard Space Flight Center (GSFC) and collaborators had proposed to observe 4U 1728–34 in order to look for neutron star r and g-modes and for the neutron star spin. I had traveled down to GSFC to perform the first RXTE observations of Sco X-1, which our group (comprising some of the same people of the RXTE PCA team as Tod's) had proposed to take advantage of RXTE's high-throughput capability which would improve the detection significance of any weak variability by an order of magnitude. In our proposal we had remarked that we suspected beat-frequency QPO phenomena near "~700 Hz" might be observed. We, too, had promised to look for the neutron star spin. While, after a bit of waiting for our data to come out of the production pipeline, I was just starting the first Sco X-1 analysis, Tod showed me his first power spectrum of 4U 1728–34 exhibiting a QPO peak near 800 Hz. This was the first "kHz" phenomenon. When I got on with the Sco X-1 data, sure enough, there was a peak near 1100 Hz! This was the first "sub-millisecond" phenomenon. Of course I discussed this result with Tod and the Sco X-1 team members. We were very excited, but cautious, as obviously the new effect might be instrumental, and our conversations focused on ways to make certain it wasn't. I analyzed some of our data on Cyg X-1 which had similar observational parameters and count rate as the 4U 1728–34 data, and showed the result to the 4U 1728–34 team: it was clear that there was no sign of any 800 Hz QPO there. For Sco X-1 I found that the frequency of the 1100 Hz QPO increased along the so-called "normal branch" (NB) in the X-ray color-color diagram, i.e., with mass accretion rate. Both teams were sufficiently confident that the effects were real, and IAU Circulars were submitted (van der Klis et al. 1996a, Strohmayer et al. 1996a). These initial results were eventually published in the Astrophysical Journal (van der Klis et al. 1996c, Strohmayer et al. 1996c; Fig. 1). Tod later told me that the 4U 1728–34 circular was submitted before the Sco X-1 circular, and that he was dismayed it came out second. Of course, officially the IAU Central Bureau for Astronomical Telegrams is not in the business of establishing priorities; their mission is to rapidly disseminate urgent information. Our discoveries were not independent, as we discussed our results and assisted each other before publication, and the two teams even had common members. But let it be noted for history: Tod saw the 800 Hz in 4U 1728–34 before I saw the 1100 Hz in Sco X-1! I think it is good that in these exciting first few days we focused on the science and freely exchanged information rather than letting the competitive spirit get the better of us. As we ventured on into the new field, this principle of a rapid and open flow of information has been maintained, much helped by that invaluable medium, the IAU Circulars. Particularly with satellite missions, with their limited life spans, the rapid distribution of scientific results is essential to make optimal use of the observational resources.

CURRENT SITUATION

Kilohertz QPO have now[1] been reported in 11 LMXBs, 3 of which are Z sources and 8 of which are atoll sources and probable atoll sources (see van der Klis 1995 for a recent review of LMXB subclasses; hereafter I shall use "atoll source" for LMXBs that probably fall in this class as well as for those that definitely do so), together covering nearly three orders of magnitude in X-ray luminosity ($\sim 10^{-3}$ to $\sim 1 L_{Edd}$). Table 1 summarizes some of these results, and provides an overview of the literature that is approximately complete

---

[1] May 1st, 1997

as of this writing. Rather than getting into an exhaustive description of the phenomenology, or following the historical line I shall concentrate on what I consider at this point to be the main clues. I refer to the Table for all kHz QPO observational references in the remainder of this section.

Table 1. Observed frequencies of kilohertz QPO.

Source (in order of RA)	Lower peak freq. (Hz)	Upper peak freq. (Hz)	Peak sepa-ration (Hz)	"Third" freq. (Hz)	References
4U 0614+091	480 ↓ 800	520 ↓ 750 ↓ 1150	327±4	328	Ford et al. 1996, 1997 van der Klis et al. 1996d Mendez et al. 1997 Vaughan et al. 1997
4U 1608−52	691 830 ↓ 890				Van Paradijs et al. 1996 Berger et al. 1996 Vaughan et al. 1997
Sco X-1	570 ↓ 800 ↓ 830	870 ↓ 1050 ↓ 1080 ↓ 1130	292±2 ↓ 247±3		van der Klis et al. 1996a,b,c, 1997b
4U 1636−53	898 ↓ 920	1147 ↓ 1183 ↓ 1193	249±13	581	Zhang et al. 1996, 1997 van der Klis et al. 1996d Wijnands et al. 1997 Vaughan et al. 1997
	835 ↓ 897				
4U 1728−34	640 ↓ 790	500 ↓ 990 ↓ 1100	355±5	363	Strohmayer et al. 1996a,b,c
KS 1731−260	898	1159 ↓ 1207	260±10	524	Morgan and Smith 1996 Smith et al. 1997 Wijnands and van der Klis 1997
4U 1735-44	1150				Wijnands et al. 1996
X 1743-29?				589	Strohmayer et al. 1996d
GX 5−1	325 ↓ 448	567 ↓ 652 ↓ 746 ↓ 895	327±11		van der Klis et al. 1996e
GX 17+2	682 ↓ 880	988	306±5		van der Klis et al. 1997a
4U 1820−30	546 ↓ 796	1065	275±8		Smale et al. 1996, 1997

Arrows indicate observed frequency variations.
Frequencies in the same row were observed simultaneously, except "third" frequencies.
Entries straddling the upper and lower peak columns are of single, unidentified peaks.

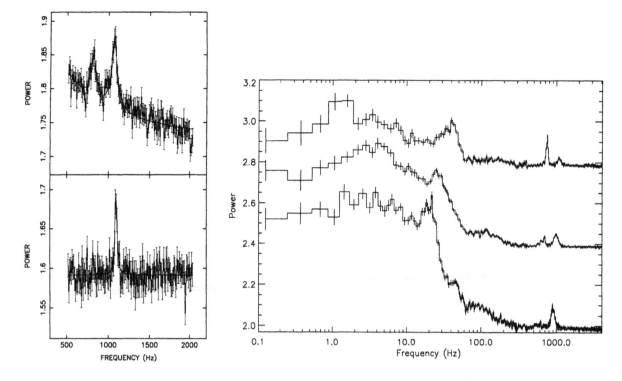

Fig. 1. Power spectra of Sco X-1 (left) showing double (top) and single (bottom) kHz QPO peaks (van der Klis et al. 1996c), and of 4U 1728−34 (right), also with double and single peaks in the kHz range (Strohmayer et al. 1996c). The sloping continuum above 1 kHz in Sco X-1 is instrumental. The peaks in the 1–10 Hz range in 4U 1728−34 are a known aspect of atoll source phenomenology (Hasinger and van der Klis 1989).

A clear pattern of systematic behaviour has emerged. In most sources (8 out of 11) *two* simultaneous kHz peaks (hereafter: twin peaks) are observed in the power spectra of the X-ray count rate variations (Fig. 1). The lower-frequency peak (hereafter the *lower peak*) has been observed at frequencies between 325 and 920 Hz, the higher-frequency peak (hereafter the *upper peak*) has been observed at frequencies between 500 and 1207 Hz. When the accretion rate $\dot{M}$ increases, both peaks move to higher frequency. In atoll sources $\dot{M}$ is inferred to correlate with X-ray count rate, and kHz QPO frequency increases with count rate. In Z sources in the so-called "normal branch" (NB), $\dot{M}$ is inferred to *anti*correlate to count rate, and indeed in Z sources in the NB kHz QPO frequency increases when the count rate drops.

In three atoll sources (4U 1728−34, 4U 1636−53 and KS 1731−260), oscillations have been seen during X-ray bursts whose frequencies (360–580 Hz) are consistent with being equal to the frequency *differences* between the twin peaks (in 4U 1728−34), or twice that (in the other two sources). In a fourth atoll source (4U 0614+09) there is marginal evidence for a third peak at the twin-peak separation frequency which corresponds to an oscillation in the persistent emission rather than in X-ray bursts.

These cases of three commensurate frequencies very strongly suggest that some kind of beat-frequency model is at work, with the "third peaks" at the neutron star spin frequencies (or twice that), the upper kHz peak at the Kepler frequency corresponding to some preferred orbital radius around the neutron star, and the lower kHz peak at the difference frequency between these two. Strohmayer et al. (1996c) suggested that this preferred radius is the magnetospheric radius. Miller, Lamb and Psaltis (1996) proposed it is the sonic radius. In models of this kind, which involve the neutron star spin as one of the frequencies participating in the beat-frequency process, the twin peak separation is predicted to be constant. However, in Sco X-1 the peak separation varies systematically with inferred $\dot{M}$, from ∼310 Hz when the upper peak is near 870 Hz to ∼230 Hz when it is near 1075 Hz: the peaks move closer together by ∼80 Hz while they both move up in frequency as $\dot{M}$ increases. This is in strong contradiction to straightforward beat-frequency models. To maintain the beat-frequency idea, in Sco X-1 a *second* unseen frequency has to be postulated in addition to the (in Sco X-1) unseen neutron-star spin frequency. See Models for a discussion of this and related issues.

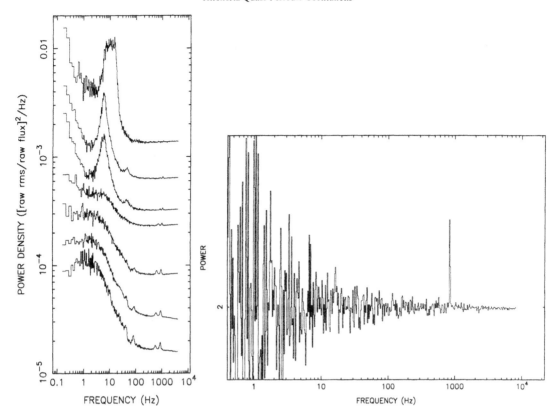

Fig. 2. *Left:* Power spectra of Sco X-1, with inferred $\dot{M}$ increasing upwards. Notice the decrease in strength and increase in frequency of the kHz peaks as a function of $\dot{M}$. The peaks near 45 and 90 Hz are identified as horizontal branch oscillations (HBO), that between 6 and 20 Hz as normal/flaring branch oscillations (N/FBO). The large width of the N/FBO peak in the top trace is due to peak motion. The sloping continua in the kHz range are instrumental (van der Klis et al. 1997b). *Right:* Power spectrum of 4U 1608−52 showing a single, narrow QPO peak near 800 Hz (Berger et al. 1996).

In the Z sources Sco X-1, GX 5−1 and GX 17+2 twin kHz QPO peaks and the so-called horizontal-branch oscillations (HBO; van der Klis et al. 1985) are seen simultaneously (Fig. 2). HBO are thought to be a product of the magnetospheric beat-frequency mechanism (Alpar and Shaham 1985, Lamb et al. 1985). If this is correct, then this model can *not* explain the kHz QPO in these sources. It is possible in principle that the kHz QPO in the Z sources is a different phenomenon from that in the atoll sources (e.g., Strohmayer et al. 1996c), but this seems unlikely: the frequencies, their dependence on $\dot{M}$, the coherencies, the peak separations and the fact that there are *two* peaks, one of which sometimes becomes undetectable at extreme $\dot{M}$, are too similar to attribute to just coincidence. If this is correct, then the variable twin peak separation detected in Sco X-1, the simultaneous presence of kHz QPO and HBO in Z sources, *and* the direct indications for a beat frequency in the atoll sources must all be explained within the same model, a formidable challenge.

One of the twin peaks sometimes (at high or low $\dot{M}$ for the source considered) drops below the detection limit while the other one is still observable. "Third peaks", whether they were seen in X-ray bursts or in the persistent emission, so far were always single. In addition to these cases of single kHz QPO peaks, there has been a number of cases where the suspicion arose that a single kHz peak might be something else than either one of the twin peaks or a third peak located near the difference frequency of the twin peaks. In particular this was the case in 4U 1608−52 (Berger et al. 1996) and 4U 1636−53 (Wijnands et al. 1997). In these cases, the frequency of the QPO peak varied erratically over a range of 830–890 Hz (4U 1608−52) and 835–897 Hz (4U 1636−53) while there were no obvious variations in the X-ray colors that indicated that $\dot{M}$ changed. The frequency variations seemed unrelated to count-rate variations as well. It is too early to say whether this behavior is just another aspect of twin-peak kHz QPO, or constitutes another phenomenon.

One of the distinguishing characteristics of kHz QPO is that they often show a relatively large coherence. The quality factor Q, defined as the QPO peak's centroid frequency ν divided by its full width at half maximum $\Delta\nu$ regularly reaches values of more than 100 in one or both of the twin peaks (although much lower Q's are also common). This provides a strong constraint on "orbiting clump" type models, as lifetime broadening considerations show that the clumps must persist over hundreds of cycles. The oscillations in bursts have shown even larger coherence. They attained a record-level Q of ~900 in a burst in KS 1731−260 (Smith, Morgan and Bradt 1997). This high Q value supports models where these oscillations are caused by the neutron-star spin. In 4U 1728−34 (Strohmayer et al. 1996c), drifts by ~1 Hz have been observed in the ~363 Hz frequency of the QPO in bursts that are suggestive of the bursting layer slightly expanding and then recontracting, changing its rotation rate to conserve angular momentum and thus modulating the QPO frequency. It is not clear at this point if this phenomenon is unique to sources where the burst QPO frequency is equal to the twin peak separation (4U 1728−34 is the only known source of this type so far) or is also seen in sources where it is twice that. It may be that in the former case we are seeing brightness contrasts in the bursting layer spinning around at its own frequency, which is slightly different from that of the star as a whole, whereas in the latter case we see the magnetic poles which are attached to the body of the neutron star and therefore do not drift in frequency.

The amplitudes of kHz QPO have, in all cases where a check was possible, shown a strong positive dependence on photon energy (e.g., Berger et al. 1996, Zhang et al. 1996). Their amplitudes when measured in a broad photon-energy band can therefore be expected to depend strongly on details of the low-energy part of the spectrum, which contributes many photons and little kHz QPO amplitude: detector cutoff and interstellar absorption will affect the overall fractional amplitude. Reported fractional amplitudes vary between 0.5 and a few percent in Z sources and 3 and 15% (rms) in atoll sources when measured over a 2–20 keV band; for higher energies amplitudes up to 40% (rms) have been observed.

Mendez et al. (1997) show that the energy spectrum of the oscillating flux can be fitted with a blackbody spectrum with a temperature of ~1.6 keV and a radius of 0.5±0.2 km. The QPO could therefore be due to modulation of emission from a region on the neutron star surface with these properties. However, the data allow many different spectral models. For example, alternative interpretations that fit the data as well are that the oscillations are caused by variations with an amplitude of 2.5% in the temperature of a ~1.1 keV blackbody with a radius of ~10km, or by ~5% variations in the optical depth of an unsaturated Comptonization spectrum.

A final strong model constraint is provided by the small magnitude of any time lags between the kHz QPO signal as observed in different energy bands (Vaughan et al. 1997). Time-lag measurements require very high signal-to-noise ratio's, and have so far only been made in the single, apparently count-rate independent peaks in 4U 1608−52 and 4U 1636−53 near 850 Hz, and in a 730 Hz peak in 4U 0614+09 which was probably an upper peak. Finite lags of 10–60μsec were discovered in 4U 1608−52; the hard photons lag the soft ones by increasing amounts as the photon energy increases. Upper limits of 30μsec and 45μsec were set in 4U 1636−53 and 4U 0614+09, respectively. These are by far the smallest lags ever measured; they correspond to light travel distances of 3–20 km. For rather general assumptions about the spectral formation mechanism, this limits the scale of any Compton scattering regions dominating the spectral shape to between a few and a few tens of km.

The great enigma in the phenomenology right now is, in my opinion, the peculiar lack of correlation between kHz QPO frequency and average source luminosity, whereas *in each individual source* a strong correlation between frequency and $\dot{M}$ is observed. In 4U 0614+09, at a luminosity of a few times $10^{-3} L_{Edd}$, similar QPO frequencies have been observed as in 4U 1820−30, which is near $10^{-1} L_{Edd}$, and in Sco X-1, which is inferred to be a near-Eddington accretor. In at least 6 sources, spread over this entire range of average X-ray luminosity, the upper peak has been observed to disappear below the detection limit when its frequency is somewhere between 1100 and 1200 Hz as the flux exceeds a certain limit, but this flux limit is widely different between sources. This must mean that another, compensating, parameter than just accretion rate is affecting the properties of the kHz QPO, most likely by directly affecting the frequency, although some kind of selection effect that leads to suppression of any QPO outside the 300–1200 Hz range is also a possibility. This latter possibility of course requires that the peaks actually observed in sources with different luminosities are in some sense "different". One would expect that in sources that go through a large decrease in accretion rate [transients] several "new" QPO peaks would successively appear near 1200 Hz, move down in frequency and disappear near 300 Hz. This has not been seen and seems somewhat unlikely, but can not be excluded at this point.

An obvious candidate for such a compensating parameter is the neutron star magnetic field strength, but neutron star mass or spin, either by their effects on the surrounding space-time or directly, might play a role as well. What would be required, specifically, is that there exists a correlation or an anti-correlation between, say, the magnetic field strength B of the neutron star and its mean accretion rate $\langle \dot{M} \rangle$, and that the QPO frequency depends on B in such a way as to approximately compensate the $\dot{M}$ effect. Interestingly, it has been concluded previously (Hasinger and van der Klis 1989, see van der Klis 1995) on the basis of comparing Z and atoll source phenomenology that $\langle \dot{M} \rangle$ and B are correlated among LMXBs, and recently spectral modeling (Psaltis 1997) has tended to confirm this. The magnetospheric beat-frequency model (Alpar and Shaham 1985), when combined with this inferred correlation, qualitatively fits the requirements sketched above, but the results on the Z sources make this model unattractive for kHz QPO. Perhaps the magnetic field strength affects the inner accretion flows in other ways than by just terminating the disk at the magnetospheric radius. If magnetic stresses could somehow slow down the (for example, orbital) motion responsible for the kHz QPO, that would do it. Of course, radiative stresses can diminish the effective gravity and slow down orbital motion (Miller and Lamb 1993), but the luminosity is not independent from $\dot{M}$, but instead is expected to vary proportionally to it, so that radiative stresses cannot fulfill this role: we know already that when in a given source $\dot{M}$ goes up so does L_x, but this does not prevent the QPO frequency from going up as well.

There is a lively discussion about the nature of the observed frequencies and their potential to constrain neutron-star masses and radii and to test general relativity. Kaaret, Ford and Kaiyou Chen (1997) have proposed that the behavior of the single, count-rate independent QPO peaks in 4U 1608−52 and 4U 1636−53 described above is related to orbital motion near the marginally stable orbit, and from this derive neutron star masses of ~$2M_\odot$. Will Zhang, Strohmayer and Swank (1997) have proposed that the narrow range of maximal frequencies (1100–1200 Hz) described above must be identified with the general relativistic marginally stable frequencies, which leads them to the conclusion that the neutron stars' masses are near $2M_\odot$ as well. An alternative possibility is that the maximal frequencies are set by the Keplerian frequency at the neutron star surface. This requires the star to be larger than the marginally stable orbit and for ~$1.4M_\odot$ neutron stars would favour the stiffest equations of state. S.N. Zhang, Wei Cui and Wan Chen (1997) have recently discussed possible observable effects of the marginally stable orbit in the related LMXB black hole systems. They mention the possibility that 300 Hz and 67 Hz QPO observed in the black hole candidates GRO J1655−40 and GRS 1915+105, respectively (Remillard 1997, Morgan et al. 1996, 1997) are radial trapped g-mode oscillations of the disk. To my knowledge, this model has not yet been applied to kHz QPO in neutron stars.

Just the assumption that the upper peak corresponds to Keplerian motion around the neutron star allows to set stringent limits on neutron star parameters, a point made by Miller et al. (1996) in their paper on a particular model that interprets the upper peak in this way (see Models). Different from the proposals just mentioned, these limits do *not* rely on identifying any of the observed frequencies with the marginally stable orbital frequency. There are two direct constraints on the neutron star mass and radius from the simple assertion that there is stable Keplerian motion at the frequency ν_u of the upper peak: (1) the radius of the star R must be smaller than the radius of this Keplerian orbit, in Schwarzschild coordinates $R < (GM/4\pi^2\nu_u^2)^{1/3}$, and (2) the radius of the marginally stable orbit must *also* be smaller than this: $6GM/c^2 < (GM/4\pi^2\nu_u^2)^{1/3}$, as no stable orbit is possible within this radius. Condition (1) is a mass-dependent upper limit on the radius of the star, and condition (2) provides an upper limit on the mass: $M < c^3/(2\pi 6^{3/2} G\nu_u)$. For $\nu_u = 1193\,Hz$ (Wijnands et al. 1997), $M < 1.9$ and $R_{NS} < 16.3$ km. Putting in the corrections for the frame dragging due to the neutron star spin requires knowledge of the spin rate (which in the sonic point model is equal to the twin peak separation, or perhaps half that; see Models). The correction also depends somewhat on the neutron star model, which determines the relation between spin rate and angular momentum, so that the limits become slightly different for each EOS. Putting in these Kerr corrections (for a spin rate of 275 Hz) changes the limits quoted above only slightly, to $M < 2.1$ and $R_{NS} < 16.5$ km for a wide range of equations of state (Wijnands et al. 1997).

MODELS

This section is intended not so much to provide an exhaustive comparative discussion of the models that have so far been proposed for kHz QPO, as to alert the reader to the various basic physical pictures that have been discussed.

Of course, the phenomenology as described in the previous section very strongly suggests that a beat-frequency model of some kins is at work. Neutron star spin and disk Keplerian motion are periodic phenomena known to be present in the system and are therefore natural candidates for providing the basic frequencies. However, it is to early to declare any proposed implementation of a beat-frequency model for kHz QPO an unqualified success. Let us first look at other models that have been put forward.

Remarkably short shrift has been given so far to *neutron star vibration models*. The short time scale variations in kHz QPO frequency and the lack of higher-frequency peaks have been cited as reasons for rejecting these models. Of course, a fundamental problem for vibration models (also for lower-frequency QPO) has always been to explain how the vibrations would be able to produce appreciable modulations of the X-ray flux, in the case of kHz QPO with amplitudes up to 15% (full band) to 40% (>10 keV). Also, it is unclear what physics would be required to pick out just 2 (or 3) frequencies from the range of modes one would expect in most of these models.

A model based on numerical radiation hydrodynamics has been proposed by Klein et al. (1996) for the case of the kHz QPO in Sco X-1. In this model accretion takes place by way of a magnetic accretion funnel. In the funnel, the mass accretion rate is locally super-Eddington, and *photon bubbles* form, which rise up by buoyancy through the accreting matter. Klein et al. find that the "bursting of the bubbles" at the top of the flow can produce kHz QPO signals that resemble the observed ones in several respects. For Sco X-1 their model assumes a total X-ray luminosity of only $3\ 10^{37}$ erg/s, which seems unlikely from the point of view of the general picture of Z-source phenomenology (e.g., van der Klis 1995) which instead suggests that these sources, including Sco X-1, are near-Eddington accretors. The model predicts that there should be QPO peaks at higher frequencies. High time-resolution observations of Sco X-1 performed in order to check on this show no evidence of higher-frequency peaks down to quite good limits (van der Klis et al. 1997b). The higher-frequency peaks may not always be strong enough to be observable (Klein 1997, private communication). The model is currently being further explored.

The dependence between the QPO frequencies observed in Sco X-1 can be explained with a model where each of the two QPO signals comes from one of two diametrically opposed *relativistic jets* emanating from the central source. In this picture a central shrouded X-ray pulsar provides the basic high-frequency signal. The observer sees the pulsar signal reflected off inhomogeneities in the two jets. Three frequency shifts affect the observed frequencies: a redshift (identical for each jet) because the inhomogeneities move away from the pulsar at relativistic speed, special-relativistic time dilation, and an additional redshift and blueshift for the receding and approaching jet, respectively. The predicted frequencies for the signals reflected off the two jets will be $\nu_{\mp} = \nu_{pulse}(1 - v/c)/(1 \pm (v/c) \cos \theta)$, where v is the jets' speed, θ their angle with the line of sight, and ν_{pulse} the unseen pulse frequency (van der Klis et al. 1997b). This model fits the Sco X-1 data remarkably well, for ν_{pulse}=1370 Hz (which could be twice the neutron star spin frequency) and θ=61°, if we assume that with increasing $\dot{M}$, the jet's speed decreases from v/c=0.48 to 0.26. However, it is hard to see how this model can account for the atoll sources' kHz QPO properties, as it can't explain frequency shifts *without* the peaks' frequency separation being affected, as are observed in those sources.

Finally, let's turn to beat-frequency models. The two versions of the model that have been discussed both identify the upper peak's frequency with the Keplerian frequency of the accretion disk at some preferred radius, and the lower peak with the beat between this Keplerian frequency and the neutron star spin frequency. In the magnetospheric beat-frequency model, which has been used previously to explain certain cataclysmic variable oscillations (Patterson 1979) and the so-called horizontal-branch oscillations (HBO) in Z sources (Alpar and Shaham 1985, Lamb et al. 1985), uses the magnetospheric radius r_M as this preferred radius. As HBO and kHz QPO have been seen *simultaneously* in all three Z sources where kHz QPO have so far been observed, at least *one* additional model is required.

According to Miller, Lamb and Psaltis (1996), applying the magnetospheric beat-frequency model to the kHz QPO leads to several difficulties. They propose the *sonic-point model* instead. This model uses the sonic radius as the preferred radius of which we observe the Kepler frequency. The sonic radius is defined as the radius where the radial inflow velocity becomes supersonic. In the absence of other stresses, the sonic radius is located near the general relativistic marginally stable orbit (at $6GM/c^2$ in a Schwarzschild geometry and closer in in a Kerr geometry). Radiative stresses may change the location of the sonic radius, as indeed is required by the observation that the kHz QPO frequencies vary with $\dot{M}$. As we shall see below, interpretations along these lines have direct consequences for the EOS of high-density matter and provide possibilities to test general relativity in the strong-field regime. See Kluzniak and Wagoner (1985) and Kluzniak, Michelson and Wagoner (1990) for pre-kHz QPO considerations of the question how to use the innermost accretion flows onto neutron stars to determine neutron star parameters and test general relativity.

In the sonic point model, the mechanism that produces the upper peak is as follows. At the sonic radius clumps form in the accretion flow that live for tens to hundreds of QPO cycles (i.e., several 0.01 to several 0.1 s). During a clump's lifetime, its matter gradually accretes onto the neutron star surface. This accreting matter follows a fixed spiral-shaped trajectory in the frame corotating with the Keplerian motion, and therefore hits the neutron star surface at a specific spot, the "footpoint"of that clump's spiral flow. Enhanced accretion at the footpoint produces enhanced emission there, and because the footpoint travels around the neutron star surface at the Keplerian angular velocity, irrespective of what the neutron star's spin is, the observer sees the hot spot change aspect and perhaps appear and disappear with the sonic radius' Keplerian frequency. The narrowness of the QPO peaks implies that all the clumps are near one precise radius, allows for relatively little fluctuations in the spiral flow, and requires the above-mentioned long clump lifetimes. In Z sources, applying this model for the kHz QPO, and the magnetospheric beat-frequency model for the HBO leads to the conclusion that the sonic radius r_S is well within the magnetosphere, so that only part of the matter is apparently "picked up" by the magnetic field lines at r_M, and the remainder must maintain its Keplerian flow to much closer in.

The modulation of the radiation caused by this mechanism is not a modulation of the total luminosity, but a modulation of the direction into which this luminosity is emitted ("beaming"). If some of these neutron stars have a magnetic field strong enough to channel some of the accreting matter to the magnetic poles (as required by the magnetospheric beat-frequency model for Z sources, and also by the sonic point model itself, see below), then *another* beaming modulation is expected at the neutron star spin frequency. However, so far *this* modulation has not been detected in any Z source, which presents the difficulty that one has to somehow get rid of the spin frequency but not of the sonic-radius Keplerian frequency. As both are beaming modulations and the frequencies are similar, this may require some finetuning of the scattering process that is smearing the pulsations.

Miller et al. (1996) predict that as the sonic radius approaches the general-relativistic marginally stable orbit the frequency of the upper peak will hit a "ceiling" and remain stable for further increases in accretion rate. There are so far no data that have shown this. Instead it has been observed that the QPO disappear above some level of inferred accretion rate, which level, however, is very different between sources (much higher in sources with a higher average luminosity), but at frequencies that are mostly in the range 1100–1200 Hz. Perhaps this is what *really* happens when the marginally stable orbital radius is reached.

Now let's turn to the question how the lower peak is produced in the sonic point model. The footpoint running over the neutron star surface will of course encounter the same point on the surface once every beat period between the neutron star spin and the sonic-point Keplerian frequencies. If there are B gradients over the surface this could affect the emission properties of the footpoints at the beat-frequency. However, this is *not* the model proposed by Miller et al. They propose instead that the physical interaction that eventually leads to the modulation of the X-ray flux at the beat-frequency takes place at the sonic radius. In their model, X-rays produced by the accretion of matter channeled onto the magnetic poles are emitted into two broad "lighthouse beams", sweeping around at the neutron star spin frequency. These beams irradiate the clumps at the sonic radius when they sweep over them, which happens once per beat period. This leads to a modulation, at the beat frequency, in the rate at which the clumps provide matter to their spiral flows. Consequently, the accretion of matter onto the footpoints, and therefore their emission, is modulated at the beat frequency, and this leads to the lower peak we see in the power spectra. This model predicts various aliases and harmonics of the observed peaks should also be present, but these have not been observed so far. No quantitative predictions have to my knowledge been made for the strengths of these additional peaks. The model requires the pulsar beams to extend out to the sonic radius with sufficient strength to affect the accretion flow there, yet to be smeared, presumably by scattering further out, to below the <1% detection levels that have so far been reached (e.g., Vaughan et al. 1994). On the other hand the footpoints' emission beams must be able to propagate to infinity in order for us to see the sometimes quite strong upper peaks. As double peaks are seen in sources between 10^{-3} and 1 L_{Edd}, all these processes must operate in a way to keep the phenomenology the same over a large range in $\dot{M}$ (but this, of course, is a problem in any interpretation that attempts to explain these phenomena by one mechanism across the board). The cases of 4U 1636−53 and KS 1731−260, where the pulsar frequency appears to be twice the twin-peak separation frequency in the Miller et al. model requires a further explanation of the question why the two observed pulsar beams are not *both* interacting with the the sonic-radius clumps, but only one of them. Both the magnetospheric and the sonic-point beat-frequency model predict the twin-peak separation to be equal to the neutron star spin frequency (or the pulsar frequency). This is in contradiction to the result on Sco X-1, where the twin peak separation varies systematically from ∼310 Hz when the upper peak is at 870 Hz to ∼220 Hz when it is at 1075 Hz. As already mentioned above, explaining this with a beat-frequency model requires introducing *another* unseen frequency in addition to the (in Sco X-1) unseen pulsar frequency. A modification of the sonic point model along these lines seems possible (Lamb, priv. comm. 1996).

Obviously, a large amount of effort is still required to make any of the models so far proposed stick. Fortunately, as it looks now the theoretical efforts that are underway at this point will be guided by a very constraining body of RXTE data. Eventually, most LMXBs will likely exhibit the new phenomenon, and many of its properties can be measured with RXTE with great precision.

ACKNOWLEDGEMENTS

This work was supported in part by the Netherlands Organization for Scientific Research (NWO) under grant PGS 78-277 and by the Netherlands Foundation for Research in Astronomy (ASTRON) under grant 781-76-017.

REFERENCES

Alpar, M.A., Shaham, J. Nature, 316, 239 (1985).
Berger, M., Van der Klis, M., Van Paradijs, J., Lewin, W.H.G., Lamb, F., Vaughan, B., Kuulkers, E., Augusteijn, T., Zhang, W., Marshall, F.E., Swank, J.H., Lapidus, I., Lochner, J.C., Strohmayer, T.E., ApJ, 469, L13, (1996).
Cook, G.B., Shapiro, S.L., Teukolsky, S.A., ApJ, 424, 823 (1994).
Ford, E., Kaaret, P., Tavani, M., Harmon, B.A., Zhang, S.N., Barret, D., Bloser, P., Grindlay, J., IAU Circ.6426 (1996).
Ford, E., Kaaret, P., Tavani, M., Barret, D., Bloser, P., Grindlay, J., Harmon, B.A., Paciesas, W.S., Zhang, S.N., ApJ, 475, L123 (1997).
Hasinger, G., Van der Klis, M. A&A, 225, 79 (1989).
Jongert, H.C., Van der Klis, M., A&A, 310, 474 (1996).
Kaaret, Ph., Ford, E., Chen, K., ApJ, 480, L27 (1997). astro-ph/9701101
Klein, R.L., Jernigan, G.J., Arons, J., Morgan, E.H., Zhang, W., ApJ, 469, L119 (1996).
Kluźniak, W., Wagoner, R.V., ApJ, 297, 548 (1985).
Kluźniak, W., Michelson, P., Wagoner, R.V., ApJ358, 538 (1990).
Lamb, F.K., Shibazaki, N., Alpar, M.A., Shaham, J. Nature, 317, 681 (1985).
Méndez, M., Van der Klis, M., Van Paradijs, J., Lewin, W.H.G., Lamb, F.K., Vaughan, B.A., Kuulkers, E., Psaltis, D., ApJ, 485, L37 (1997).
Miller, M.C., Lamb, F.K., ApJ, 413, L43 (1993).
Miller, M.C., Lamb, F.K., Psaltis, D., ApJ, submitted (1996). astro-ph/9609157
Morgan, E.H., Smith, D.A., IAU Circ. 6437 (1996).
Morgan, E.H., Remillard, R., Greiner, J., IAU Circ.6392 (1996).
Morgan, E.H., Remillard, R., Greiner, J., ApJ, 482, 993 (1997).
Patterson, J., ApJ, 234, 978 (1979).
Psaltis, D., PhD Thesis, UIUC (1997).
Remillard, R., Proc. Texas Conf. *Texas in Chicago*, Chicago, December 1996 (1997).
Smale, A.P., Zhang, W., White, N.E., IAU Circ. 6507 (1996).
Smale, A.P., Zhang, W., White, N.E., ApJ, 483, L119 (1997).
Smith, D.A., Morgan, E.H., Bradt, H., ApJ, 479, L137 (1997).
Strohmayer, T., Zhang, W., Swank, J., IAU Circ.6320 (1996a).
Strohmayer, T., Zhang, W., Smale, A., Day, C., Swank, J., Titarchuk, L., Lee, U., IAU Circ.6387 (1996b).
Strohmayer, T., Zhang, W., Smale, A., Day, C., Swank, J., Titarchuk, L., Lee, U., ApJ, 469, L9 (1996c).
Strohmayer, T., Lee, U., Jahoda, K., IAU Circ. 6484 (1996d).
Van der Klis, M. NATO ASI C262: *Timing Neutron Stars*, Ögelman and van den Heuvel (eds.), Kluwer, p. 27 (1989).
Van der Klis, M. in: *X-Ray Binaries*, Lewin, Van Paradijs and Van den Heuvel (eds.), Cambridge University Press, p. 252 (1995).
Van der Klis, M., Jansen, F., Van Paradijs, J., Lewin, W.H.G., van den Heuvel, E.P.J., Trümper, J.E., Sztajno, M. Nature, 316, 225 (1985).
Van der Klis, M., Swank, J., Zhang, W., Jahoda, K., Morgan, E., Lewin, W., Vaughan, B., Van Paradijs, J., IAU Circ.6319 (1996a).
Van der Klis, M., Wijnands, R., Chen, W., Lamb, F.K., Psaltis, D., Kuulkers, E., Lewin, W.H.G., Vaughan, B., Van Paradijs, J., Dieters, S., Horne, K., IAU Circ.6424 (1996b).
Van der Klis, M., Swank, J.H., Zhang, W., Jahoda, K., Morgan, E.H., Lewin, W.H.G., Vaughan, B., Van Paradijs, J., ApJ, 469, L1 (1996c).
Van der Klis, M., Van Paradijs, J., Lewin, W.H.G., Lamb, F.K., Vaughan, B., Kuulkers, E., Augusteijn, T., IAU Circ.6428 (1996d).
Van der Klis, M., Wijnands, R., Kuulkers, E., Lamb, F.K., Psaltis, D., Dieters, S., Van Paradijs, J., Lewin, W.H.G., Vaughan, B., IAU Circ. 6511 (1996e).
Van der Klis, M., Homan, J., Wijnands, R., Kuulkers, E., Lamb, F.K., Psaltis, D., Dieters, S., Van Paradijs, J., Lewin, W.H.G., Vaughan, B., IAU Circ. 6565 (1997a).
Van der Klis, M., Wijnands, R., Chen, W., Horne, K., ApJ, 481, L97 (1997b).
Van Paradijs, J., Zhang, W., Marshall, F., Swank, J.H., Augusteijn, T., Kuulkers, E., Lewin, W.H.G., Lamb, F., Lapidus, I., Lochner, J., Strohmayer, T., Van der Klis, M., Vaughan, B., IAU Circ.6336 (1996).
Vaughan, B.A., Van der Klis, M., Wood, K.S., Norris, J.P., Hertz, P., Michelson, P.F., Van Paradijs, J., Lewin, W.H.G., Mitsuda, K., Penninx, W., ApJ, 435, 362 (1994).
Vaughan, B.A., Van der Klis, M., Van Paradijs, J., Wijnands, R.A.D., Lewin, W.H.G., Lamb, F.K., Psaltis, D., Kuulkers, E., Oosterbroek, T., ApJ, 483, L115 (1997).
Wijnands, R.A.D., Van Paradijs, J., Lewin, W.H.G., Lamb, F.K., Vaughan, B., Kuulkers, E., Augusteijn, T., IAU Circ. 6447 (1996).
Wijnands, R.A.D., Van der Klis, M., ApJ, 482, L65 (1997).
Wijnands, R.A.D., Van der Klis, M., Van Paradijs, J., Lewin, W.H.G., Lamb, F.K., Vaughan, B., Kuulkers, E., ApJ, 479, L141 (1997).
Zhang, S.N., Cui, W., Chen, W., ApJ, 482, L155 (1997).
Zhang, W., Lapidus, I., White, N.E., Titarchuk, L., ApJ, 469, L17 (1996).
Zhang, W., Lapidus, I., Swank, J.H., White, N.E., Titarchuk, L., IAU Circ. 6541 (1997).
Zhang, W., Strohmayer, T.E., Swank, J.H., ApJ, 482, L167 (1997).

Pergamon

Adv. Space Res. Vol. 22, No. 7, pp. 935–938, 1998
© 1998 COSPAR. Published by Elsevier Science Ltd. All rights reserved
Printed in Great Britain
0273-1177/98 $19.00 + 0.00

PII: S0273-1177(98)00144-6

THE COMPLEX CONTINUUM MODEL FOR THE LMXB DIPPING SOURCES

M. J. Church[1,2], M. Bałucińska-Church[1,2], T. Dotani[1], K. Mitsuda[1], H. Inoue[1], and T. Takahashi[1]

[1] *Institute of Space and Astronautical Science, Yoshinodai 3-1-1, Sagamihara, Kanagawa 229-8510, Japan*
[2] *School of Physics and Space Research, University of Birmingham, Edgbaston, Birmingham B15 2TT, UK*

ABSTRACT

We report the results of a programme of analysis of the LMXB dipping sources, recently concentrating on the spectral changes taking place on rapid timescales during dip ingress and egress in XB 1916-053 using the high quality ASCA data. We tested the hypothesis that the source can be fitted by the same model that we have previously shown gives very good explanations of the two very different sources X 1755-338, the energy-independent dipper and X 1624-490, the Big Dipper. This complex continuum model consists of a blackbody identified with emission from the surface of a neutron star, plus a power law seen as Comptonised emission from an ADC. Analysis of the ASCA observation of XB 1916-053 shows this source to be remarkable in that all emission components are completely absorbed in dips, and that the rapid spectral evolution during dip ingress and egress is well fitted by our model. On the basis that the model fits these 3 very different sources we now propose that the complex continuum model will be able to fit all of the dipping sources.   ©1998 COSPAR. Published by Elsevier Science Ltd.

INTRODUCTION

There are ∼ 10 Low Mass X-ray Binary sources that show periodic dipping in X-ray intensity at the orbital period of the binary, generally accepted as being caused by occultation of the central emission regions by the bulge in the outer accretion disc where the accretion flow from the companion impacts (White & Swank 1982). It was expected that dipping would show the characteristics of photoelectric absorption in the bulge, since this region is relatively cool because of its distance from the central neutron star, ie that there would be an increase in column density in dipping so that the low energy cut-off of the spectrum shifts to higher energies and there is a marked hardening of the spectrum. However this was not found to be generally the case. Some sources show a hardening, but others show no change in hardness at all in the band 1 - 10 keV (eg X 1755-338) which is difficult to understand, and in X 1624-490, the spectral evolution is complicated: first a hardening and then a marked softening in deeper dipping.

On the basis of our analysis of the *Exosat ME* data on these two sources X 1755-338 and X 1624-490, we have proposed a model for the dipping sources (Church & Bałucińska-Church 1995). This model requires two emission regions (at least): the boundary layer at the surface of the neutron star emitting blackbody radiation as a point source, and an extended region, probably the Accretion Disc Corona, emitting a Comptonised component which we model by a power law at energies below the expected Comptonisation break. In this model, dipping consists primarily of absorption of the point source, whereas the extended region will only be occulted to a small extent by the absorbing region.

More recently, we have analysed the high quality ASCA observation of the dipping source XB 1916-053 made in 1993 May 2nd and lasting 18 hr (Church *et al.* 1997). This source is very different from both X 1755-338 and X 1624-490. Firstly, dipping has been seen at several different levels (Smale *et al.* 1988), varying from relatively shallow dipping as seen in X 1755-338, to much deeper dipping without any obvious dip variability that is normally seen. Moreover previous analysis of this source has been carried out using the "absorbed plus unabsorbed" approach to spectral modelling. This model has been applied in several sources in which it appeared that part of the emission did not have an increased column density during dipping, but was nevertheless attenuated as revealed by the normalisation of the unabsorbed component decreasing sharply. This decrease in normalisation has generally been ascribed to electron scattering in the absorber of this part of the emission.

RESULTS

During the ASCA observation, the dipping reached 100% in all dips as can be seen in the folded light curve shown in Figure 1; ie during the deepest part of each dip, all source emission was totally absorbed - very remarkable behaviour. Secondly, the duration of dipping was very long, equal to 32.7% of the orbital cycle, showing that the absorbing region subtends the very large angle of 120° at the neutron star. The immediate consequence of this is that as the absorbing region is very extended azimuthally, it is very likely that it must therefore also be extended in height above the accretion disc. Thus the 100% dipping in this source is directly due to this very large size of the absorber in that not only point-source emission is covered, but also the extended emission originating in our model in the Accretion Disc Corona.

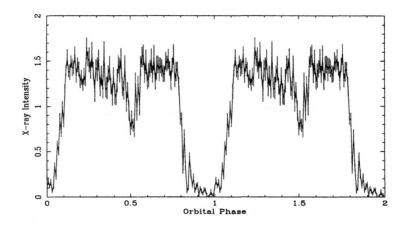

Fig. 1. The ASCA GIS2 light curve folded on the 3009 s best-fit orbital period showing that dipping reaches 100%.

Spectral analysis of the ASCA observation was carried out by dividing the data into 5 intensity bands corresponding to non-dip and dip emission, specifically 4.0 - 5.0 c/s (non-dip) and 0.0 - 1.0, 1.0 - 2.0, 2.0 - 3.0, and 3.0 - 4.0 c/s. Spectral fitting of the non-dip data showed that simple models such as absorbed bremsstrahlung and absorbed blackbody could be rejected. An absorbed power law fitted non-dip data well, but could be rejected for dip data. We therefore applied the two-component model consisting of a blackbody and a power law and found a good fit to the non-dip data with $kT_{bb} = 1.8$ keV and power law photon index $\Gamma = 2.6$. In order to fit this model to the dip data, it is clear that we have to allow both emission regions to be absorbed, since at the deepest parts of dips, the count rate falls to zero. The point-source blackbody will be covered instantaneously, however it is clear that we should allow the extended power law component to be covered progressively as the large absorbing bulge passes across the extended emission region. A model of the form: $AB_1 * BB + AB_2 * [AB_3 * f + (1 - f)] * PL$ will achieve this, where AB are variable absorption terms, f is the covering fraction, BB the blackbody and PL the power law terms. The ASCA dip data could be fitted very well by this model, with the values of kT_{bb} and Γ held constant at the non-dip values. The normalisations were also held fixed in order to prevent an "absorbed plus unabsorbed" approach being used (which results in a decreasing normalisation), and because we know that the actual emission parameters of the source regions cannot change during dipping. Dipping was seen to be due to very large increases in N_H for the blackbody, with large increases of N_H for the power law, and a covering factor that rose from 0 to 1.0 at the deepest parts of dipping where the count rate was small. Thus we have been able to show that the two-component model is able to fit this very different member of the dipping class (Church *et al.* 1997), and we now proceed to discuss the implications of this.

DISCUSSION

Spectral Evolution During Dipping. Firstly, we are able to explain the different spectral changes during dipping in X 1755-338, X 1624-490 and XB 1916-053. Typical non-dip and dip parameters for these sources are shown in Table 1. (For XB 1916-053, N_H for the power law is equal to $AB_2 + AB_3$ above; for the other sources, a progressive covering term was not needed.)

Table 1. Typical spectral fitting parameters for the dipping sources. N_H is in units of 10^{21} H atom cm^{-2}

source	state	kT_{bb}	N_H(BB)	Γ	N_H(PL)
X 1755-338	non-dip	0.88	10.7	2.67	3.6
	dip	0.88	19.4	2.67	4.3
X 1624-490	non-dip	1.39	56.0	2.41	56.0
	dip	1.39	>3500	2.41	56.0
XB 1916-053	non-dip	1.82	4.8	2.62	4.9
	dip	1.82	> 3000	2.62	> 400

In X 1755-338, the spectrum is dominated by the power law and the peak of the blackbody is at 2.5 keV. As we previously demonstrated (Church & Bałucińska-Church 1993), the two components combine in such a way that partial absorption of the blackbody resulting in an overall decrease in count rate of about 20%, has the effect that within the band 1 - 10 keV, the flux of the source decreases approximately independently of energy. In X 1624-490, the blackbody is 1.6 times hotter, (Church & Bałucińska-Church 1995) so that the blackbody emission is increased by 6.2 times relative to X 1755-338. The blackbody also peaks at a higher energy of $\sim$ 4.2 keV. Thus when absorption takes place in dipping, there is first of all the natural hardening of the spectrum expected. However dipping can be saturated in this source, corresponding to complete absorption of the blackbody, and in this deepest dipping the spectrum becomes much softer than in the non-dip state, since the harder blackbody part of the spectrum has been completely removed, leaving the relatively soft power law.

XB 1916-053 represents an interesting further variation on this theme. Figure 2 shows light curves in two energy bands and the associated hardness ratio for the ASCA observation. It is clear that in dip ingress and egress, there is a hardening which can also be seen in interdips. However in the main part of the dip the count rate is so close to zero that the hardness ratio cannot be determined. The original work of White and Swank on this source (1982) also showed a hardening in dipping in *Einstein MPC* data. In this source, spectral changes in dipping are not determined by the relative contributions of the blackbody and the power law or by kT_{bb}, since both components are absorbed. As in X 1624-490, the lower energy part of the spectrum at $\sim$ 1 keV is dominated by the power law, and the blackbody forms the minor part of the spectrum (as in X 1755-338) and peaks at about 5.5 keV. Thus as the major part of the spectrum, the power law, is absorbed in dipping, the spectrum becomes harder. This is the reverse of what happens when a hot blackbody determines the change in hardness as in X 1624-490.

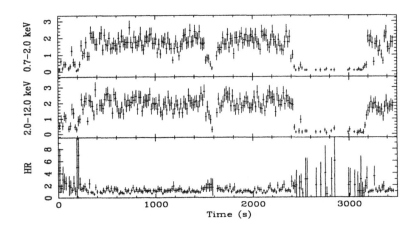

Fig. 2. ASCA light curves in the bands 0.7 - 2.0 and 2.0 - 12.0 keV and the hardness ratio derived from these, showing hardening in dip ingress and egress and in the interdip.

It is interesting to consider why the absorber is so extended in XB 1916-053. The geometry of this binary system is markedly different compared with X 1755-338 because of the low mass of the companion of 0.1 M$_\odot$ or less (Walter *et al.* 1982), so that the separation of the stars is 0.5 R$_\odot$ or less, about 3.5 times smaller, and the radius of the accretion disc in XB 1916-053 will also be smaller (by $\sim$60%). Thus an absorbing bulge of given size will subtend

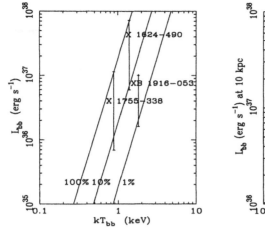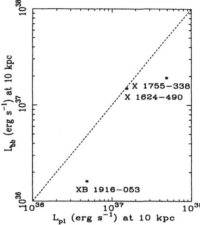

Fig. 3. (a) Luminosity of the blackbody component as a function of kT_{bb}; (b) comparison of blackbody and power law luminosities.

a larger angle at the neutron star. Moreover the distance of the outer edge of the disc from the Lagrangian point from which mass injection originates is very small, less than $0.2R_{\odot}$, and this may effect the size of the bulge.

Comparison of the Blackbody and Power Law Fluxes of the Sources. In Figure 3a we show the unabsorbed luminosities of the 3 sources plotted as a function of blackbody temperature kT_{bb}. Also plotted are lines expected for blackbody emission from a neutron star of radius 12 km according to the relation $L = f' 4\pi R_{bb}^2 \sigma T^4$ where f' is the fraction of the surface area emitting. The 3 cases shown have f = 1.0, 0.1 and 0.01. The data plotted consist of our best-fit results for the 3 sources, the fluxes having been converted to luminosity using possible source distances. For X 1755-338 and X 1624-490, the upper limits correspond to the intersection with the f=1 line; ie corresponding to distances of 7.7 kpc and 21.6 kpc respectively. The lower limit distance for X 1755-338 is 6.0 kpc, since the radio values of N_H decrease below this value, and our spectral fitting results are most consistent with the largest radio value. For XB 1916-053, we fix the limits at 10 and 25 kpc, since the data suggest that the source is relatively distant. It is clear that the sources differ in more than temperature, and in particular XB 1916-053 appears to have an emitting area substantially less than the other 2 sources. In Figure 3b we compare the unabsorbed luminosities of the blackbody and power law components evaluated at 10 kpc, ie we are comparing the fluxes and so are not subject to the large distance uncertainties. The power law flux is evaluated in the band 0.5 - 100 keV. It can be seen that in all 3 sources the blackbody contribution is substantial, equal to that of the power law for X 1624-490 (although the choice of band for the power law is somewhat arbitrary). By extending this work to include more sources we hope to elucidate the relation between the two spectral components.

REFERENCES

Church, M. J., and M. Bałucińska-Church, "Cosmic abundance absorption dips in X 1755-33", Mon. Not. R. Astr. Soc. **260**, 59 (1993).

Church, M. J., and M. Bałucińska-Church, "A complex continuum model for the low-mass X-ray binaries: application to X 1624-490", Astron. & Astrophys. **300**, 441 (1995).

Church, M. J., M. Bałucińska-Church, H. Inoue, T. Dotani, K. Mitsuda, et al., "Simple photoelectric absorption in the LMXB dipping source XB 1916-053 from spectral analysis of the ASCA observation", Proc. Conf. on Imaging and Spectrsocopy of Cosmic Hot Plasmas, Tokyo (1997).

Smale, A. P., K. O. Mason, N. E. White, and M. Gottwald M., "X-ray observations of the 50-min dipping source XB 1916-053", Mon. Not. R. astr. Soc. **232**, 647 (1988).

Walter F. M., S. Bowyer, K. O. Mason, J. T. Clarke, J. P. Henry, J. Halpern, and J. E. Grindlay, Ap. J. **253**, L67 (1982).

White, N. E., and J. H. Swank, "The discovery of 50 minute periodic absorption events from 4U 1915-05", Ap. J. **253**, L66 (1982).

 Pergamon

Adv. Space Res. Vol. 22, No. 7, pp. 939–942, 1998
© 1998 COSPAR. Published by Elsevier Science Ltd. All rights reserved
Printed in Great Britain
0273-1177/98 $19.00 + 0.00

PII: S0273-1177(98)00141-0

TEMPORAL AND SPECTRAL PARAMETERS OF SLOW X-RAY PERIODIC SOURCES OBSERVED DURING PROGNOZ 9 MISSION

M.I. Kudryavtsev[1], S.I. Svertilov[2], and V.V. Bogomolov[2]

[1]*Space Research Institute, Russian Academy of Science, Profsoyznaya st. 84/32, Moscow 117810, Russia*
[2]*Skobeltsyn Institute of Nuclear Physics, Moscow State University, Vorob'evy Gory, Moscow 119899, Russia*

ABSTRACT

Periodic components in hard X-rays from some objects near the Galactic Center were discovered in the experiment during *Prognoz-9* mission. The corresponding period values lay in the range from several hours to several days. Some of these periodic processes were identified with the X-ray binaries which were not known before as eclipse systems: *H1705-25 (Nov Oph 1977)* - 152 h, *Sco X-1* - 62 h, *Cen X-4* -8.2 h, *4U1755-33* - 4.4 h. Other periods are still unidentified: 98 h, 69 h, 13 h, 9.4 h, 7.8 h, 3.4 h, 1.96 h. The common property of all these processes is the very complicated form of mean light curves similar to those pulse profiles which were obtained for ordinary "more fast" (with periods less than 1000 s) X-ray pulsars. The energy spectra of these sources are characterized by the values of the effective temperature kT in the range from ~5 (1.96 h) to >40 keV (152 h).

©1998 COSPAR. Published by Elsevier Science Ltd.

INSTRUMENTATION AND OBSERVATIONS

The observations of galactic sources in hard X-rays (10-200 keV) were made in 1983-84 during the complex experiment on a high-apogee (~720000 km) satellite *Prognoz-9* with the wide-field of view (FOV) (~45° *FWHM*) scintillator spectrometer (~40 cm^2 effective area) (Kudryavtsev and Svertilov, 1985). The X-ray instrument was arranged in such a way that the center of its field of view, averaged over the satellite's rotation period (~120 s), coincided with the spin axis which pointed in the solar direction every 5-7 days. According to the experiment conditions sky areas adjacent to the ecliptic plane (±25° - for instrument beam *FWHM*) were observed and slow (1°/day on average) scanning along the ecliptic was made. The count rates for X-ray photons were measured over the energy ranges 10-50, 25-50, 50-100, 100-200 keV.

From November 1983 to February 1984, the Galactic-center region, where the Galactic X-ray sources concentration is the highest, was within the instrument's field of view. Since each source in that region can be observed as long as 100 days, while virtually continuous measurements of count rates averaged over 10 s were made, the experiment provided favourable conditions for the study of periodic events over a wide range of periods.

DATA ANALYSIS AND RESULTS

The number of periodic events in hour and day ranges of periods: 152, 98, 82, 69, 13, 9.4, 8.2, 8.0, 4.4, 3.4 h were detected as the result of the epoch-folding technique analysis of the data time sets (Kudryavtsev *et al.*, 1988; Kudryavtsev and Svertilov, 1992). To identify some of these periodicities we used information about period, flux, spectrum and location of possible source. The 82 h period was identified naturally with orbital period of the eclipsing binary *4U1700-37*. The 8.2 h period was discovered by Kaluzienskii *et al.*, (1980) in *Cen X-4* X-rays during outburst in May 1979. The 8.2 h periodic process was observed in *Prognoz 9* experiment during the limited time interval when except for *Cen X-4* there were no sources with similar periodicities in X-rays in the instrument FOV. Thus we suppose that this period can be connected just with *Cen X-4* although we have no information about brightness of *Cen X-4* in that time. The 4.4 h periodicity was identified in the same way with the equivalent period of dips in *4U1755-33* X-ray flux (White *et al.*, 1984). The other periodicities mentioned above were hitherto not observed.

Due to the triangle-like form of the instrument beam the error box of the observed periodic source positions could be determined in some cases by the time of the full visibility of the source and the correlation of its amplitude with pointing direction. The error box obtained for the source of 62 h periodicity allowed us to conclude that it may be identified with *Sco X-1* (Kudryavtsev *et al.*, 1989). Recently we have analysed the new data about X-ray sources (including X-ray novas) in *Scorpius* and nearest constellations. As the result of this analysis we have obtained that 62 h periodicity can be alternatively identified with *GRO J1655-40* (*XN Sco 1994*) - an eclipsing binary with an orbital period of 2.62 days which is equal to 62 h in error limits. Basing on all data about 152 h periodicity obtained in the experiment, we concluded that it can be identified most probably with *H1705-25* (*XN Oph 1977*). In spite of that as we know there were no other observations of this source in hard X-rays in 1983, November - 1984, January, we assume that it could be active at that time.

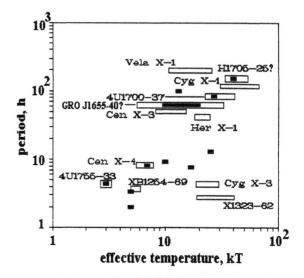

Fig. 1. Period versus *kT* for the periodic sources detected in *Prognoz-9* experiment (filled rectangles) and some other binaries (empty rectangles).

The Period and *kT* Values

The values of periods were determined by the maximums of the peaks in χ^2 versus period periodogramms. The resolution on revealed periods defining by the peak width depends on the attitude of period value and the time of corresponding periodicity observation. This resolution change from ~10% (82 h, 98 h, 152 h) to ~1% (8.0 h). The values of the effective temperature *kT* (for thermal bremsstrahlung spectra) were evaluated for each periodic process. This parameter characterizes the spectral hardness of the periodic flux component (only in the case of eclipsing binary *4U1700-37* it corresponds to the full flux spectrum). The detected periodic source period and *kT* values are presented in Figure 1. The corresponding parameters of some other binaries (only with *kT* ≥5 keV spectra - which can be detected in *Prognoz-9* experiment), for which periodicities were observed only in X-rays, are also shown on the Figure. (The long-term as well as the "short" - with period less than ~2 h periodicities not discussed here.) The detected periodic sources can be separated into two groups: with relatively short (<15 h) and rather long (>60 h) periods (see Figure 1). The sources with rather soft spectra (*kT* <~20 keV), typical for low-massive binaries, dominate in the first group, while the objects of X-ray pulsars type, with more hard spectra (*kT* >25 keV), - in the second.

The Mean Light Curves

The mean light curves of the most intensive detected periodic sources are presented by solid lines in Figure 2. To confirm the reliability of revealed periodicities itself and to distinguish temporal (stochastic) and phase peculiarities (peaks) on mean light curves, we considered besides the mean phase profiles corresponding to the main periods the mean phase profiles corresponding to the "double" periods. The latter ones were divided into two "halves". From the first "halfs" the "odd" profiles (upper dotted lines in Figure 2) and from the second "halves" - the "even" ones (lower dotted lines in Figure 2) were composed. Thus we had folded separately data for non-overlapping time subsets with the same reference point. The similarity of "even" and "odd" phase profile forms confirm the reliability of periodic processes. It is necessary to note, that except the 82 h light curve all other curves are not typically eclipsing. Because most of the small details at mean light curves can be caused by the temporal stochastic variability, we can definitely conclude that they correspond to complicated phase structure only if such details are presented on both "odd" and "even" profiles. Such "repeated" details can be seen at light curves corresponding 62 h, 82 h and 152 h periodicities. The subharmonic of about 1.8 h at 13 h light curve is not confirmed because it is revealed clearly only on the "odd" profile. However, the quasisymmetrical structure of 82 h light curve consisting of the 4 local maximums (at ~0.2, ~0.3 and ~0.7, ~0.8 of the full phase), which can be seen also on the non-overlapping intervals of observations, should be especially noted. Because this light curve was obtained as the result of continuous observations of X-ray binary *4U1700-37* during the time interval containing 8 full orbital cycles, its structure can reflect the peculiarities of compact companion orbital motion.

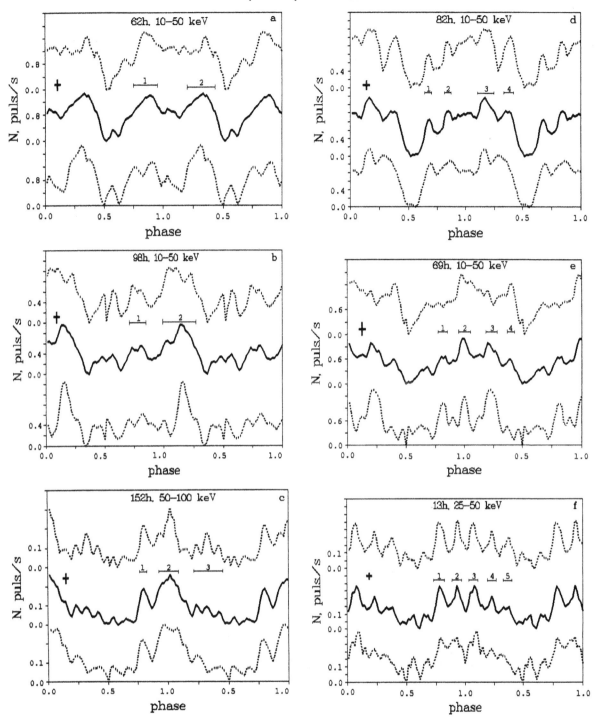

Fig. 2. The mean light curves corresponding to the periodic sources observed during *Prognoz-9* mission: a - 62 h (time of observation 13.11-20.12.1983), b- 98 h (07.12.1983-06.01.1984), c - 152 h (31.10.1983-12.01.1984), d - 82 h (7.12.1983-06.01.1984), e -69 h (07.12.1983-06.01.1984),, f - 13.3 h (13.12.1983-06.01.1984). The curves are presented for the energy ranges where the corresponding periodicities are the most contrast. The lines marked with numbers are enumerating the most prominent peaks at mean light curves.

Possible Identification of 152 h Periodicity with *H1705-25* (*XN Oph 1977*)

The source of 152 h periodicity is the most hard and long-periodic between the bright objects observed in *Prognoz 9* experiment. The periodic component flux value at the energies ~100 keV is no less than 100 mCrab. Only two objects in the error box of this source position can be characterized by such fluxes. They are the X-ray novas of 1977 (*H1705-25* from *HEAO 1 A4* catalogue (Cook *et al.*, (1984)) and 1993. The curves, which are presented in Figure 3, confirm that the 152 h periodicity may be associated with *Nov Oph 1977*. These phase dependencies were obtained by means of epoch-folding technique, and by this the *SSI* initial data (the flux values J in 2-18 keV energy range averaged over ~1.7 h time intervals) was normalised on that averaged flux values J', which present only the trend-like intensity monotonous decrease during the transient decline phase.

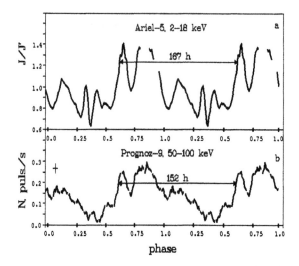

Fig. 3. The mean phase dependencies obtained on the base of *Prognoz 9* (152 h period) - a and *Ariel 5 SSI* (Watson *et al.*, 1978) (167 h period) - b data.

The value 167 h is equal to the 6.7 days period of quasiperiodic component in *Nov Oph 1977* 2-18 keV X-rays, which was supposed by the authors of *SSI* experiment on the base of their own experiment data analysis. The outstanding similarity of the curves on the Figure 3 (especially in view of its complicated structure) should be noted. Because the difference of 152 (±6) h and167 h period values exceeds the errors, it can be concluded, that during the time between both experiments (more than 5 years) a ~10% period drift took place.

DISCUSSION

The various forms of slow periodic sources mean light curves, detected in *Prognoz 9* experiment, testify that slow regular variations of X-ray binaries fluxes can be caused, not only by the companion's orbital motion, but by other factors such as a periodic disturbances inherent to these systems. The accretion disk precession is the most often discussed cause of such periodicities. But other reasons connected with the rotation of binary system companions also can be taken into account for the explanation of light curves shapes. It should be noted, in this way, the similarity of presented light curves and the mean pulse profiles of the fast X-ray pulsars (Nagase, 1989).

REFERENCES

Cook, B.A., A.M. Levine, F.L. Lang, F.A. Primini, and W.H.G. Lewin, HEAO 1 High-energy X-ray Observations of Three Bright Transient X-ray Sources H1705-25 (Nova Ophiuchi), H1743-322, and H1833-077 (Scutum X-1), *Astrophys. J.*, 285, 258 (1984).

Gottlieb, E.W., E.L. Wright, and W. Liller, Optical Studies of Uhuru Sources. XI. A Probable Period for Scorpius X-1 = V818 Scorpii, *Astrophys. J. (Lett.)*, 195, L33 (1975).

Kaluzienski, L.J., S.S. Holt, and J.H. Swank, The 1979 X-ray Outburst of Centaurus X-4,*Astrophys. J.*, 241, 779 (1980).

Kudryavtsev, M.I. and S.I. Svertilov, The X-ray Experiment on Prognoz 9 Satellite, *Mosk. Univ. Phys. Bull.*, 39, 83 (1985).

Kudryavtsev, M.I. and S.I. Svertilov, Periodic Sources of Hard X-rays near the Galactic Center: Periods of the order of Hours in Prognoz 9 Data, *Sov. Astron. Lett.*, 18, 235 (1992).

Kudryavtsev, M.I., Yu.I. Logachev, and S.I. Svertilov, Hard X-ray Sources near the Galactic Center with Periods of several Days: the Prognoz 9 Experiment, *Sov. Astron. Lett.*, 14, 379 (1988).

Kudryavtsev, M.I., N.A. Mamontova, S.I. Svertilov, and E.D. Tolstaya, An Indication of 62-hour Periodicity in Sco X-1 based on Prognoz 9 X-ray Experiment, *Sov. Astron. Lett.*, 15, 466 (1989).

Nagase, F., Accretion-Powered X-ray Pulsars, *Space Sci. Rev.*, 41, 1 (1989).

Watson, M.G, M.J. Ricketts, and R.E. Griffiths, The X-ray Light Curve of Nova Ophiuchi 1977 (H1705-25), *Astrophys. J. (Lett.)*, 221, L69 (1978).

Pergamon

Adv. Space Res. Vol. 22, No. 7, pp. 943–950, 1998
© 1998 COSPAR. Published by Elsevier Science Ltd. All rights reserved
Printed in Great Britain
0273-1177/98 $19.00 + 0.00

PII: S0273-1177(98)00126-4

TIME VARIABILITIES OF X-RAY BINARY STARS

Sigenori Miyamoto

Osaka University of Health and Sport Sciences, Noda 1558-1, Kumatori-cho, Sennangun, Osaka, 590-04, Japan

ABSTRACT

The hard and the soft power law components and the disk blackbody component of the black hole candidate X-ray binaries have their inherent normalized power spectrum densities and phase lags. These short term variabilities are also similar to those of the X-ray energy spectral components of the non-pulsating neutron star X-ray binaries, although X-ray energy spectra of these two kinds of X-ray binaries are different in their high X-ray intensity state. There are two states in the black hole candidate X-ray binaries; the power law hard state and the power law soft state. Some of the black hole candidate-X-ray binaries show large hysteretic behavior between the states and the X-ray intensity. There seems to be no observation of large hysteretic behavior of this sort in the non-pulsating neutron star X-ray binaries. A model is assumed that in the X-ray emitting region of the accretion disk, the power law hard state corresponds to the advection dominated optically thin accretion disk and the power law soft state corresponds to the cooling dominated optically thick accretion disk, and this model is compared with observations and discussed. It is concluded that to describe long term variabilities of X-rays from the black hole candidate-X-ray binaries and the non-pulsating neutron star-X-ray binaries, other factors such as magnetic field seem to be necessary. ©1998 COSPAR. Published by Elsevier Science Ltd.

INTRODUCTION

The black hole candidate X-ray binaries (BHC-XBs) have been known to have two states; the high state and the low state (for the review, see for example Oda, 1977; Liang & Nolan, 1984; Tanaka, 1989). Figure 1 shows schematic summary of the X-ray energy spectra in these two states of the BHC-XBs and the non-pulsating neutron star X-ray binaries (NS-XBs) together with their short term variabilities (Miyamoto, 1993; 1994; van der Klis, 1995).

In the high state, X-rays consist of two components: the disk blackbody component and the soft power law (soft PL) component. The disk blackbody component is emitted from the optically thick accretion disk, and its energy spectrum is almost similar to the blackbody radiation of the temperature (kT) of about 1 keV. The soft power law component has a power law energy spectrum in the X-ray energy region. Its photon index is about -2.2~ 2.7. In the low state, X-ray consists of one energy spectral component: the hard power law component, although this energy spectrum can not be represented by a power law above 500 keV. The photon index of the power law component in the high state is about -2.2 ~ 2.7, and the photon index in the low state is about -1.6 ~ 1.7. These indexes are different from each other and the disk blackbody component usually coexists with the soft power law component in the high state.

As mentioned later, X-rays of some of the BHC X-ray transients increase with the hard power law (hard PL) X-ray energy spectrum, i.e. in the low state. Thus sometimes X-ray flux in the low state becomes larger than that in the high state. So, hereafter sometimes we call the power law hard state (the PL hard state) and the power law soft state (the PL soft state) instead of the low state and the high state, respectively.

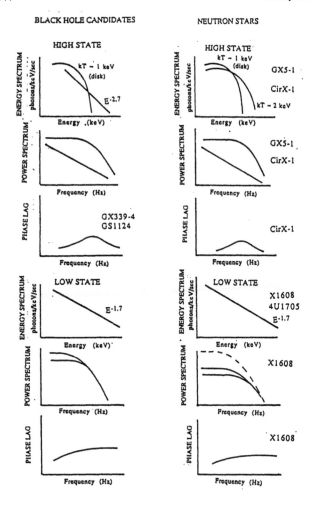

Fig. 1.   A schematic summary of the energy spectra and the short term variability of X-rays from BHC-XBs and NS-XBs (Miyamoto, 1993 & 1994).

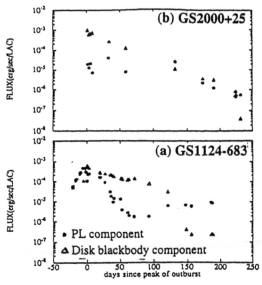

Fig. 2.   Long term variability of the X-ray component of (a) Nova Muscae 1991 or GS1124-683, and (b) GS2000+25 observed with GINGA (Miyamoto, 1993;1994; Terada et al. 1996).

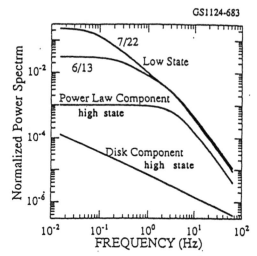

Fig. 3. Typical values of the normalized power spectrum densities of the X-ray energy spectral components of a BHC-XB (GS1124-683) (Miyamoto et al. 1994a).

SHORT TERM VARIABILITY OF X-RAY COMPONENTS

Figure 2-a shows long term variability of the X-ray component of Nova Muscae 1991 (GS1124-683) observed with GINGA (Miyamoto et al., 1993). This nova increased its X-ray flux in the PL hard state for about 10 days, and near its X-ray flux maximum, the hard PL component changed to the soft PL component, and the disk blackbody component increased rapidly. Then the soft power law component decreased and the disk blackbody component became the main component and showed the first peak. After the soft power law component reached its bottom intensity, the disk blackbody component made the second peak. Several tens days afterward, the soft power law component increased again. After some time, the disk blackbody component decreased rapidly and the soft power law component was replaced with the hard power law component. This was the transition from the high state (the PL soft state) to the low state (the PL hard state).

X-rays in the PL hard state and the PL soft state consist of many triangular shots, and larger and longer shots are numerous in the PL hard state (e.g. Miyamoto, 1993;1994). To see variability of X-rays, the power spectrum density normalized to the X-ray intensity; the Normalized Power Spectrum Density (NPSD) has been used (Miyamoto et al., 1991; 1994a).

We found that the NPSDs in the high state can be separated into two components: the flat top noise and the power law noise, and that the flat top noise is due to the soft power law component and the power law noise is due to the disk blackbody component. The PSDs normalized to the corresponding energy spectral components separately have their own inherent values as well as their shapes. We also found that the hard power law component has its inherent NPSD. Typical values of the NPSDs of these three components are shown in Figure 3 (Miyamoto et al., 1994a). These values fluctuate less than a factor of about 4.

Even if the X-ray energy range is different, the NPSDs of the soft power law component are similar. This is also true for the hard power law components. These suggest that high and low energy X-rays of the soft power law components are produced in a small part of the accretion disk at a time, and this is also valid in the hard power law component. However, the NPSDs of the disk blackbody component become larger if the X-ray energy becomes higher. This suggests that the higher energy X-rays of the disk blackbody component are produced in the inner part of the accretion disk than the lower energy X-rays (Miyamoto et al., 1994b).

The phase lag or the time lag of variability between different energy X-rays gives us information about the X-ray production processes in the PL soft state and the PL hard state (Miyamoto et al., 1988; 1991). In the high state, on 1991 January 22, a few days after the first peak of the X-ray flux of a X-ray nova of GS1124-683, we observed a large peak in the phase lag as shown in Figure 1 (Miyamoto et al., 1993). This is the same large peaked phase lag observed in GX339-4, just after the peak of its X-ray outburst in 1988 (Miyamoto et al., 1991). This large peak of the phase lag corresponds to the time lag of about 0.1 sec, and can be explained by a model that soft X-rays are Compton scattered in a large (about 109 cm) hot electron cloud of the temperature (kT) of about 30 keV of the Thomson scattering depth of about 0.5 (Miyamoto et al., 1991).

The phase lag in Nova Muscae 1991 (GS1124-683) increased gradually from 1991 January 11 to 22 (Miyamoto et al., 1994b). Almost simultaneously, on 1991 January 17, a radio burst was observed in GS1124-683 by Ball et al., (1995), and also on 1991 January 20~21, a broad Gaussian line of 474 keV, the positron annihilation line, was observed with GRANAT by Sunyaev et al. (1992). Thus, the production of high energy electrons were confirmed in radio waves, in X-rays and in γ-rays on around 1991 January 22, when the large phase lag was observed in the Nova Muscae 1991.

Very recently, Pound et al. (1995) discovered that the PL soft state existed also in a Seyfert galaxy. As many of active galactic nuclei (AGNs) have the hard PL spectrum, the PL hard state and the PL soft state also exist in the AGNs: the X-ray energy spectra and the variabilities of X-rays seem to be universal. Thus observation of the galactic BHC and the NS X-ray binaries is very important to understand the phenomena not only in the galactic X-ray stars but also in the AGNs.

Non-pulsating neutron star X-ray binaries (NS-XBs) are believed to contain a weak magnetic field neutron star.

X-ray energy spectra and short term variabilities are shown in Figure 1. The energy spectra of X1608 -522 (Mitsuda et al., 1989) and 4U1705-44 (Langmeier et al., 1987) are also shown in Figure 4. In this figure, there are two kinds of the X-ray spectra. One consists of the two thermal spectra: the disk blackbody component and a hard black body component (the NS surface component), when the X-ray flux is high. The other is represented by the hard power law, whose photon index is about -1.7, when the X-ray flux is low (Mitsuda et al., 1989). These are consistent with the observations that in the weakly magnetized NS-XBs, the brightest soft X-ray sources have no hard X-rays larger than 30 keV, while hard power law tails extending up to about 100~200 keV (or more) are likely to arise in X-ray bursters, if they reach sufficiently low intensity state (Barret & Vedrenne, 1994).

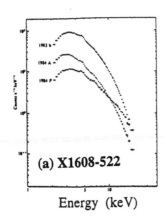

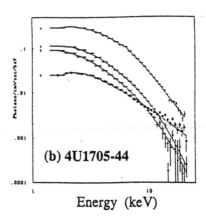

Energy (keV)   Energy (keV)

Fig. 4.   The energy spectra of (a) X1608-522 (Mitsuda et al. 1989) and (b) 4U1705-44 (Langmeier et al., 1987)

The NPSDs are the flat top type when the NS surface component is relatively large. When the NS surface component decreases, the NPSD becomes the power law type. Thus in the high X-ray intensity state of GX5-1, although the energy spectrum is different from that of BHC-XBs, the NPSDs are similar to those of BHC-XBs (Miyamoto, 1993; 1994; Miyamoto et al., 1993; 1994b). The phase lag of GX5-1 was difficult to obtain because of large statistical errors.

Cir X-1 is considered to be a NS-XB. Energy spectra of Cir X-1 is represented by the two component model similar to the case of X1608-522.   Cir X-1 showed the same large peaked phase lag just after its flaring up (Miyamoto et al., 1994b).   At that time, the energy spectrum of Cir X-1 was similar to those of X1608-522 in the high intensity state and the NPSDs was the flat top type (Miyamoto, 1993; 1994). Thus short term variabilities of X-rays from Cir X-1 are quite similar to those of the BHC-XBs in the high state including their phase lag, although the energy spectrum of their higher energy spectrum component (the NS surface component) is different from that of the soft power law component in the BHC-XBs.

The NPSDs of X1608-522 in the low intensity state is shown in Figure 1.   The NPSDs are a little smaller at less than 8 Hz and a little larger at above 8 Hz than those of the NPSDs of BHC-XBs in the low state (Yoshida et al., 1993; Miyamoto et al., 1994b).   The phase lag of X1608-522 in the low intensity state is also similar to those of BHC-XBs in the low state.   The values are about a factor of two smaller than those of BHC-XBs (Miyamoto, 1993;1994).

Comparison of energy spectra and short term variabilities of X-rays from the BHC-XBs and the weak magnetic field NS-XBs is shown in Figure 1.   In the high intensity state, X-ray energy spectra are different between the BHC-XBs and the NS-XBs.   However, the NPSDs and the phase lags are similar in this state.   In the low intensity state, on the other hand, the energy spectrum, the NPSD and the phase lag are all similar between the BHC- and the NS-XBs, and it is curious that there is no X-ray component from the neutron star surface in this state. There is clear correspondence of the energy spectral components between the BHC-XBs and the NS-XBs.

LONG TERM VARIABILITY OF X-RAY COMPONENTS

Until recently, terminology of the high state and the low state has been used to represent the states of BHC-XBs. However, this terminology is not suitable, because GX339-4, Cyg X-1 and GS1124-683 increased their X-ray intensity with a X-ray energy spectrum similar to those in the low state, and near their X-ray flux maximum, these XBs changed their energy spectrum to those in the high state and decreased their X-ray flux in the high state. These X-ray stars showed large hysteretic behavior between the states, which was recognized firstly by Miyamoto et al. (1995).

One example is an old observation of Cyg X-1 with HEAO-3, ARIEL-V and HAKUCHO in 1979 (Figure 2 in Ling et al., 1987). Cyg X-1 increased X-ray flux in the PL hard state (the low state) and the state changed to the PL soft state (the high state) near its X-ray maximum flux ($\sim 7 \times 10^{-8}$ erg/s/cm^2 in 2-100 keV). Then Cyg X-1 increased its disk blackbody component, and decreased its X-ray flux in the PL soft state (the high state). Thus, Cyg X-1 shows a large hysteretic behavior (Miyamoto et al., 1995).

Similar transition was also observed in GX339-4 with HAKUCHO in 1981 (Figure 1 in Maejima et al. 1984). X-ray (2-100 keV) energy flux increased to 1.1×10^{-8} erg/s/cm^2 in the PL hard state, and after the energy spectrum changed to the soft one, soft X-rays increased rapidly. Thus, the large hysteresis was also observed in GX339-4 (Miyamoto et al., 1995).

The long term variation of GS1124-683 or Nova Muscae 1991 observed with GINGA is shown in Figure 2. The increase with the PL hard state observed with GINGA ASM is shown by the dotted line.

As for the BHC X-ray transients, two types of the rise time (the fast-rise-type and the slow-rise-type) (Harmon et al., 1994b) and three ways of transition between the two states have been observed with BATSE (GRO) and GINGA.

In the slow-rise-type, X-ray transits increase their X-ray flux (2~100 keV) with the time constant of about 30 ~ 60 days. GX339-4, Cyg X-1, and GRS 1915+105 (Harmon et al., 1994a;b) belong to this type. In this type, X-ray flux increases in the PL hard state, and at near X-ray maximum flux the state changes to the PL soft state and the disk black body component increase rapidly. Then the X-ray flux decreases gradually and the state changes to the PL hard state again. Thus, the large hysteretic behavior was observed in these X-ray transients.

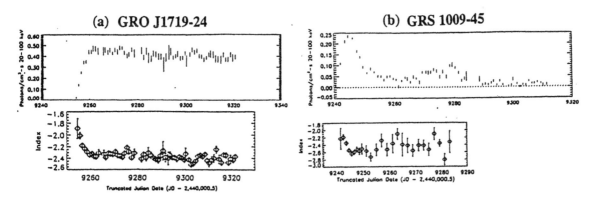

Fig. 5.   Examples of fast rise type X-ray transients.   (a) Fast rise transients which increase the X-ray flux in the PL hard state, change to the PL soft state near the X-ray flux maximum and decrease the X-ray flux and then change to the PL hard state (GRO J1719-24). (b) Fast transients which are always in the PL soft state (GRS1009-45).

In the fast-rise-type, X-ray transits increase their X-ray flux (2~100 keV) with the time constant of about 2~6 days. There seems to be three ways of the transition between the states in this type.   The first is to increase its X-ray flux in the PL hard state, changes to the PL soft state near its X-ray flux maximum and decreases its X-ray flux and then changes to the PL hard state, whose transition is similar to that of GX339-4 and Cyg X-1.   GRO J1719-24 (Harmon et al., 1994b) (see Figure 5-a), GS1124-683, and probably GS2000+25 belongs to this.   The second is the BHC transients which are always in the PL hard state. They increase their X-ray flux in the PL hard state, decrease their X-ray flux also in the PL hard state. GS2023+338 (Tanaka, 1989; Terada et al., 1991; Miyamoto, 1993; 1994) and GRO J0422+32 (Harmon et al., 1994b) belong to this.   GS2023+338 was observed to show X-ray flux saturation (probably the Eddington limit) of 13~16 minutes duration several times in its flare maximum, although its mean X-ray flux was about 1/50 of the saturated X-ray flux (Terada et al., 1994).   The third is those which are always in the PL soft state.   GRS1009-45 belongs to this kind (Harmon et al., 1994b) (see Figure 5-b).

It is of interest to note that there is no BHC X-ray transients which increase their X-ray flux in the PL soft state and change their state to the PL hard state near their X-ray flux maximum.

COMPARISON WITH THE ADVECTION DOMINATED ACCRETION DISK MODEL.

Let us compare long term variabilities of BHC-XBs with a AD & CD disk model that in the X-ray emitting region of the accretion disk, the cooling dominated accretion disk (CD disk) (see Frank, King & Raine 1992 for a review) corresponds to the PL soft state and the advection dominated optically thin accretion disk (AD disk) (e.g. Abramowicz et al., 1988; 1995; Narayan & Yi, 1994, 1995a,b) corresponds to the PL hard state.   The log $(\dot{M}/\dot{M}_{Edd})$ vs log Σ relation of the accretion disk is shown schematically in Figure 6, where Σ is the matter density of the accretion disk and $\dot{M}/\dot{M}_{Edd}$ is the ratio of the matter accretion rate to that at the Eddington limit.

Two rise types of the BHC X-ray transients may be explained as follows. The slow rise transients may be induced by the instability of the accretion disk near the inner edge. On the other hand, most of the fast-rise-type transients may be induced by the accretion disk instability started at larger radii (e.g. Cannizzo, 1998). Some of these may occur when accreting matter falls directly from the companion star to the central part of the accretion disk.   One example of this may be GS2023+338 (Miyamoto, 1993; 1994).

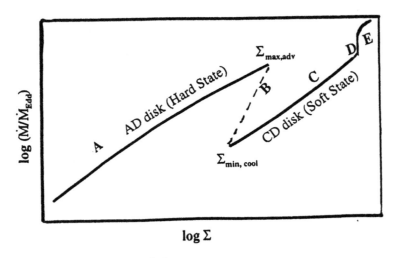

Fig.6.   Schematic relation between log $(\dot{M}/\dot{M}_{Edd})$ and log Σ in the X-ray emitting region of the accretion disk of the black hole.

The hysteretic behavior of BHC-XBs by the AD & CD disk model might be explained as follows. The X-ray intensity of the BHC X-ray transients increases in the AD disk, if the initial matter density of the disk is below the lower surface density limit $\Sigma_{min,cool}$ (see Fig. 6). With increase of the accreting rate, when the surface density becomes higher than the upper density limit $\Sigma_{max,adv}$, the disk changes to the CD disk. The low energy X-rays emitted from the inner edge of the accretion disk are Compton up-scattered in the hot electron clouds existed at the central part of the disk, and become the soft PL component (Miyamoto et al., 1991; 1992). When the matter density of the accreting disk decreases less than $\Sigma_{min,cool}$, the CD disk changes to the AD disk, which is the start of the PL hard state. If the density of the accretion disk does not exceed $\Sigma_{max,adv}$, the X-ray transients are always in the PL hard state. X-ray transients are always in the PL soft state, if these have transited to the CD disk at their high X-ray intensity, and since then the density of the accretion disk has not decreased less than $\Sigma_{min,cool}$.

However, there is one large defect in this model. According to the Narayan and Yi (1995), the value of $\dot{M}/\dot{M}_{Edd}$ corresponding to the point of $\Sigma_{min,cool}$ is estimated to be less than 10^{-10}, which is too small to be compared with the observations mentioned previously. Narayan and Yi (1995) mentioned that there seems to be some reason that the CD disk goes into the AD disk automatically in the region where Σ is less than $\Sigma_{max,adv}$.

In the followings, let's estimate several parameters, assuming that the value of $\Sigma_{min,cool}$ is less than $\Sigma_{max,adv}$, but much larger than the value estimated by Narayan and Yi (1995). We define the parameters f_{AD} and f_{CD}, the fractions of the gravitational energy which are advected with the flow onto the black hole and not emitted as the X-ray luminosity in the AD disk and the CD disk respectively. The value of f_{AD} is expected to be almost one, and f_{CD} is small. We also define the parameters A_{AC} and A_{CA}, the multiplication factors of the accretion rate after the transitions from the AD disk to the CD disk and from the CD disk to the AD disk, respectively. If the CD disk changes to the AD disk, the luminosity will change by a multiplication factor of $A_{CA}(1- f_{AD})/(1- f_{CD})$. In the reverse, the luminosity will change by the factor of $A_{AC}(1- f_{CD})/(1- f_{AD})$. From observed results of the long term variability of X-ray binary stars, luminosities corresponding to the values of $\Sigma_{max,adv}$, $\Sigma_{min,cool}$, and the values of $f_{CD}, f_{AD}, A_{CA}, A_{AC}$ will be estimated.

In the case of GS1124-683 (see Figure 3a), the transition from the PL soft state to the PL hard state occurred at the luminosity of about a factor of 100 less than the Eddington limit in the CD disk. The value of $A_{CA}(1- f_{AD})/(1- f_{CD})$ is estimated to be about 0.2. In the case of GS2000+25 (Fig. 3b), the transition occurred at the luminosity of about a factor of 10000 less than the Eddington limit in the CD disk, and the value of $A_{CA}(1- f_{AD})/(1- f_{CD})$ is estimated to be about 0.4.

The transition from the PL hard state to the PL soft state occurred at the luminosity of 7×10^{-8} erg/s/cm^2 in the case of Cyg X-1 (Ling et al., 1987) and 1.1×10^{-8} erg/s/cm^2 in the case of GX339-4 (Maejima et al., 1984; Miyamoto et al., 1995). Thus the luminosities which correspond to the density $\Sigma_{max,adv}$ in the AD disk are about a factor of 24 less than the Eddington limit luminosity (LEDD) in both Cyg X-1 and GX339-4. The X-ray luminosities did not change largely before and after these transitions. Thus the value of $A_{AC}(1- f_{CD})/(1- f_{AD})$ is of the order of one at the transition from the AD disk to the CD disk at $\Sigma_{max,adv}$.

In the NS-XBs case, the transition from the high intensity state (the CD disk) to the low intensity state (the AD disk) and its reverse were observed in X 1608-522 (Mitsuda et al. 1989) and in 4U1705-44 (Langmeier et al. 1987), respectively. The X-ray energy spectra of these NS-XBs are the two components thermal type in the high intensity state. Transitions in both directions occurred when the X-ray luminosity was about 10^{37} erg/sec, i.e. about 1/10 of L_{Edd}, and there seems to be no large hysteretic behavior such as observed in BHC-XBs. In the low intensity state, the energy spectrum has only one component: the hard power component and there is no X-ray component from the neutron star surface. Thus f_{AD} seems to be quite small, and A_{AC} in BHC-XBs seems to be of the order of one.

Can we explain the quite small value of f_{AD} with the existing AD disk model ? Why is the neutron star surface component not observed in the AD disk of the NS-XBs, if f_{AD} is about one ? Can we explain short term variability by the AD & CD disk model ? Why does the disk blackbody component increase its X-ray flux after the X-ray energy spectrum has changed from the PL hard one to the soft one ? The large peaked phase lags observed in BHC-XBs and NS-XBs can be explained by large hot electron clouds, which are produced by rapid rotation of the magnetic field trapped in the CD disk (Miyamoto and Kitamoto, 1991).   Why does the magnetic field in the AD disk not play an important role to solve above questions ?

REFERENCES

Abramowicz, M.A., B. Czerny, J.P. Lasota, & E. Szuszkiewicz, *APJ*, **332**, 646 (1988).
Abramowicz, M.A., X. Chen, & R.E. Taam, *APJ*, **452**, 379 (1995).
Ball, L., M.J. Kesteven, D. Campbell-Wilson, A.J. Turtle, and R.M.Hjellming, *MNRAS* **273**, 722 (1995).
Barret, D & Vedrenne, G, *ApJS,* **92,** 505 (1994).
Cannizzo, J.K., *APJ,***494**, 366 (1998).
Frank, J., A.R. King, & D.J. Raine, *Accretion Power in Astrophysics*, Cambridge University Press, Cambridge, (1992).
Harmon, B. A., et al., *ApJ Lett.* **425**, L17 (1994a).
Harmon,B. A. et al., *AIP Conf.* **304**, 210 (1994b).
Langmeier, A., M. Sztajno, G. Hasinger, J. Truemper, and M. Gottwald, *ApJ*, **323**, 288 (1987).
Liang, E.P. and P.L. Nolan, *Space Sci. Rev.* **38,** 353 (1984).
Ling, J. C., W.A. Mahoney, Wm.A. Wheaton, & A.S. Jacobson, *ApJ. Lett.*, **321**, L117 (1987).
Maejima,Y., K. Makishima, M. Matsuoka, Y. Ogawara & M. Oda, *ApJ*, **285**, 712 (1984).
Mitsuda, K., H. Inoue, N. Nakamura and Y. Tanaka, *PASJ* **41**, 97 (1989).
Miyamoto, S., *Proceedings of the IIAS workshop on Mathematical Approach to Fluctuations* (vol. **II**) (held in Kyoto, Japan,13-23 September, 1993 by International Institute for Advanced Study, Kyoto, Japan), ed. T.Hida, pp.254-298, World Scientific Publish Co.; Singapore, New Jersey, London, Hong Kong (1995).
Miyamoto, S., *ISAS RN* **548** (1994).
Miyamoto, S., S. Kitamoto, K. Mitsuda and T. Dotani, *Nature* **336**, 450 (1988)
Miyamoto, S. and S. Kitamoto, *Nature* **342**, 773 (1989).
Miyamoto, S. and S. Kitamoto, *ApJ*, **374**, 741 (1991).
Miyamoto, S., K. Kimura, S. Kitamoto, T. Dotani and K. Ebisawa, *ApJ*, **383**, 784 (1991)..
Miyamoto, S., S. Kitamoto, S. Iga, H. Negoro, K. Terada, *APJ Lett.*, **391**, L21 (1992).
Miyamoto, S., S. Iga, S.Kitamoto and Y. Kamado, *APJ Lett.*, **403**, L39 (1993).
Miyamoto, S., S. Kitamoto, S. Iga, K. Hayashida & K. Terada, *ApJ*, **435**, 398 (1994a).
Miyamoto, S., S. Kitamoto, Y. Kamado, K. Hashimotodani, K. Hayashida, K. Terada, & H. Negoro, in *New Horizon of X-ray Astronomy*, ed.Makino, F. & Ohashi, T. , p47, Tokyo, Universal Academy Press, Inc, (1994b).
Miyamoto, S., S. Kitamoto, K. Hayashida, W.Egoshi, *APJ Lett.* **442**, L13 (1995).
Narayan, R., & I.Yi, *ApJ Lett.*, **428**, L13 (1994).
Narayan, R., & I. Yi, *ApJ*, **444**, 231 (1995a)
Narayan, R., & I. Yi, *ApJ,* **452,**710 (1995b).
Oda, M. *Space Sci. Rev.* **20,** 757 (1977)..
Pounds, K.A., C. Done and J.P. Osborne, *MNRAS*, **277**, L5 (1995).
Sunyaev, R.A. et al., *ApJ Lett.*, **389**, L75 (1992).
Tanaka Y. 1989, in *Proc. 23rd ESLAB Symp.on Two Topics in X-ray Astronomy*, ed. J. Hunt, B. (1989).
Terada, K., S. Miyamoto, S.Kitamoto, H. Tsunemi, K. Hayashida, *Proc. Frontiers of X-ray Astronomy*, ed Y.Tanaka & K. Koyama (Universal Academy Press, Tokyo) p323 (1991).
Terada, K., S. Miyamoto, S.Kitamoto, W. Egoshi, *PASJ*, **46**, 677 (1994).
van der Klis, M., in *X-ray Binaries*, ed.W.H.G.Lewin, J.Van Paradijs, & E.P.J. Van der Heuvel, Cambridge University Press (1995).
Yoshida, K., K. Mitsuda, K.Ebisawa, Y.Ueda, R. Fujimoto, T. Yaqoob, *PASJ*, **45**, 605 (1993).

Pergamon

Adv. Space Res. Vol. 22, No. 7, pp. 951–960, 1998
© 1998 COSPAR. Published by Elsevier Science Ltd. All rights reserved
Printed in Great Britain
0273-1177/98 $19.00 + 0.00

PII: S0273-1177(98)00127-6

HYDRODYNAMICS OF ACCRETION ONTO BLACK HOLES

M.-G. Park[1,2] and J. P. Ostriker[1]

[1] *Princeton University Observatory, Princeton, NJ 08544, USA*
[2] *Kyungpook National University, Department of Astronomy and Atmospheric Sciences, Taegu 702-701, KOREA*

ABSTRACT

Spherical and axisymmetric accretion onto black holes is discussed. Physical processes in various families of solutions are explained and their characteristics are summarized. Recently discovered solutions of axisymmetric flow provide us with various radiation efficiency and spectrum, which may successfully model diverse accretion systems. Possible role of preheating is also speculated. The various families of solutions can be plotted as trajectories on the plane of (l,e) [or $(l,\dot{m})$] where l is the luminosity in units of the Eddington luminosity, $\dot{m}$ is the similarly defined mass accretion rate and e is the efficiency defined by $e \equiv L/\dot{M}c^2 = l/\dot{m}$. We discuss the domains on these planes where solutions are known or expected to be unstable to either spherical or non-spherical perturbations. Preliminary analysis indicates that a preheating instability will occur along the polar direction of the advection dominated flow for $e \gtrsim 10^{-2}$.
©1998 COSPAR. Published by Elsevier Science Ltd.

INTRODUCTION

The generic accretion flow can be broadly classified as (quasi) spherical or axisymmetric one. The angular momentum of the flow and its transfer processes will determine whether the flow is spherical or not. True spherically symmetric flow is possible only when the angular momentum is rigorously zero. However, when the angular momentum is small enough and there exists a process that can effectively remove or transport the angular momentum, then the flow maintains a roughly spherical shape (Loeb and Laor, 1992; Narayan and Yi, 1995a). If the angular momentum is significant, the flow becomes rotationally supported and is flattened to disk.

Until recently, most works on spherical accretion flow have focused on zero angular momentum flow and those on axisymmetric flow on thin flat disk. Purely spherical flow is physically well defined and has high degrees of symmetry, which make it possible to treat hydrodynamics, radiative transfer and various gas-radiation microphysics rather accurately. Yet it might be too ideal for real astrophysical situations. The disk flow produces more radiation for a given mass accretion and the solution is well understood in some parameter ranges except the angular momentum transport process remains uncertain. However, the true solution of the disk flow requires solving the two-dimensional hydrodynamic and radiation transport equations, a quite daunting challenge even with today's enormous computing power.

In this review, we use the dimensionless luminosity $l \equiv L/L_E$ and the dimensionless mass accretion rate $\dot{m} \equiv \dot{M}/\dot{M}_E$ where L is the luminosity of the accretion flow, L_E the Eddington luminosity, $\dot{M}$ the mass accretion rate and $\dot{M}_E$ the Eddington mass accretion rate, $\dot{M}_E \equiv L_E/c^2$. In some literature, the Eddington accretion rate divided by the radiation efficiency $e \equiv L/(\dot{M}c^2) = l/\dot{m}$ is used as the basic unit. In accretion onto a black hole, e can be arbitrary unlike in accretion onto a neutron star.

SPHERICAL FLOW

The first study of spherical accretion onto compact objects dates back more than forty years (Bondi, 1952). In this classic work, hydrodynamics of polytropic flow is studied within the Newtonian dynamics, and it is found that either a settling or transonic solution is mathematically possible for the gas accreting onto compact objects. Especially, the accretion rate is highest for the transonic solution. Relativistic version of the same problem was solved by Michel (1972) twenty years later, after the discovery of celestial X-ray sources. He showed that accretion onto the black hole should be transonic.

However, the real accretion necessarily involves radiative processes, which make the flow non-adiabatic and, at the same time, more diverse and interesting. In general, compression of the infalling gas raises the temperature of the accreting gas and then radiation is produced by the radiative cooling of heated gas, which in turn interacts with the other part of the flow via scattering and absorption, thereby heats or cools the upstream gas. So understanding the accretion flow involves simultaneously solving for the gas and radiation field for given outer boundary conditions. This is particularly important for accretion onto black holes because the lack of the hard surface in a black hole makes the radiation efficiency unknown a priori. In accretion onto neutron star, the surface potential of the star determines how much luminosity is generated for a given accretion rate, regardless of the details in radiative and hydrodynamic processes. In accretion onto black hole, the radiation efficiency is determined by the state of the gas at all radii. If the gas is kept at low temperature, most of the gravitational potential energy would be lost to the hole in the form of gas kinetic energy. If the gas is kept at high temperature by the radiative heating or other processes, the luminosity is increased and less kinetic energy is lost to the hole.

Spherical accretion onto a black hole is generally specified by the mass accretion rate and by the luminosity and the spectrum of the radiation field. As long as two body processes are important, the solutions become scale free and are applicable to black holes of arbitrary mass (Chang and Ostriker, 1985). These scale-free solutions then depend primarily on the dimensionless parameters l and $\dot{m}$ or l and e (Ostriker et $al.$, 1976). However, if we require the self-consistency between the gas and the radiation field, only certain combinations of the accretion rate and the luminosity (plus spectrum) are allowed, that is a given type of solution is described by a line on the $(l, \dot{m})$ plane. Now we describe these solutions one by one in the order of increasing $\dot{m}$.

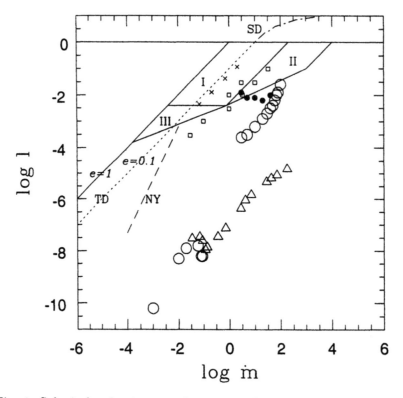

Fig. 1. Spherical and axisymmetric accretion flow solutions in luminosity vs mass accretion rate plane.

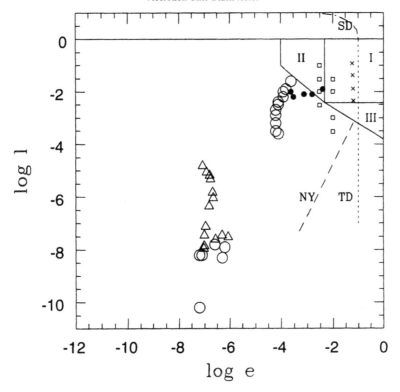

Fig. 2. Spherical and axisymmetric accretion flow solutions in luminosity vs radiation efficiency plane.

Solutions with $\dot{m} \ll 1$

Since the electron scattering optical depth from infinity down to the horizon is roughly given by $\tau_{es} = \dot{m}(r/r_s)^{-1/2}$ where r_s is the Schwarzschild radius, $\dot{m} \ll 1$ flow is optically thin to scattering, and whatever photons produced escape without difficulty. Any kind of interaction between photons and gas particles is minimal. Gas is freely falling so $v \propto (r/r_s)^{-1/2}$ and its temperature has the virial value, $T \propto (r/r_s)^{-1}$, until electrons become relativistic (Shapiro, 1973). In Newtonian approximation, $\gamma = 5/3$ accreting flow has a constant Mach number, because the sound speed is increasing as fast as the infall velocity (Bondi, 1952).

But in the relativistic regime, the sound speed or the electron temperature is increasing less rapidly and the flow becomes supersonic with increasing Mach number for $T \gtrsim 10^9$ K (Shapiro, 1973; Park, 1990a). The temperature can reach as high as few 10^{10} K, producing a hard relativistic bremsstrahlung spectrum. Yet the luminosity is quite small due to the very low density. Any radiative heating, especially Compton heating, for this family of solutions can be ignored because of the low luminosity. Typical solutions are marked as *large empty circles* in $\dot{m} < 1$ region of Figures 1 and 2 and their radiation efficiency e is less than 10^{-8} (Park, 1990a).

Low-Temperature Solutions with $\dot{m} \gtrsim 0.1$

As $\dot{m}$ approaches 0.1, bremsstrahlung and atomic cooling becomes more efficient and gas cools down $\sim 10^4$ K, the equilibrium temperature of the surrounding gas. Depending on the mass of the hole, the flow can form an effectively optically thick[1] core with blackbody radiation field inside. This regime represents the transition from adiabatic to non-adiabatic flow (Park, 1990a; Nobili *et al.* , 1991).

---

[1] $\tau^* \equiv (3\tau_{es}\tau_{abs})^{1/2} \gg 1$ where τ_{abs} is the absorption optical depth.

954   M.-G. Park and J. P. Ostriker

As the gas density at the boundary increases further, so does the mass accretion rate, and the flow becomes effectively optically thick out to a larger radius. The flow is maintained at a low temperature. These solutions are quite similar to those in a stellar interior except that the flow is infalling by the gravity and PdV work is the source of the luminosity (Flammang, 1982, 1984; Soffel, 1982; Blondin, 1986; Nobili et al., 1991). Typical solutions are shown in Figures 1 and 2 as *large triangles* (Nobili et al., 1991).

An important and interesting radiation transport process called radiation trapping happens inside certain radius where $v > c/(\tau_{es} + \tau_{abs})$ is satisfied. When the diffusion speed of photons is slower than the flow's bulk velocity, most of the photons are carried inward with the flow (Begelman, 1978). The correct treatment of this process requires relativistic radiative transport equations, at least up to the order $(v/c)^1$ (Flammang, 1982, 1984; Park, 1990a; Nobili et al., 1991; Park, 1993). In relativistic flow, especially in an optically thick one, the flux seen by the observer comoving with the flow and the flux seen by the stationary observer (relative to the coordinate) should be carefully distinguished. The momentum transferred to the gas from the radiation is closely linked to the former, a comoving-frame flux, and the luminosity seen by the observer at infinity to the latter, a fixed-frame flux. (Mihalas and Mihalas, 1984; Park, 1993). For an accretion flow, the comoving flux is always larger than the fixed frame flux and should be less than the Eddington flux for steady inflow (Park and Miller, 1991). This radiation trapping and the low temperature of the flow make this family of solutions very inefficient radiators with $e \sim 10^{-7}$.

High-Temperature Solutions with $\dot{m} \gtrsim 0.1$

All this would change if gas can be preheated by outcoming hard radiation produced at smaller radii. The existence of higher temperature, higher luminosity accretion solution due to this preheating was first proved by Wandel et al. (1984) in simplified treatment. A more accurate and relativistic treatment show that these higher luminosity and higher efficiency steady-state solutions exist only for $3 \lesssim \dot{m} \lesssim 100$ (Park and Ostriker, 1989; Park, 1990a,b; Nobili et al., 1991). So there exist two different families of solutions for given $\dot{m}$, low temperature and high-temperature one (Figure 1).

The gas temperature can reach up to $10^9 \sim 10^{10}$ K; the flow is optically thick to scattering, and the radiation trapping occurs near the hole. However, it is optically thin to absorption, $\tau^* \ll 1$. Bremsstrahlung photons produced in the inner region are upscattered by hot electrons around, which subsequently heat cool electrons in the outer part by Comptonization. These solutions are plotted in Figures 1 and 2 as *large empty circles* (Park, 1990a; Nobili et al., 1991). These solutions are much more luminous, $l \simeq 10^{-4} - 10^{-2}$, and produce much harder photons than the low-temperature solutions of the same mass accretion rate. Still, they are much less efficient radiators, $e \sim 10^{-4}$, than the thin disk.

There are ways to increase the radiation efficiencies. The dissipational heating of turbulent motion and the magnetic field reconnection (Mészáros, 1975; Maraschi et al., 1982) and self-consistently sustained high electron-positron pair production (Park and Ostriker, 1989) are two examples. Solutions of the former type are shown as *small squares* (Maraschi et al., 1982) with e as high as ~ 0.1 and the latter type as *small filled circles* (Park and Ostriker, 1989) in Figures 1 and 2. They constitute yet another families of solutions for given $\dot{m}$. Though these solutions are quite attractive because of the high efficiency and high temperature, they are quite likely to be subject to the preheating instability described below. The reason that these solutions are found in steady-state calculation at all is that either preheating is ignored (Mészáros, 1975; Maraschi et al., 1982) or only the inner part of the flow is considered (Park and Ostriker, 1989). It is possible to maintain the steady flow under significant preheating by having a shock (Chang and Ostriker, 1984). However, no self-consistent solutions with a steady shock is yet constructed, although there is some indication that such solution might exist around $\dot{m} \sim 1$ or $\dot{m} \sim 100$ (Park, 1990b; Nobili et al., 1991).

Preheating

When radiation and gas interact, both momentum and energy are transferred from one to the other. The luminosity of a given spherical accretion flow has a well known upper limit above which photons would give the gas particles too much momentum to infall steadily. When gas is fully ionised and opacity is dominated by the Thomson scattering, this limit is the well-known Eddington luminosity.

Ostriker et al. (1976) found another limit based on the energy transfer. They found that if the gas is preheated too much around the sonic point, it gets too hot to accrete in steady-state. When the gas temperature at the sonic point is 10^4 K and the Compton temperature of the preheating radiation is 10^8 K, the disruption of the accretion can occur at l less than 0.01. This is why Park (1990a,b) and Nobili *et al.* (1991) could not find high-temperature self-consistent solutions for $\dot{m} \lesssim 3$ and $\dot{m} \gtrsim 100$: in the former parameter region, the photon energy of the preheating radiation is too high and in the latter, the luminosity is too high.

This disruption of flow by preheating happens in a square region labeled I in Figures 1 and 2. The exact location of the boundary depends on various conditions of the gas and the radiation field as well as the shape of the cooling curve. Under certain conditions, the region could be much smaller than the one shown in Figures 1 and 2 (Ostriker *et al.* , 1976; Cowie *et al.* , 1978; Bisnovatyi-Kogan and Blinnikov, 1980; Stellingwerf and Buff, 1982; Stellingwerf, 1982; Krolik and London, 1983).

The time-dependent behavior of the preheated flow was investigated by Cowie *et al.* (1978) with analytic analysis and hydrodynamic simulations (see also Grindlay, 1978). They found two distinct type of time dependent behaviors both inside and near the boundaries of region I. High luminosity, low efficiency flow (region II in Figs. 1 and 2) develops preheating inside the sonic point, which produces recurrent flaring in short time intervals, $\sim 100(M/M_\odot)$ sec, with its average luminosity similar to the steady-state solutions. On the other hand, in higher efficiency, lower luminosity flow (region III in Figs. 1 and 2) preheating outside the sonic point induces longer time scale, $\sim 10^9(M/M_\odot)$ sec, changes in accretion rate and luminosity. Even the solution outside the preheating regime appears to have short time variability due to the preheating (Zampieri *et al.* , 1996).

The fact that high luminosity, high efficiency steady-state solutions would suffer the preheating instability implies the time-dependent nature of these solutions. As the luminosity and efficiency of the accretion increase, we expect more variability in the flow and in the emitted radiation in general.

AXISYMMETRIC ACCRETION

Thin Disk

Until recently most works on non-spherical accretion have focused on the thin disk accretion (see Pringle, 1981, Treves *et al.* , 1988, Chkrabarti, 1996 for reviews). Since the pioneering work by Shakura and Sunyaev (1973), Pringle and Rees (1972), and Novikov and Thorne (1973), thin disk models with so-called α viscosity have been successfully applied to many astronomical objects including various X-ray sources (see Frank *et al.* , 1985 for reviews).

Thin disk accretion model has merits of being simple: the equations become one-dimensional (in radius) and all physical interaction can be described by local quantities. Gas is rotating with Keplerian angular momentum, which is transported radially by viscous stress. The gravitational potential energy is locally converted to heat by this viscous process. The gas cools by radiating in the vertical direction, thereby not affecting any other part of the flow. The radiation efficiency depends only on the location of the inner disk boundary and is generally ~ 0.1. So any thin accretion disk model lies on the *dotted line* $e = 0.1$, denoted as TD, in Figures 1 and 2. However, its application cannot be extended to high luminosity systems due to the instability at a high mass accretion rate (Lightman and Eardley, 1974; Shakura and Sunyaev, 1976; Pringle, 1976; Piran, 1978) nor to hard X-ray sources due to the low temperature of the disk. Although Shapiro *et al.* (1976) discovered another family of thin disk solutions that have very high electron temperature, they are also found to be thermally unstable (Pringle, 1976; Piran, 1978).

Slim Disk

If the dimensionless mass accretion rate $\dot{m}$ approaches e^{-1}, the vertical height of the disk becomes comparable to the radius and the disk is not thin any more. In this regime, the radial motion of the gas becomes important and the angular momentum of the gas is below the Keplerian value unlike in thin disk. Abramowicz *et al.* (1988) improved over the thin disk approach by incorporating the effect of pressure in the radial motion to find new type of solutions in high $\dot{m}$ regime. They are stable against the viscous and thermal instabilities. Viscously dissipated energy can now be advected into the hole along with the gas in addition to being radiated away through the surface of the disk. They named these solutions 'slim disk' because the disk is not thin, yet not so thick that vertically integrated equations are valid. This work shows that disk accretion too can have efficiency other than ~ 0.1. These solutions are represented in Figures 1 and 2 as the *dot-dashed curve* (SD) at the end of $e = 0.1$ line (Szuszkiewicz *et al.* , 1996).

Self-Similar Flow

To get the slim disk solutions, Abramowicz *et al.* (1988) had to explicitly integrate the hydrodynamic equations that have a critical point. However, recently Narayan and Yi (1994) found that the equations admit self-similar solutions if the advective cooling is always a constant fraction of the viscous dissipation. These solutions are very simple and at the same time very restrictive. The density, velocity, angular velocity, and the total pressure are simple power laws in radius: $\rho \propto r^{-3/2}$, $v \propto r^{-1/2}$, $\Omega \propto r^{-3/2}$, and $P \propto r^{-5/2}$. Unfortunately the temperature of these solutions is always close to the virial value and the vertically integrated equations may not be valid.

They addressed this question in the next work (Narayan and Yi, 1995a). By assuming the self-similar form of physical quantities in radius, the axisymmetric two-dimensional hydrodynamic equations are reduced into one-dimensional equations in polar angle only, and easily solved. The solutions show a radial velocity of a few percent of free-fall value at the equatorial plane and zero radial velocity on the pole. So the gas is preferentially accreted along the equatorial plane. However, the flow is always subsonic because the pressure is close to the virial value. Since only the transonic accretion is allowed, this makes the solution inapplicable close to the hole[2].

One important result of this work is that physical quantities calculated from the vertically integrated equations generally agree with the polar angle averaged quantities. This justifies using vertically integrated one-dimensional equations, i.e., slim disk equations, even to the two-dimensional accretion flow. But rigorously, this convenience applies only to the self similar solutions.

Due to the simplicity of the solutions, many diverse microphysics can be incorporated into these solutions without much difficulty (Narayan and Yi, 1995b). The typical solutions with low $\dot{m}$ have efficiency much smaller than the standard thin disk (dashed line in Figs. 1 and 2). These advection-dominated accretion models with low $\dot{m}$ successfully explain various low luminosity sources which have been hard to model with highly efficient thin disk, like Sgr A* (Narayan *et al.* , 1995), NGC 4258 (Lasota *et al.* , 1996), and X-ray transients (Narayan *et al.* , 1996; Chen and Taam, 1996). Typical solutions are denoted as NY in Figures 1 and 2 (Narayan and Yi, 1995b). All these models have most of the gravitational potential energy of the accreted matter advected into the hole without ever being converted to the radiation. So these flows are quite similar to the almost adiabatic, high temperature, low $\dot{m}$ spherical solutions of Shapiro (1973). In a way it should be because the above self-similar forms of various physical quantities are exactly those in the non-relativistic adiabatic spherical accretion flow. In fact, optically thin two-temperature spherical accretion with a magnetic field also produces a spectrum that roughly agrees with that of Sgr A* (Melia, 1992, 1994). Another interesting feature of these models is that because the total pressure is a constant fraction of the virial value, the flow has to be two-temperature and/or some magnetic field should exist, otherwise the radiation spectrum will be simply that of the free-fall, virial temperature plasmas.

Unified Description

All these seemingly different solutions are actually just different families of solutions for the axisymmetric accretion flows with angular momentum. In each family, appropriate assumptions are made to ease the difficulty of solving the relevant equations. Specific name is attached to each family to describe the physical characteristics of the flow, e.g., thin disk, slim disk, and advection dominated flow. Although slim disk solutions and self-similar flow solutions are the results of different approach—the former from the one-dimensional approach and the latter from the self-similarity in two-dimensional flow—their final results agree in general, albeit not in detail, and the description of these axisymmetric flows in consistent manner has become finally possible (Chen *et al.* , 1995).

A very revealing way of presenting these solutions is to look at the relation between the mass accretion rate $\dot{m}$ and the surface mass density Σ at a given radius. Different families of solutions appear as specific curves in $\dot{m}$ vs Σ plane (Figure 3). [3] Since the shape of the curves depends on the mass of the hole and the value of the viscosity parameter α, we describe a specific case where $M = 10\,\mathrm{M}_\odot$ and $\alpha = 0.01$ (Chen *et al.* , 1995). Solutions can be broadly separated into (vertically) effectively optically thin and thick flows. The left *dotted*

---

[2] However, recent global solutions of slim disk equations show that self-similar solutions describe the flow reasonably well at intermediate radii (Narayan *et al.* 1997; Chen *et al.* , 1997).

[3] The solutions are for $\alpha = 0.01$, $M = 10\,\mathrm{M}_\odot$, and $r = 20GM/c^2$. The surface mass density Σ is in g/cm^3.

curve (SLE,NY) in Figure 3 represents the former and the right **S** shape *solid curve* (GTD,RTD,SD) the latter. The former exists only for $\dot{m} \ll 1$ and the latter for any value of $\dot{m}$.

The lower (GTD) and middle (RTD) branch of **S** curve are the classic gas pressure dominated and radiation pressure dominated thin disk solutions of Shakura and Sunyaev (1973). In both, radiative cooling dominates over the advective one, and the lower branch is thermally and viscously stable while the middle branch is unstable to both type of instability. The upper (SD) branch of **S** curve is the original slim disk solution in which the advective cooling dominates and stabilizes the radiation pressure dominated flow (Abramowicz *et al.*, 1988). This solution is not geometrically thin. The disk scale height can be comparable to the cylindrical radius, especially when $\dot{m} > e^{-1}$. The temperature of these **S** curve solution is not high, $\lesssim 10^8$ K, and the radiation spectrum will be the superposition of modified blackbodies at different temperatures. However, they are very efficient radiators, $e \sim 0.1$.

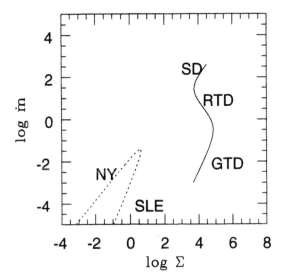

Fig. 3. Various axisymmetric accretion solutions in accretion rate vs surface mass density plane.

The right (SLE) branch of the effectively optically thin solutions represent the two-temperature hot disk solutions of Shapiro *et al.* (1976), which are viscously stable yet thermally unstable (Pringle, 1976; Piran, 1978) and radiative cooling dominated. The left (NY) branch is the advection dominated flow of Narayan and Yi (1994) and Abramowicz *et al.* (1995), in which most of the viscously generated heat is advected into the hole stabilizing the flow. The gas temperature is very high, $\sim 10^9$ K, and the radiation spectrum will be that of Comptonized synchrotron and bremsstrahlung. The radiation efficiency of the right branch is always ~ 0.1 and that of the left branch has roughly $l \propto \dot{m}^2$ and $e \propto \dot{m}$ as in optically thin spherical accretion (Narayan and Yi, 1995b).

As in spherical accretion, there can be other interesting family of solutions if we consider the pair production (Kunsunose and Mineshige 1992 and references therein) or various type of shocks (Chakrabarti, 1996 and references therein), which we will not discuss further.

Preheating and Outflows

One interesting aspect of axisymmetric accretion unexplored so far is the effect of preheating. In thin disk flow, any part of the disk is free from the radiation produced in the other parts of the disk either due to the geometry (vertical direction) or due to the high absorption optical depth (radial direction). But in true two-dimensional flow, hard radiation generated at the inner part of the flow should go through the outer part. This radiative heating will be more complex than in a spherical flow. For example, the self-similar flow of Narayan and Yi (1995a) has zero radial velocity along the pole, around which viscous heating, radiative cooling, and advective cooling are very small compared to those in the equatorial region. So this part of the flow can be very easily heated by Comptonization to a high temperature, possibly higher than the virial value. It is quite plausible that this could develop some interesting phenomena, like relaxational oscillations seen in spherical flow (Cowie et al. , 1976) or simply outflow. Two-dimensional flow with steady accretion onto the equatorial plane and time-dependent accretion or outflow in the polar direction will be quite useful in explaining some high energy sources. We know that time-dependence and outflow are the norm rather than the exception in these sources. Thus the potential instability of the self-similar solutions to preheating in the polar region should be seen as a virtue of the solution. A preliminary analysis indicates that a preheating instability will first occur along the polar direction for $e \gtrsim 10^{-2}$. Two-dimensional numerical hydro simulations show that accretion and outflow can coexist (Chakrabarti and Molteni, 1993; Molteni et al. , 1994; Ryu et al. , 1995; Molteni et al. , 1996). Preheating might be yet another ingredient in the outflow and the time variabilities in accreting flows.

CONCLUSION

Rapid and exciting recent developments in the theory of accretion, especially on the viscous axisymmetric flow, are the significant steps toward understanding various astronomical sources that are believed to be powered by the accretion onto black holes. Although the theory is far from complete and we may have to wait many years to understand the real three-dimensional hydrodynamics and radiative transfer, it already is rather successful in explaining most X-ray sources. In addition, the incomplete part of our understanding could be the key to the remaining mysteries since the most attractive current quasi spherical solutions are likely to be unstable to the formation of (unsteady) jets.

ACKNOWLEDGMENTS

We thank Xingming Chen and Insu Yi for useful discussions, especially Xingming Chen for kindly providing the data for Figure 3. This work is partly supported by the KOSEF grant 971-0203-013-2, Korea Research Foundation, and NSF grant AST 94-24416.

REFERENCES

Abramowicz, M. A., X. Chen, S. Kato, J.-P. Lasota, and O. Regev, Thermal Equilibria of Accretion Disks, Astrophys. J., **438**, L37 (1995).

Abramowicz, M. A., B. Czerny, J. P. Lasota, and E. Szuszkiewicz, Slim Accretion Disks, Astrophys. J., **332**, 646 (1988).

Begelman, M. C., Black Holes in Radiation-dominated Gas - An Analogue of the Bondi Accretion Problem, Mon. Not. R. Ast. Soc., **184**, 53 (1978).

Bisnovatyi-Kogan, G. S., and S. I. Blinnikov, Spherical Accretion onto Compact X-ray Sources with Preheating - No Thermal Limit for the Luminosity, Mon. Not. R. Ast. Soc., **191**, 711 (1980).

Blondin, J. M., Hypercritical Spherical Accretion onto Compact Objects, Astrophys. J., **308**, 755 (1986).

Bondi, H., On Spherically Symmetrical Accretion, Mon. Not. R. Ast. Soc., **112**, 195 (1952).

Chang, K. M., and J. P. Ostriker, Standing Shocks in Accretion Flows onto Black Holes, Astrophys. J., **288**, 428 (1985).

Chakrabarti, S. K., Accretion Processes on a Black Hole, Phys. Rep., **266**, 229 (1996).

Chakrabarti, S. K., and D. Molteni, Smoothed Particle Hydrodynamics Confronts Theory: Formation of Standing Shocks in Accretion Disks and Winds around Black Holes, Astrophys. J., **417**, 671 (1993).

Chen, X., M. A. Abramowicz, and J.-P. Lasota, Advection-dominated Accretion: Global Transonic Solutions, Astrophys. J., **476**, 61 (1997).

Chen, X., M. A. Abramowicz, and J.-P. Lasota, R. Narayan, and I. Yi, Unified Description of Accretion Flows around Black Holes, Astrophys. J., **443**, L61 (1995).

Chen, X., and R.E. Taam, The Spectral States of Black Hole X-Ray Binary Sources, Astrophys. J., **466**, 404 (1996).

Cowie, L. L., J. P. Ostriker, and A. A. Stark, Time-Dependent Spherically Symmetric Accretion onto Compact X-Ray Sources, Astrophys. J., **226**, 1041 (1978).

Flammang, R. A., Stationary Spherical Accretion into Black Holes. II - Theory of Optically Thick Accretion, Mon. Not. R. Ast. Soc., **199**, 833 (1982).

Flammang, R. A., Stationary Spherical Accretion into Black Holes. III Optically Thick Accretion in Particular Cases, Mon. Not. R. Ast. Soc., **206**, 589 (1984).

Frank, J., A. R. King, and D. J. Raine, Accretion Power in Astrophysics, Cambridge U. Press, Cambridge (1985).

Grindlay, J. E., Thermal Limit for Spherical Accretion and X-Ray Bursts, Astrophys. J., **221**, 234 (1978).

Krolik, J. H., and R. A. London, Spherical Accretion onto Quasars, *Astrophys. J.*, **267**, 18 (1983).

Kusunose, M., and S. Mineshige, Hot Accretion Disks with Nonthermal Pairs, *Astrophys. J.*, **440**, 100 (1995).

Lasota, J.-P., M. A. Abramowicz, X. Chen, J. Krolik, R. Narayan, and I. Yi, Is the Accretion Flow in NGC 4258 Advection Dominated?, *Astrophys. J.*, **462**, 142, (1996).

Lightman, A. P., and D. M. Eardley, Black Holes in Binary Systems: Instability of Disk Accretion, *Astrophys. J.*, **187**, L1 (1974).

Loeb, A., and A. Laor, Accretion Flows near Black Holes Mediated by Radiative Viscosity, *Astrophys. J.*, **384**, 115 (1992).

Maraschi, L., R. Roasio, and A. Treves, The Effect of Multiple Compton Scattering on the Temperature and Emission Spectra of Accreting Black Holes, *Astrophys. J.*, **253**, 312 (1982).

Melia, F., An Accreting Black Hole Model for Sagittarius A*, *Astrophys. J.*, **387**, L25 (1992).

Melia, F., An Accreting Black Hole Model for Sagittarius A*. II: A Detailed Study, *Astrophys. J.*, **426**, 577 (1994).

Mészáros, P., Radiation from Spherical Accretion onto Black Holes, *Astr. Astrophys.*, **44**, 59 (1975).

Michel, F. C., Accretion of Matter by Condensed Objects, *Astrophys. Space Sci.*, **112**, 195 (1972).

Mihalas, D., and B. W. Mihalas, *Foundations of Radiation Hydrodynamics*, Oxford University Press, Oxford (1984).

Molteni, D., G. Lanzafame, and S. K. Chakrabarti, Simulation of Thick Accretion Disks with Standing Shocks by Smoothed Particle Hydrodynamics, *Astrophys. J.*, **425**, 161 (1994).

Molteni, D., D. Ryu, and S. K. Chakrabarti, Numerical Simulations of Standing Shocks in Accretion Flows around Black Holes: A Comparative Study, *Astrophys. J.*, **470**, 460 (1996).

Narayan, R., J. E. McClintock, and I. Yi, A New Model for Black Hole Soft X-Ray Transients in Quiescence, *Astrophys. J.*, **457**, 821 (1996).

Narayan, R., S. Kato, and F. Honma, Global Structure and Dynamics of Advection-Dominated Accretion Flows Around Black Holes, *Astrophys. J.*, **476**, 49 (1997).

Narayan, R., and I. Yi, Advection-dominated Accretion: A Self-similar Solution, *Astrophys. J.*, **428**, L13 (1994).

Narayan, R., and I. Yi, Advection-dominated Accretion: Self-similarity and Bipolar Outflows, *Astrophys. J.*, **444**, 231 (1995a).

Narayan, R., and I. Yi, Advection-dominated Accretion: Underfed Black Holes and Neutron Stars, *Astrophys. J.*, **452**, 710 (1995b).

Narayan, R., I. Yi, and R. Mahadavan, Explaining the Spectrum Of Sagittarius-A* with a Model of an Accreting Black-Hole, *Nature*, **374**, 623 (1995)

Nobili, L., R. Turolla, and L. Zampieri, Spherical Accretion onto Black Holes - A Complete Analysis of Stationary Solutions, *Astrophys. J.*, **383**, 250 (1991).

Novikov, I. D., and K. S. Thorne, Astrophysics of Black Holes, in *Black Holes*, ed. C. DeWitt and B. DeWitt, Gordon and Breach, New York (1973).

Ostriker, J. P., R. McCray, R. Weaver, and A. Yahil, A New Luminosity Limit for Spherical Accretion onto Compact X-Ray Sources, *Astrophys. J.*, **208**, L61 (1976).

Park, M.-G., Self-consistent Models of Spherical Accretion onto Black Holes. I. One-temperature Solutions, *Astrophys. J.*, **354**, 64 (1990a.)

Park, M.-G., Self-consistent Models of Spherical Accretion onto Black Holes. II. Two-temperature Solutions with Pairs, *Astrophys. J.*, **354**, 83 (1990b).

Park, M.-G., Relativistic Theory of Radiative Transfer: Time-dependent Radiation Moment Equations, *Astr. Astrophys.*, **274**, 642 (1993).

Park, M.-G., and G. S. Miller, Near-critical Spherical Accretion by Neutron Stars - General Relativistic Treatment, *Astrophys. J.*, **371**, 708 (1991).

Park, M.-G., and J. P. Ostriker, Spherical Accretion onto Black Holes - A New, Higher Efficiency Type of Solution with Significant Pair Production, *Astrophys. J.*, **347**, 679 (1989).

Piran, T., The Role of Viscosity and Cooling Mechanisms in the Stability of Accretion Disks, *Astrophys. J.*, **221**, 652 (1978).

Pringle, J. E., Thermal Instabilities in Accretion Discs, *Mon. Not. R. Ast. Soc.*, **177**, 65 (1976).

Pringle, J. E., Accretion Discs in Astrophysics, *Ann. Rev. Astron. Astrophys.*, **19**, 137 (1981).

Pringle, J. E., and M. J. Rees, Accretion Disc Models for Compact X-Ray Sources, *Astr. Astrophys.*, **21**, 1 (1972).

Ryu, D., G. L. Brown, J. P. Ostriker, and A. Loeb, Stable and Unstable Accretion Flows with Angular Momentum near a Point Mass, *Astrophys. J.*, **452**, 364 (1995).

Shakura, N. I., and R. A. Sunyaev, Black Holes in Binary Systems. Observational Appearance, *Astr.*

Astrophys., **24**, 337 (1973).

Shakura, N. I., and R. A. Sunyaev, A Theory of the Instability of Disk Accretion on to Black Holes and the Variability of Binary X-Ray Sources, Galactic Nuclei and Quasars, *Mon. Not. R. Ast. Soc.*, **175**, 613 (1976).

Shapiro, S. L., Accretion onto Black Holes: The Emergent Radiation Spectrum, *Astrophys. J.*, **180**, 531 (1973).

Shapiro, S. L., A. P. Lightman, and D. N. Eardley, A Two-temperature Accretion Disk Model for Cygnus X-1 - Structure and Spectrum, *Astrophys. J.*, **204**, 187 (1976).

Soffel, M. H., Stationary Spherical Accretion into Black Holes - The Transition from the Optically Thin to the Optically Thick Regime, *Astr. Astrophys.*, **116**, 111 (1982).

Stellingwerf, R. F., X-radiation Limited Accretion Flow. II. Stability and Time Dependence, *Astrophys. J.*, **260**, 768 (1982).

Stellingwerf, R. F., and J. Buff, X-radiation Limited Accretion Flow. I. Steady Flow Regime, *Astrophys. J.*, **260**, 755 (1982).

Szuszkiewicz, E., M. A. Malkan, and M. A. Abramowicz, The Observational Appearance of Slim Accretion Disks, *Astrophys. J.*, **458**, 474 (1996).

Treves, A., L. Maraschi, and M. Abramowicz, Basic Elements of the Theory of Accretion, *Pub. Ast. Soc. Pac.*, **100**, 427 (1988).

Wandel, A., A. Yahil, and M. Milgrom, Nonadiabatic Self-consistent Spherical Accretion as a Model for Quasars and Active Galactic Nuclei, *Astrophys. J.*, **282**, 53 (1984).

Zampieri, L., J. C. Miller, and R. Turolla, Time-dependent Analysis of Spherical Accretion onto Black Holes, *Mon. Not. R. Ast. Soc.*, **281**, 1183 (1996).

Adv. Space Res. Vol. 22, No. 7, pp. 961–964, 1998
© 1998 COSPAR. Published by Elsevier Science Ltd. All rights reserved
Printed in Great Britain
0273-1177/98 $19.00 + 0.00

Pergamon

PII: S0273-1177(98)00134-3

DISCOVERY OF THE CESSATION OF FLICKERING DURING DIPS IN CYGNUS X-1

M. Bałucińska-Church[1,2], M. J. Church[1,2], Y. Ueda[1], T. Takahashi[1], T. Dotani[1], K. Mitsuda[1], H. Inoue[1]

[1] *Institute of Space and Astronautical Science, Yoshinodai 3-1-1, Sagamihara, Kanagawa 229-8510, Japan*
[2] *School of Physics and Space Research, University of Birmingham, Edgbaston, Birmingham B15 2TT, UK*

ABSTRACT

We report the discovery of the cessation of flickering in dips in the black hole candidate Cyg X-1, detected for the first time in the ASCA observation of May 9, 1995. During this observation, particularly deep dipping took place resulting in strong changes in hardness ratio corresponding to absorption of the power law spectral component. The deadtime corrected light curve with high time resolution clearly shows a dramatic decrease in the extent of flickering in the band 0.7 - 4.0 keV during dipping, but in the band 4.0 - 10.0 keV, there is relatively little change. We show that the rms flickering amplitude in the band 0.7 - 4.0 keV is proportional to the x-ray intensity in this band which changes by a factor of almost three. This is the first direct evidence that the strong low state flickering consists of fluctuations in the power law emission.   ©1998 COSPAR. Published by Elsevier Science Ltd.

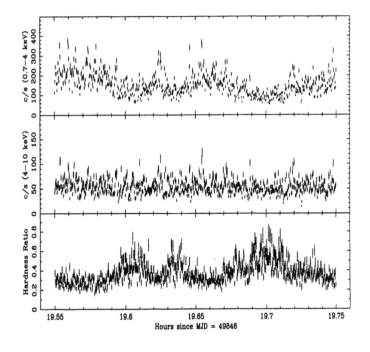

Fig. 1. GIS light curves in 1 s timebins for the ASCA observation of Cyg X-1 of May 9, 1995 in two energy band: 0.7 - 4.0 keV and 4.0 - 10.0 keV, together with the hardness ratio formed by dividing the light curves.

INTRODUCTION

Cygnus X-1 is well known as one of the best Galactic black hole candidates. Its X-ray spectrum is complex, consisting in the Low State of a hard, underlying power law, a reflection component (Done *et al.* 1992), and a soft excess (Bałucińska & Hasinger 1991). It was previously shown that in a *Rosat* observation, kT_{bb} for the soft excess agreed well with the characteristic temperature of the inner accretion disc for the low luminosity state of the

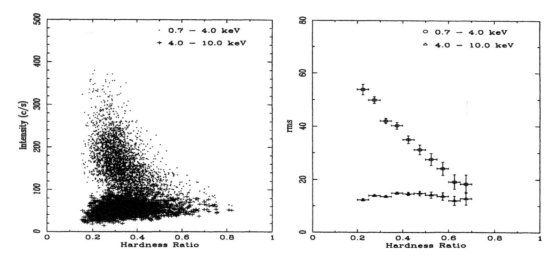

Fig. 2. (a) Intensity versus hardness ratio in the same two energy bands as in Figure 1, and (b) rms flickering amplitude in c/s versus hardness ratio.

source, indicating that the soft excess was thermal disc emission (Bałucińska-Church *et al.* 1995). Rapid aperiodic variability was discovered in Cygnus X-1 by Oda *et al.* (1971). Since then there have been extensive studies of such variability in various black hole candidates and in LMXB, and it is generally accepted that the flickering phenomenon originates in the inner accretion disc although the mechanism is not clear. The strength of variability may be quantified in terms of the *fractional rms amplitude* σ/I_x, and values of 30% - 50% are typical of the low state of Cygnus X-1 (Belloni & Hasinger 1990). The evidence is that this strong variability arises in the hard power law spectral component that dominates the Low State of the source, whereas the soft spectral component that dominates the High State shows little variability (van der Klis 1995). In Cygnus X-1, previously no major variation of the flickering has been detected during a particular observation as far as we are aware. In the present paper we present analysis of the May 9, 1995. ASCA observation of Cygnus X-1 in which strong dipping took place, and we show that the amplitude of flickering decreased dramatically in dips.

RESULTS

Spectral Analysis of Dip Evolution. Cyg X-1 has been observed several times with ASCA; we present results for the 7 hours observation made on May 9, 1995. At $\sim 19^h$ UT dipping took place lasting approximately one hour. The dead time corrected light curve of the strongest dips are shown in Figure 1 in an expanded view with 1s binning. The details of the dip light curves are given in two bands: 0.7 – 4.0 keV and 4.0 – 10.0 keV, together with the hardness ratio (HR) formed by dividing these. Three dips can be seen lasting 2 to 3 min in each of which there is a strong increase in hardness ratio associated with the photoelectric absorption taking place. We have analysed the spectral changes in dipping by dividing the data into 7 intensity bands including non-dip emission, selected in a time interval including the strong 3rd dip in Figure 1, and non-dip data on each side of the dip. Spectral analysis of the intensity-selected spectra in the band 0.7 – 4.0 keV were carried out using a blackbody to represent the soft excess, and a power law to represent the hard component, since in this band the reflection component makes little contribution. The data required a model in which partial covering of the power law took place. Dipping was seen to be due primarily to absorption of the major power law component with a N_H increasing from $8.9^{+18.5}_{-0.8} \cdot 10^{21}$ H atom cm^{-2} in non-dip to $72^{+11}_{-9} \cdot 10^{21}$ H atom cm^{-2} in the deepest part of dipping. We were not able to constrain very well possible changes in absorption of the blackbody taking place at energies below the ASCA band.

Phase of Dipping. The value of phase that we calculated for the observed dipping using the ephemeris of Gies & Bolton (1982) was 0.70 ±0.02. It was the case, that dipping has previously been seen at around phase zero, consistent with inferior conjunction with the Companion. However, we cannot be definite about the actual phase because of the errors involved in extrapolating this ephemeris to the date of the observation. This is especially the case because of the possibility that the orbital period is changing, as suggested by Ninkov *et al.* (1987) based on analysis of all data available at that time. This will not be resolved until there is a new determination of the

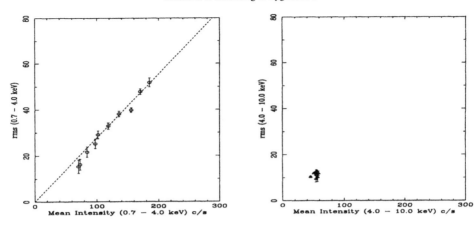

Fig. 3. Rms amplitude versus x-ray intensity (a) in the band 0.7 - 4.0 keV and (b) in the band 4.0 - 10.0 keV.

ephemeris. It is important to resolve this point, since dipping at phase ~0.7 would imply absorption in the stellar wind of the Companion.

Cessation of Flickering. The striking effect that flickering essentially stops in the deepest part of the dipping, is obvious in Figure 1, particularly in the 3rd dip which is the strongest. In the lower energy band 0.7 - 4.0 keV, individual low frequency flickering has an amplitude of 200 c/s in non-dip emission, i.e. the soft X-ray intensity rises from 200 – 400 c/s. However, in dipping it can be seen that this amplitude decreases considerably. The effect is less obvious in the higher energy band 4.0 - 10.0 keV, although there may be some small decrease in amplitude at the third dip.

To demonstrate the effect more clearly, the data are replotted in Figure 2a and 2b. In Figure 2a the X-ray intensities in the soft band and the hard band are plotted against hardness ratio, so that dipping corresponds to the high values of hardness ratio. It is clear that the peak-to-peak variation in the X-ray intensity falls sharply as dipping takes place. This is shown more clearly in Figure 2b in which the rms amplitude of intensity variation is plotted against hardness ratio for the two energy bands. This was produced by evaluating the rms deviation of the variability of the data selected within a narrow band of hardness ratio. Firstly, it can be seen that the non-dip value of the amplitude of 60 c/s with a non-dip count rate of 200 c/s in the low band gives a fractional rms amplitude of 30% which is quite typical for the low state of Cygnus X-1. There is a strong decrease of the amplitude at low energy, but with little change at high energies. To show this more clearly, we plot rms amplitude against X-ray intensity for the two energy bands in Figure 3. Poisson noise is subtracted from the rms, and also an approximate correction to the intensity is made for the dust-scattered halo in the band 0.7 - 4.0 keV based on the work on Predehl and Schmitt (1995). The scattered component is not expected to show flickering as this will be smoothed out by the variable time delays. The data in the low band are all consistent with a simple relationship between the variability and intensity which changes by almost a factor of 3 due to photoelectric absorption. The points at lowest dip intensity fall a little below a linear relationship but these have larger errors. In the high energy band the amplitude remains constant and there is little change in intensity since the increase in N_H in the dip has little effect in this band. Thus, all of the data are consistent with the idea that the variability takes place in the power law component, and that the reduction in rms amplitude simply reflects the decreasing power law intensity in the dips due to photoelectric absorption which is strong in the low energy band, but has little effect in the high band. Furthermore, if in the band 0.7 - 4.0 keV, the rms amplitude is divided by the intensity, the fractional rms amplitude is constant, as expected from the linearity of Figure 3a, again indicating the simple relation between amplitude and intensity.

We have also plotted the data as a function of time in the observation by evaluating the rms amplitude in a running 85 s timebin, and also evaluating a running mean of the intensity in this length of timebin. This is shown in Figure 4 for the lower energy band 0.7 - 4.0 keV. The middle panel shows the unintegrated light curve with 1 s binning, and the lower panel shows the integrated average values in 85 s timebins. There is some underestimation of the depth of dipping in the averaged light curve, and also of the depth of dipping in the rms amplitude plot as a result of smoothing over a long timebin. However, the good correspondence between amplitude and intensity is clear.

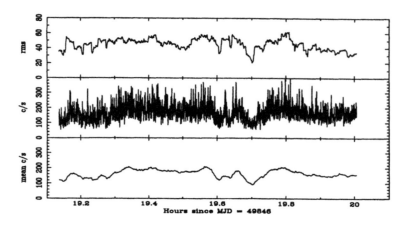

Fig. 4. (a) rms flickering amplitude evaluated in an 85 s timebin running through the observation; (b) the unsmoothed X-ray light curve in the band 0.7 - 4.0 keV; (c) running 85 s means of the light curve.

DISCUSSION

We have demonstrated for the first time the cessation of flickering during deep dipping in Cyg X-1 and we may now consider the implications of this in terms of the spectral components present in the low state source. The fractional rms amplitude of the variability is about 30% in the non-dip emission, a typical value for the low state, in which it is thought that the variability is associated with the hard power law component. In the energy band 0.7 - 4.0 keV, there are two spectral components the soft excess blackbody and the power law. Taking typical parameters for the soft excess, as determined from ASCA and *Rosat* (Bałucińska-Church *et al.* 1995), we can estimate that this contributes only 8% of the count rate in the band 0.7 - 4.0 keV. Thus it is clear that the power law *must* be involved in the variability.

Secondly, we have shown that the change in the strength of variability is simply related to the intensity of the power law spectral component which changes by almost a factor of three in the energy band 0.7 - 4.0 keV due to photoelectric absorption. All of the data are consistent with a linear relation between amplitude and intensity; however, there may be some evidence that the lowest intensity points deviate slightly from this relation. We have not attempted to correct the data for the contribution of the blackbody, as required if it was assumed that this has a fractional rms amplitude significantly smaller than the power law; however, the contribution of this component to the intensity is small.

Thus our results support the general understanding that the strong variability in the low state of Cygnus X-1 takes place in the hard spectral component. It is likely that the reduction of the flickering during dipping will provide a useful diagnostic of the variability process.

REFERENCES

Bałucińska, M., and G. Hasinger, A&A **241**, 439 (1991).

Bałucińska-Church, M., T. Belloni, M. J. Church, and G. Hasinger, A& A **302**, L5 (1995).

Belloni, T., and G. Hasinger, A&A **227**, L33 (1990).

Done, C., J. S. Mulchaey, R. F. Mushotzky, and K. A. Arnaud, Ap. J. **395**, 275 (1992).

Gies, D. R., and C. T. Bolton, Ap J **260**, 240 (1982)

Ninkov, Z., G. A. H. Walker, and S. Yang, Ap J **321**, 425 (1987).

Oda, M., P. Gorenstein, H. Gursky, E. Kellogg, and E. Schreiber, *et al.*, Ap. J. **166**, L1 (1971).

Predehl, P., and J. H. M. Schmitt, A & A **293**, 889 (1995).

Van der Klis, M., in " X-Ray Binaries", Cambridge Astrophysical Series vol. 26, Cambridge University Press (1995).

Adv. Space Res. Vol. 22, No. 7, pp. 965–968, 1998
© 1998 COSPAR. Published by Elsevier Science Ltd. All rights reserved
Printed in Great Britain
0273-1177/98 $19.00 + 0.00

Pergamon

PII: S0273-1177(98)00138-0

PCA OBSERVATIONS OF CYG X-1 FROM RXTE USING FAST TIMING AND HIGH TELEMETRY RATES

A. B. Giles [1], K. Jahoda, and T. Strohmayer [1]

Code 662, NASA Goddard Space Flight Center, Greenbelt, MD 20771, USA

[1] also Universities Space Research Association

ABSTRACT

The Rossi X-ray Timing Explorer (RXTE) spacecraft contains a very large area, high resolution timing experiment provided by the NASA's Goddard Space Flight Center. This Proportional Counter Array (PCA) has an area of 7000 cm^2 and a time resolution down to one micro second. A 450 Mbit memory on board the spacecraft allows science data acquisition at rates up to 512 kbps but a more typical high rate of ~256 kbps can be continuously supported for up to 30 minutes. This performance far surpasses previous rocket and satellite timing experiments. Since the launch of RXTE several observations have been made of Cyg X-1 in this special high telemetry rate mode to specifically search for, and attempt to confirm, the existence of millisecond bursts. Such features were reported in several observations of Cyg X-1 in the mid 1970's. This paper briefly summarizes the PCA performance compared to earlier large timing experiments and then presents early results from the analysis of these data sets. Individual millisecond bursts do not seem to exist in the data examined so far.
©1998 COSPAR. Published by Elsevier Science Ltd.

INTRODUCTION

The earliest observations of very short time scale variability or millisecond (ms) bursts from Cyg X-1 were those reported by Rothschild *et al.* (1974 & 1977). The analysis method employed revealed a total of 12 bursts. Further ms burst observations with a much larger detector were reported by Giles (1981). This paper also lists references to earlier observations of rapid variability from Cyg X-1. A number of authors (Weisskopf and Sutherland, 1978; Press and Schechter, 1974) reported statistical difficulties with the analysis methods employed to locate ms bursts, particularly given the longer time scale variability or shot noise seen from Cyg X-1 first described by Terrell (1972). A complete description of the mathematical theory for the shot noise model has been given by Sutherland *et al.* (1978). The first experiment to individually time tag all photons from Cyg X-1 to μs resolution was that flown in 1976 (Giles, 1981). Unfortunately only a short observation was made due to an attitude control failure but several bursts of similar significance to those found by Rothschild *et al.* (1974, 1977) were located.

The first satellite data obtained on Cyg X-1 with rapid sampling was a HEAO A1 observation by Meekins *et al.* (1984). No bursts were seen but this data has recently been reanalyzed, taking into account dead time corrections, and suggests an excess in the bin count distribution around the few milliseconds range (Wen *et al.* 1996). The Medium Energy (ME) array (Turner *et al.* 1981) on EXOSAT observed Cyg X-1 on 13 occasions (Belloni and Hasinger, 1990). Although EXOSAT could in principle have sampled much faster, only a few of the observations were made at ms resolution, apparently due to telemetry limitations. GINGA observed Cyg X-1 on four occasions, the principal periods being in August 1987 (Miyamoto and Kitamoto, 1989) and in May 1990 (Negoro *et al.* 1991). A description of the Large Area Counter (LAC) on GINGA can be found in Turner *et al.* (1989). Several papers have appeared discussing Cyg X-1 shot profiles and power density spectra obtained from GINGA but apparently none report ms bursts. Finally, during the AO1 period RXTE supported 8 programs on Cyg X-1, and also performed an extended set of public TOO observations when the source went into its rare high state in May 1996. Table 1 summarizes the key features of the previous significant medium energy experiments that have observed Cyg X-1. The key requirements when searching for rapid individual bursts are to simultaneously have a large detector area, fast electronics with minimal dead time and a high telemetry rate.

Table 1.   The PCA & Previous Large Detectors used to Observe Cyg X-1

Parameter	GSFC	SL-1306	HEAO 1 NRL	EXOSAT ME	GINGA LAC	XTE PCA
Date	Oct 73 & Oct 74	Nov 76	Aug 77 to Jan 79	May 83 to May 86	Feb 87 to Nov 91	Dec 95 to + 5 years ?
Area cm^2	1360	4000	1650	1800 [a]	4500	7000
Energy range keV	1.5 - 35	1.5 - 12	1 - 20	1 - 50	1.5 - 30	2 - 60
No. of E channels	128 [b]	8	1 (128 max)	128	2 (48 max)	256
Time resolution μs	320 / 160	2	10 [c]	7.6 [d, e]	980	1
Telemetry kbps	50	128	128	< 7.2 [f]	16 (2)	32 - 512
"Orbit"	Rocket	Rocket	LEO	97 hr	LEO	LEO
Reference	(1)	(2)	(3)	(4)	(5)	(6)

[a] often only 1/2 on source
[b] first event in 320 μs bin in 73 (160 μs in 74)
[c] single bit mode
[d] virtually never used
[e] dead time calibration problems, severe telemetry constraints
[f] all other experiments off

(1) Rothschild *et al.* (1977)
(2) Giles (1981)
(3) Meekins *et al.* (1984)
(4) Turner *et al.* (1981)
(5) Turner *et al.* (1989)
(6) Swank *et al.* (1994)

THE PCA OBSERVATIONS

Of the 3 observations made in AO1 only data from the 3rd run, on 17 February 1996, is presented in this paper. The light curve for this observation is shown in Figure 1 and the mean rate is 7325 counts per second. The PCA Very Large Event (VLE) window was set to 60 μs for this observation.

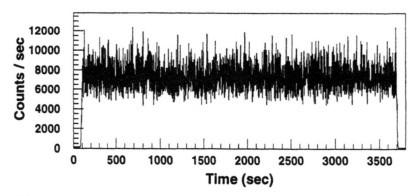

Fig. 1. Cyg X-1 light curve. The counts are binned into one second samples.

The light curve shown in Figure 1 can be processed with an FFT to provide the power spectrum shown in Figure 2. This reveals that the power spectrum is completely flat and consistent with the poisson noise level of 2 past ~300 Hz. The normalization used is that of Leahy *et al.* (1983). The power spectrum at lower frequencies, down to 0.01 Hz, is not shown here but is typical of Cyg X-1 in the low state (Belloni and Hasinger, 1990).

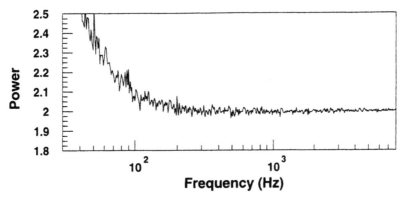

Fig. 2. Cyg X-1 power spectrum reveals no high frequency periodic or

Quasi Periodic Oscillations (QPO) features.

We have examined the bin count distribution over a 2816 second period for seven small bin sizes. These are plotted in Figure 3 with the largest bin curve being to the right. The low count bins for all seven curves are confused in this plot but this is unimportant since we are concerned with possible bursts or bins with high count values. A striking feature is that all the curves are "smooth" and there are no prominent outliers beyond the right hand end of each distribution, indicating large isolated bursts are not present in the data. The old millisecond burst observations typically had increases above the local mean by factors of 10 - 12. Such enhancements, if plotted on Figure 3, would fall beyond the observed distributions.

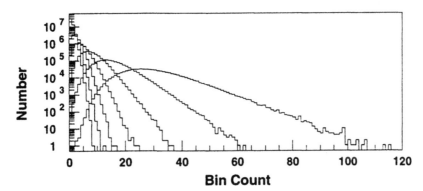

Fig. 3. Bin count distributions for bin sizes 0.061, 0.12, 0.24, 0.49, 0.98, 1.95, 3.9 ms.

To compare the curves in Figure 3 with those expected from a poisson distribution we must allow for the fact that the data in Figure 1 is typical of Cyg X-1 and varies substantially on short time scales. We have divided the data set into 2816 one second intervals and summed the predicted poisson bin count distributions for each interval. The differences between this integrated prediction and the measured data are shown in Figure 4 for bin sizes of 0.49 and 0.98 ms. Note that the Y axis has a log scale so negative values go to zero. Taking the mean in 2816 pieces results in a broader predicted distribution (with reduced height) than assuming it is constant at 7350 c/s but the measured Cyg X-1 distribution is considerably broader. Further analysis is needed to see if the differences in Figure 4 can be reduced by taking finer intervals than one second. We have also used the Burst Expectation Search (BES) method described by Giles (1996) to search the data for bursts. This analysis proved inconclusive with no prominent features being present but further work is required. The BES analysis locates burst excesses in terms of a probability search that is count rate independent. This method is capable of finding "small" bursts that have a low but real significance which get swamped in an integrated bin count distribution approach. Finally an autocorrelation analysis reveals no "fast" variability below ~3 milliseconds.

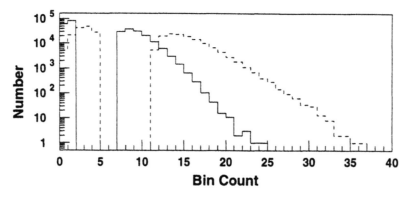

Fig. 4. Cyg X-1 minus poisson prediction. Solid trace 0.49 ms, dotted trace 0.98 ms.

SUMMARY

• Millisecond bursts of the size expected from the earlier reported observations do not exist in this data

There may be evidence of an excess of higher count bins in the bin count distribution over a span of time scales around the millisecond range. We note that this excess, if present, forms part of a continuous distribution. Further work is required on the data. This excess may relate to that reported by Wen *et al.* (1996) for the reanalyzed HEAO-1 data

• There are no high frequency periodic or QPO features

• The autocorrelation function is "flat' below ~3 milliseconds

REFERENCES

Belloni, T., and G. Hasinger, A&A, 227, L36 (1990).

Giles, A.B., MNRAS, 195, 721 (1981).

Giles, A.B., submitted ApJ (1996).

Leahy, D.A., W. Darbro, R. F. Elsner, M. C. Weisskopf, P. G. Sutherland, *et al.*, ApJ, 266, 160 (1983).

Meekins, J.F., K. S. Wood, R. L. Headler, E. T. Byram, D. J. Yentis, *et al.*, ApJ, 278, 288 (1984).

Miyamoto, S., and S. Kitamoto, Nature, 342, 773 (1989).

Negoro, H., S. Miyamoto, and S. Kitamoto, *Frontiers of X-ray Astronomy*, Universal Academy Press Inc, Tokyo, 313 (1991).

Ogawara,Y., K. Doi, M. Matsuoka, and M. Oda, ISAS RN 26, University of Tokyo (1977).

Press, W.H., and P. Schechter, ApJ, 168, 437 (1974).

Priedhorsky, W., G.P. Garmire, R.E. Rothschild, E.A. Boldt, P.J. Serlemitsos, *et al.*, ApJ, 233, 350 (1979).

Rothschild, R.E., E.A. Boldt, S.S. Holt, and P.J. Serlemitsos, ApJ, 189, L13 (1974).

Rothschild, R.E., E.A. Boldt, S.S. Holt, and P.J. Serlemitsos, ApJ, 213, 818 (1977).

Sutherland, P.G., M.C. Weisskopf, and S.M. Kahn, ApJ, 219, 1029 (1978).

Swank, J.H., K. Jahoda, W. Zhang, A.B. Giles, F.M. Marshall, *et al.*, eds. M.A.Alphar *et al.*, *The Lives of the Neutron Stars*, NATO ASI Series, Kluwer Academic Publishers, Dordrecht, pp. 525 (1994).

Terrell, N. J., ApJ, 174, L35 (1972).

Turner, M.J.L., A. Smith, and H.U. Zimmerman, Space Sci. Rev., 30, 513 (1981).

Turner, M.J.L., R.D. Thomas, B.E. Patchett, D.H. Reading, K. Makishima, *et al.* , Publ. Astr. Soc. Japan, 41, 345 (1989).

Weisskopf, M.C., and P.G. Sutherland, ApJ, 221, 228 (1978).

Wen, H., E. Bloom, G. Godfrey, J. Scargle, T. Crandel, *et al.*, AAS HEAD Meeting, San Diego (1996).

Adv. Space Res. Vol. 22, No. 7, pp. 969–972, 1998
© 1998 COSPAR. Published by Elsevier Science Ltd. All rights reserved
Printed in Great Britain
0273-1177/98 $19.00 + 0.00

Pergamon

PII: S0273-1177(98)00129-X

THE NEW INTERMEDIATE POLARS FROM THE ROSAT ALL-SKY SURVEY

F. Haberl

Max-Planck-Institut für extraterrestrische Physik, Postfach 1603, D-85740 Garching, Germany

ABSTRACT

An overview of the soft X-ray temporal and spectral properties of the 7 new intermediate polar (IP) type cata-clysmic variables, discovered up to now in the ROSAT all-sky survey, is presented. Their nature as pulsating X-ray sources with spin periods between 300 and 1400 s was revealed by follow-up ROSAT pointed and optical observations. RE 0751+14, RX J0558.0+5353, RX J0512.2-3241 and probably RX J1914.4+2456 belong to the new class of soft IPs, characterized by soft X-ray spectra which are dominated by a variable blackbody-like component with kT around 60 eV. New ROSAT HRI observations of RX J1914.4+2456 confirm the spin period at 569.4 s.

RX J0028.8+5917, RX J1712.6-2414 and RX J0153.3+7446 have X-ray spectra consistent with hot thermal Brems-strahlung as it was observed from IPs known before the launch of ROSAT. These ROSAT-discovered "hard" IPs show on average however lower photo-electric absorption which varies only marginally with pulse phase. The light curve of RX J1712.6-2414 at soft X-ray energies is modulated at the beat frequency between spin and orbital period, indicating that most of the material is accreted directly via the accretion stream. New RXTE observations of RX J1712.6-2414 during a higher intensity state of the source find the X-ray modulations a factor of 5 reduced.
©1998 COSPAR. Published by Elsevier Science Ltd.

PROPERTIES OF THE ROSAT DISCOVERED INTERMEDIATE POLARS

About two dozens of cataclysmic variables have been discovered in the ROSAT Galactic Plane Survey project to optically identify X-ray sources from the ROSAT all-sky survey between galactic latitudes of ±20°. From six of these periodic modulation of the X-ray flux was found in follow-up observations, which has clearly identified them as inter-mediate polars (IPs), asynchronous magnetic CVs. Three of the systems show a ROSAT PSPC spectrum consistent with photo-electrically absorbed hot (kT a few tens of keV) thermal Bremsstrahlung, a model used to characterize the 1-20 keV spectra of IPs before the launch of ROSAT.

The soft X-ray spectrum in the ROSAT band of the other three IPs however is dominated by a soft blackbody-like component with kT around 60 eV (Haberl & Motch, 1995) in addition to a hard tail. This is reminiscent to the spectra of polars, CVs with a highly magnetized white dwarf in synchronous rotation with a late-type dwarf star. Recently another "soft" IP, RX J0512.2-3241, at higher galactic latitude was discovered by Burwitz et al. (1996).

The seven new IPs have already significantly increased the number of known CVs of this kind (about 13 before ROSAT, Ritter 1990) and the current knowledge of their properties is summarized as follows:

Hard Intermediate Polars

A1) RX J1712.6-2414
      X-ray Period: 1003 s
      polarimetry: spin period 927 s, periodic polarization up to -5%
          only 2 other IPs with polarization: BG CMi (-0.25%) and RE0751+14
      B > 5 MG
      orbital Period: 3.4 h
      modulation in X-rays at ω-Ω; disc-less accretion?

A2) RX J0028.8+5917
>   X-ray Period: 312.8 s

A3) RX J0153.3+7446
>   X-ray Period: 1414 s ?

RX J1712.6−2414, a "classical" IP with hard spectrum is a particularly interesting case. It exhibits the largest degree of polarization yet observed in an IP indicating a magnetic field of about 8 MG (Buckley et al. 1995) and is together with RE 0751+14 the only IP where the degree of polarization is modulated periodically with the white dwarf spin period (and uniquely determining it). From polarimetry a white dwarf spin period of 927 s is derived, while the soft X-rays are modulated with a period of 1003 s (Buckley et al. 1997). The modulation at the beat frequency is the strongest evidence for disc-less accretion (e.g. Norton et al. 1996) in an IP system. There is some indication for a modulation in X-rays at 927 s in the ROSAT PSPC data. Folding the data yields a ∼10% variation while the pulsed fraction at the 1003 s period is about 30%.

Soft Intermediate Polars

B1) RE 0751+14
>   X-ray Period: 834 s, orbital Period: 5.2 h
>   polarimetry: periodic polarization up to -2.6%, B > 5 MG
>   large spin-down rate

B2) RX J0558.0+5353
>   X-ray and optical Periods: 545.5 s, but most power in 1st harmonic at 272 s
>   orbital Period: 4.15 h

B3) RX J1914.4+2456
>   X-ray Period: 1139 s

B4) RX J0512.2-3241
>   X-ray Period: 863.5 s, orbital Period: 3.45 h

The soft IPs in group B have apart from the soft X-ray spectrum further similarities to polars. Polarimetry revealed periodically modulated circular polarization in RE 0751+14 (Rosen et al. 1993; Pirola et al. 1993), suggesting a magnetic field strength of >5 MG. Except RX J1914.4+2456, where no hard component was detected (due to low statistics?), the F_{br}/F_{bb} ratio of the two spectral components in the ROSAT band of the soft IPs fit into the relation found by Beuermann & Schwope (1994) for polars, also indicating a magnetic field strength at the lower end of the field distribution of polars. This supplies strong evidence that the soft IPs are evolutionary progenitors of polars with relatively low magnetic field strength which have not synchronized orbital and white dwarf rotation yet. The high spin-down rate of RE 0751+14 is also consistent with this picture.

NEW X-RAY OBSERVATIONS

RXTE Observations of RX J1712.6−2414

RXTE observed RX J1712.6−2414 on March 12-13, 1996 with an average PCA (2-20 keV) count rate of ∼85 counts s^{-1}. A combined fit of ROSAT PSPC and RXTE PCA spectra yields a kT of about 16 keV for a thermal Bremsstrahlung model, absorbed by a column density of 6.5 10^{20} H cm^{-2}. Single and more-component Bremsstrahlung fits of the PCA spectrum with different N_H and/or kT are however formally not acceptable, suggesting spectral variations with spin and/or orbital phase. The flux in the two spectra shows that the overall intensity increased by a factor of 6 between the ROSAT and the RXTE observations under the assumption of constant spectral shape.

A folding analysis of the PCA data shows a peak at 1008 s in the χ^2 test confirming the major modulation at the beat frequency. The pulse profile for a period of 1008 s is shown in Figure 1. The pulsed fraction (amplitude relative to the mean) is reduced by about a factor of 5 compared to the ROSAT observation. A comparison of different PCA energy bands shows that there is little energy dependence and in particular no absorption changes. The different modulation is therefore unlikely being caused by the different energy bands of ROSAT and PCA. On the other hand an intensity dependence of the X-ray modulation is expected when at higher accretion rate the matter penetrates deeper into the magnetosphere, resulting in a larger accretion area around the magnetic pole. The light curve folded with the 927 s period is plotted in Figure 2.

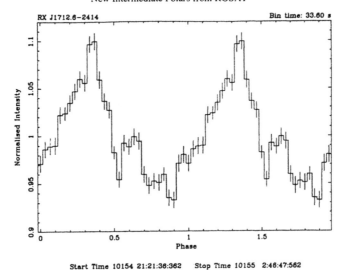

Start Time 10154 21:21:36:362   Stop Time 10155 2:46:47:562

Fig. 1. Folded RXTE PCA light curve with a period of 1008 s (2-20 keV)

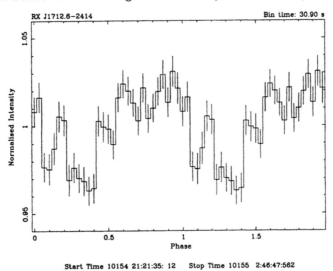

Start Time 10154 21:21:35: 12   Stop Time 10155 2:46:47:562

Fig. 2. As Figure 1 but folded with the white dwarf spin period of 927 s

ROSAT HRI Observations of RX J1914.4+2456

ROSAT re-observed RX J1914.4+2456 with the HRI in 1996, April 30 – May 5 for 26 ksec. The power spectrum shows a strong peak at 1.8×10^{-3} Hz, consistent with the frequency expected from the pulse period discovered in the PSPC observation in 1993 (Figure 3). The pulse profile shows a modulation of 100% with no X-ray emission for about 0.4 of the 569.4 s period (Figure 4). In this respect RX J1914.4+2456 is different from the other soft intermediate polars.

CONCLUSIONS

ROSAT has discovered the new class of soft intermediate polars, perhaps the long-sought progenitors of polars. The white dwarfs in these systems probably possess magnetic fields comparable in strength to the lower end of the field distribution of polars. However the magnetic field strength can not be the only parameter which rules the production of the soft spectral component in their X-ray spectra as the case of RX J1712.6-2414 shows. RX J1914.4+2456 shows some differences in its X-ray properties to the established soft intermediate polars which may indicate a different kind of object: The X-ray modulation is 100% and no hard spectral component has been detected in the PSPC spectrum.

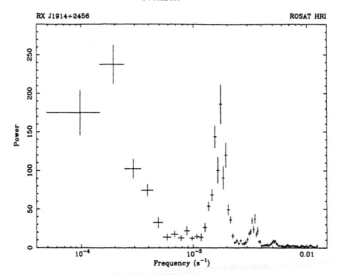

Fig. 3. Power spectrum from the ROSAT HRI observation from April 30 – May 5, 1996

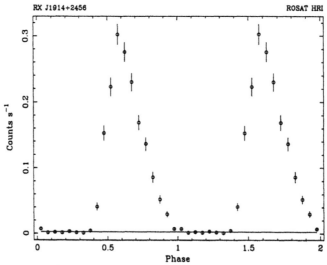

Fig. 4. ROSAT HRI light curve folded with a period of 569.41 s. The background level is indicated as horizontal line

REFERENCES

Beuermann, K., and A.D. Schwope, in *Interacting binary stars*, ASP Conference Series, Vol. **56**, 119 (1994).
Buckley, D.A.H., K. Sekiguchi, C. Motch, D. O'Donoghue, A.-L. Chen, *et al.*, MNRAS, **275**, 1028 (1995).
Buckley, D.A.H., F. Haberl, C. Motch, K. Pollard, A. Schwarzenberg-Czerny, and K. Sekiguchi, MNRAS, **287**, 117 (1997).
Burwitz, V., K. Reinsch, K. Beuermann, and H.-C. Thomas, A&A, **310**, L25 (1996).
Haberl, F., and C. Motch, A&A, **297**, L37 (1995).
Norton, A.J., A.P. Beardmore, and P. Taylor, MNRAS, **280**, 937 (1996).
Pirola, V., P.J. Hakala, and G.V. Coyne, ApJ, **410**, L107 (1993).
Ritter, H., A&AS, **85**, 1179 (1990).
Rosen, S.R., J.P.D. Mittaz, P.J. Hakala, MNRAS, **264**, 171 (1993).

Pergamon

Adv. Space Res. Vol. 22, No. 7, pp. 973–980, 1998
© 1998 COSPAR. Published by Elsevier Science Ltd. All rights reserved
Printed in Great Britain
0273-1177/98 $19.00 + 0.00

PII: S0273–1177(98)00133–1

DISC-OVERFLOW ACCRETION IN INTERMEDIATE POLARS

Coel Hellier

Department of Physics, Keele University, Staffordshire, ST5 5BG, U.K.

ABSTRACT

We have recently realised that intermediate polars accrete in a range of modes. While accretion is predominantly through an accretion disc, in some systems the accretion stream overflows the disc and interacts with the magnetic field directly. In RX 1712–24, discless accretion dominates. I review the current status of these observations, concentrating on the information from the X-ray lightcurves.

BACKGROUND

As early as 1976 Lubow and Shu pointed out that the accretion stream in cataclysmic variables was likely to be thicker than the outer rim of the accretion disc, so that the stream would not be entirely stopped by its initial collision with the disc. This idea was used intermittently afterwards, such as Frank *et al.*'s (1987) model for accretion structure in low-mass X-ray binaries, and Shafter *et al.*'s (1988) suggestion that it could explain peculiarities of some novalike variables. In 1989 Lubow revisited the theory and showed that the parts of the stream furthest from the plane would continue flowing above and below the disk, in free-fall. These streams would follow the free-particle trajectory, and re-impact the disc where they crossed the plane, near the point of closest approach to the white dwarf.

In the same year, Hellier *et al.* (1989) published spectroscopy of EX Hya in outburst showing high velocity line emission coming from near the re-impact point. Suggesting that the stream was responsible, we used the term 'disc-overflow accretion' to describe this phenomenon (Figure 1). In such magnetic systems the overflowing stream is likely to hit the magnetosphere, if it is large enough, rather than re-impact the disc. Hellier *et al.* (1989) predicted that it would result in X-rays modulated at the beat frequency (ω–Ω; where ω is the spin frequency and Ω the orbital frequency). This arises since the white dwarf is effectively rotating at the beat period, as seen by the orbitally-locked stream. Thus the stream will feed the two magnetic poles alternately, flipping between them on the beat cycle. (The connection between beat periods and stream-fed accretion had also been realised the previous year by Mason, Rosen and Hellier 1988.)

Later (Hellier 1991) I pointed out that such beat pulses could diagnose the accretion mode in intermediate polars. Pure disc-fed accretors should produce spin pulses only. Disc-overflow would produce both spin and beat pulses (with, to zeroth order, their relative amplitudes indicating the proportion of material accreting via each mode). Systems with no disc (which had been suggested by Hameury *et al.* 1986) should produce beat periods only. My systematic analysis of the *EXOSAT* archive (Hellier 1991) showed that spin pulses predominated. There were, though, smaller beat period pulsations, suggesting that some disc-overflow occurred in some stars. In FO Aqr, the strength of the beat period changed over time, which could be explained by variations in the amount of disc-overflow. The strongest beat pulse was in TX Col, where it was comparable in strength to the spin pulse, implying strong disc-overflow or discless accretion.

Wynn and King (1992) investigated the lightcurves from the various accretion geometries thoroughly for the first time. While confirming the basic picture above, they showed that for some geometries a discless accretor could produce a 2ω–Ω pulsation, instead of ω–Ω, although one of these would always be present.

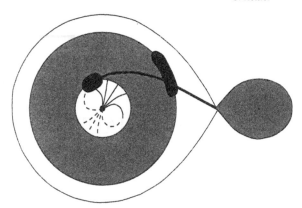

Fig. 1. The depiction of disc-overflow accretion from Hellier *et al.* (1989). The stream over-flows the initial impact with the disc, to interact directly with the magnetosphere further in.

Further, by considering asymmetric dipoles they could produce spin pulsations even in a discless accretor, although the asymmetries required to overturn the dominant pulsation were large (e.g. offsetting the dipole by half the white dwarf radius).

Stream-fed accretion will also produce a strong orbital modulation. With the stream flipping between two accretion sites at the magnetic poles, feeding the one pointing towards it, both sites will be on the hemisphere facing the stream, and thus there will be a large change in their visibility over the orbital cycle. An orbital modulation, though, is not a secure indicator of discless accretion since, in high inclination systems, accretion disc structure can obscure the line-of-sight to the white dwarf, again producing an orbital modulation (this is a common phenomenon in intermediate polars; Hellier *et al.* 1991).

The idea of the stream flipping between the magnetic poles is complicated in the disc-overflow model by the fact that material will flow both over and under the disc. Material above the disc might be prevented by the disc from flowing to the lower pole, and vice versa. Instead, the stream above the disc might readily flow to the upper pole when it points towards the stream but be inhibited from accreting, or simply join disc material, for the rest of the beat cycle. Since the opposite will be occurring below the disc, the overall effect is still, though, a flipping between poles.

A further consideration is whether the overflowing stream gets far enough in to connect to the magnetic field. Lubow (1989) calculated the stream re-impact radius as (for typical mass-ratios) 0.066 of the stellar separation, which, for typical 3–5 hr orbital periods, corresponds to 0.5–0.8×10^8 m. For comparison, assuming equilibrium rotation, the magnetosphere of a 1000-s spin period, 0.7 $M_\odot$ white dwarf is 1.3×10^8 m. Thus the stream would hit the magnetosphere for all the systems discussed here, except for GK Per and XY Ari, which have long orbital periods and short spin periods. Note, though, that such calculations are uncertain, and particle simulations (e.g. Armitage & Livio 1996) give differing results.

Before turning to discuss the individual systems, there are two more points to be borne in mind. First, it is not always obvious whether an observed pulsation is a spin or a beat pulse, or maybe even a 2ω–Ω pulse. A reliable identification occurs if both are seen, in either the X-ray or the optical. Seeing both in X-rays indicates disc-overflow accretion. In the optical, light from the magnetosphere can be spin-pulsed (if disc-fed) or beat-pulsed (stream-fed). The optical light from X-rays reprocessed from the secondary or hot-spot will always be beat-pulsed, whether the illuminating X-rays are spin-pulsed or beat-pulsed (in the latter case the reprocessing site follows the accretion site round the orbit and simply reflects the beat pulse). An additional problem is that many of the newer discoveries have not been studied sufficiently to decide this issue; I therefore concentrate here on the better studied stars.

Lastly, as pointed out by Warner (1986), modulating the amplitude of a pulse (say the spin pulse) at a different frequency (say the orbital cycle) produces power equally at both upper and lower sidebands ($\omega + \Omega$ and $\omega - \Omega$). If the modulations are sinusoidal the sidebands will be at most half the amplitude of the main pulse. If the mean flux is reduced along with the pulse amplitude (as results, for instance, from simple absorption) the sideband power will be proportional to the power at the orbital frequency.

Table 1: Amplitude of the X-ray beat (ω–Ω) pulse, as a percentage of the X-ray spin pulse, for different stars at different times.

Star	Year				Refs
AO Psc	1985	1990	1994	1995	
	13	<4	11	<3	1,2,3
FO Aqr	1985	1988	1990	1993	
	<12	49	<14	33	1,2,4
TX Col	1985	1994	1995		
	116	<30	82		1,5,6
BG CMi	1985	1988	1996		
	<17	<30	25		1,2,7
V1223 Sgr	1984	1991	1994		
	25	<15	<14		1,2
RX 1712–24	1995	1996			
	300	>180			8, 9

Refs: 1 Hellier 1991; 2 Norton *et al.* 1996; 3 Hellier *et al.* 1996; 4 Beardmore *et al.* 1996; 5 Wheatley 1996; 6 Norton and Beardmore, private communication; 7 Mukai, private communication; 8 Buckley *et al.* 1995b; 9 Hellier, unpublished.

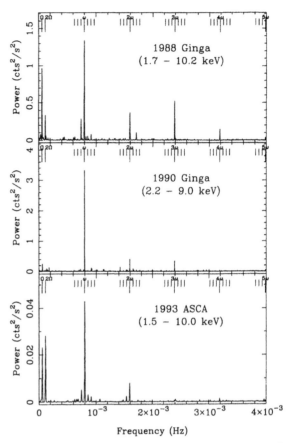

Fig. 2. Fourier transforms of X-ray lightcurves of FO Aqr showing variable beat pulse strengths (from Beardmore *et al.* 1996). This is explained by variable disc-overflow accretion.

THE INDIVIDUAL INTERMEDIATE POLARS

With the above background I turn to the individual systems. Since for disc-overflow the important diagnostic is the relative height of the spin and beat pulses, I collect these ratios for different stars at different times in Table 1. Many of the entries are from the previous compilations dealing with more than one star: Hellier (1991, hereafter H91) and Norton *et al.* (1996, NBT). The values, though, do not all have the same energy ranges; for example some are from *Ginga* 2–10 keV data and some from *Rosat* 0.1–2 keV data. Since the beat and spin pulses will have different energy dependences, a more correct comparison would use one energy range throughout — unfortunately there are not yet enough observations in the same ranges to allow this. So far no 2ω–Ω X-ray pulsations have been seen (e.g. Hellier 1992), so values for these are not given.

AO Psc *Spin 805 s, orbit 3.59 hr*
The strong optical beat pulse in AO Psc secures the dominant 805-s X-ray pulsation as the spin pulse. Some X-ray datasets, for instance the 1995 *ASCA* data reported by Hellier *et al.* (1996), show no orbital sidebands, suggesting pure disc-fed accretion. Twice, though, we've seen a beat pulse with an amplitude $\sim 10\%$ of the spin pulse. In a *Rosat* HRI observation there was also power at the upper sideband, ω+Ω (NBT). Since there are strong orbital dips in AO Psc's lightcurve, these sidebands probably result from amplitude modulation of the spin pulse. The fact that the two sidebands have different amplitudes, however, requires that disc-overflow is also occurring. The resulting modulations will interfere and produce, in this case, a weaker ω–Ω pulse. The conclusion is thus a small degree of disc-overflow occurring on 2 out of 4 occasions.

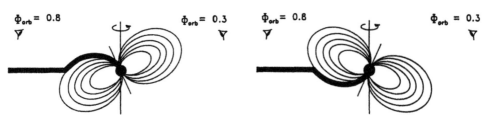

Fig. 3. These diagrams of a discless accretor, half a beat cycle apart, show the stream feeding onto the nearest magnetic pole (from Norton *et al.* 1992a). Both accretion sites are in the white dwarf hemisphere facing the stream, so their visibility will be a strong function of orbital phase. This produces an orbital modulation and amplitude-modulates the beat pulse, putting power into its $\omega-2\Omega$ and ω sidebands.

FO AQR *Spin 1254 s, orbit 4.85 hr*

FO Aqr is the star around which much of the discussion of disc-overflow has focussed, following Norton *et al.*'s (1992) report of a strong beat pulse in a 1988 *Ginga* observation, with an amplitude of 50% of the spin pulse, and Mukai *et al.*'s (1994) confirmation of the beat pulse from a 1993 *ASCA* observation. Although Norton *et al.* (1992) interpreted their data using a pure discless geometry, Hellier (1991; 1993) and Mukai *et al.* (1994) preferred the disc-overflow idea. This has the following advantages. First, the stronger spin pulse would, in the discless model, have to be attributed to an offset or asymmetric dipole, and it is unclear whether a large enough asymmetry could be accommodated before becoming unrealistic. Secondly, the beat pulse was not present in the 1985 *EXOSAT* observation (H91) and (as is now known) the 1990 *Ginga* observation (Beardmore *et al.* 1996; Figure 2). Variable disc-overflow is more likely than a disc disappearing and reforming on that timescale. Thirdly, there is clear evidence for a disk from optical observations, and also an absorption line feature which seems to originate in disc-overflow (Hellier *et al.* 1990). Again, therefore, this star has shown disc-overflow in 2 out of 4 X-ray observations.

TX COL *Spin 1911 s, orbit 5.72 hr*

TX Col was the first star to show an X-ray beat period (Buckley and Tuohy 1989), with an *EXOSAT* ME observation recording it slightly stronger than the spin pulse in hard X-rays, and it lead to the first suggestion that beat periods are linked to stream-fed accretion (Mason *et al.* 1988). Since then, an *ASCA* observation has shown no beat pulse (Wheatley 1996), and a *Rosat* observation both a $\omega-\Omega$ pulse (this time smaller than the spin pulse) and a $\omega-2\Omega$ pulse (Norton and Beardmore, private communication). Again, variable disc-overflow is the diagnosis. This last detection (a first) is interesting. As mentioned above, the accretion sites from stream-fed accretion will be on the hemisphere of the white dwarf facing the stream (Figure 3), and so we expect their visibility, and hence the amplitude of the beat pulse, to be a strong function of orbital phase. This will put power into the upper and lower sidebands of the $\omega-\Omega$ pulse (i.e. ω and $\omega-2\Omega$), explaining the detection (Figure 5).

BG CMI *Spin 913 s, orbit 3.23 hr*

In this more controversial star the identity of the spin period (913 secs or 847 secs) is currently disputed. Norton *et al.* (1992b) found a strong X-ray pulsation at 913 s (previously regarded as the spin period) and a smaller peak in the Fourier transform at 847 s (previously $\omega+\Omega$). They therefore proposed that $847^{-1} = \omega$ and $913^{-1} = \omega-\Omega$, and hence that BG CMi accreted predominantly through a stream. I personally favour the original identification: the main evidence is optical sideband periods which are very hard to explain if $\omega = 847^{-1}$, but straightforward if $\omega = 913^{-1}$. There is also strong optical evidence for an accretion disc (the full argument is too long to present here, but will be given in Hellier, in preparation).

Sticking with $\omega = 913^{-1}$, I thus have to explain the 847-s peak. This appears similar to the *Rosat* observation of AO Psc discussed above: a beat pulse caused by disc-overflow interferes with power put into the upper and lower sidebands by amplitude modulation of the spin pulse. This produces a smaller $\omega-\Omega$ pulse (in this case below the noise level) leaving the $\omega+\Omega$ pulse larger. The fact that several noise peaks have an amplitude comparable to the $\omega+\Omega$ pulse in that dataset makes this plausible. A recent *ASCA*

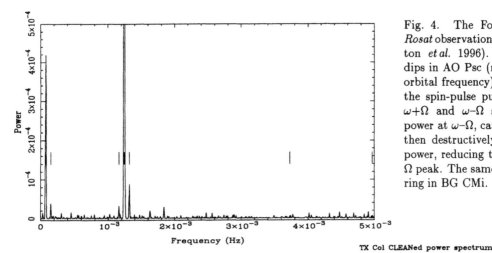

Fig. 4. The Fourier transform of a *Rosat* observation of AO Psc (from Norton *et al.* 1996). The strong orbital dips in AO Psc (note the power at the orbital frequency) amplitude-modulate the spin-pulse putting power into the $\omega+\Omega$ and $\omega-\Omega$ sidebands. Intrinsic power at $\omega-\Omega$, caused by disc-overflow, then destructively interferes with this power, reducing the strength of the $\omega-\Omega$ peak. The same appears to be occurring in BG CMi.

Fig. 5. The Fourier transform of a *Rosat* observation of TX Col, showing spin, $\omega-\Omega$ and $\omega-2\Omega$ frequencies (Norton & Beardmore, unpublished). The spin pulse arises from disc-fed accretion and the $\omega-\Omega$ pulse from stream fed accretion, while varying visibility of the stream-fed accretion sites causes power at $\omega-2\Omega$.

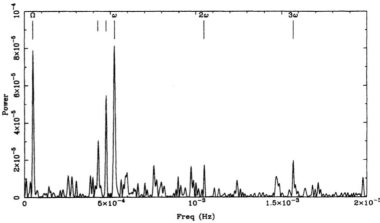

TX Col CLEANed power spectrum

observation gives a similar result, detecting both $\omega-\Omega$ and $\omega+\Omega$ sidebands, again with $\omega+\Omega$ stronger (Mukai, private communication). The strong orbital modulation needed for amplitude modulation is also seen. If this explanation for the BG CMi and AO Psc Fourier transforms is correct, it requires that the intrinsic beat pulse is anti-phased with the beat pulse from amplitude modulation. While there isn't yet enough information to establish this, there are good arguments for it being true (see Appendix).

V1223 SGR *Spin 746 s, orbit 3.37 hr*
Again the spin period identification is secured by a strong optical beat pulse. An *EXOSAT* observation showed a detection of an X-ray beat pulse (H91), while two further observations yield only upper limits (NBT) — another case of transient disc-overflow.

RX 1712–24 *Spin 927 s, orbit 3.41 hr*
This important star has a spin period identified by polarisation pulsed at 927 s (Buckley *et al.* 1995a). Pulsations in other observations, however, including the dominant X-ray pulse, are instead at the 1003-s beat period (Figure 6). Thus Buckley *et al.* propose that RX 1712–24 is the first predominantly stream-fed accretor. At this stage it is unclear whether it is completely discless, or whether a disc exists, but disc-overflow causes the accretion to bypass the disc to the extent that it is relegated to 'non-accretion disc' status, a possibility suggested by King and Lasota (1991). Recent XTE and *ASCA* observations (Haberl 1996; Hellier, unpublished) confirm that the beat pulse dominates the X-ray lightcurves, but in the *ASCA* observation most of the power is in the first harmonic. In a discless accretor it might be expected that the lightcurves are more chaotic and variable than in a disc-fed system.

EX Hya *Spin 67 mins, orbit 98 mins*

In EX Hya the signs of disc-overflow have been seen
only in the optical — a high velocity S-wave feature
from the re-impact point near the white dwarf —
and then only during its short, rare outbursts (Hellier *et al.* 1989; Buckley and Schwarzenberg-Czerny
1993). So far there has been no X-ray observation
of an outburst to test the prediction that a beat period would be seen. If it were it would be a welcome
confirmation of many of the ideas in this review.

AE Aqr *Spin 33 s, orbit 9.88 hr*

This oddity has an extremely fast spinning white
dwarf, which appears to be ejecting most of the
transfered material rather than accreting it. The
white dwarf is spinning down rapidly, and the energy lost greatly exceeds the observed accretion luminosity (e.g. De Jager 1995). There may be no
disc, so blobs of material interact directly with the
magnetic field, being accelerated out of the system
(Wynn *et al.* 1995). If the centrifugal barrier slows
down the accreting material sufficiently, the small
proportion that does accrete might circularize to
produce spin-period pulsations, rather than beat-
pulsed X-rays.

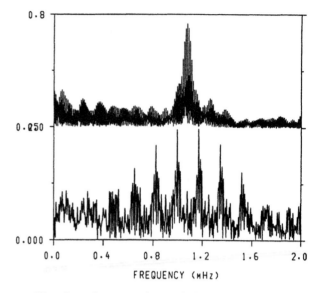

Fig. 6.   A comparison of Fourier transforms of
polarimetry (top) and X-rays (bottom) from RX 1712–
24. The polarimetry varies with the spin period,
while the X-ray is dominated by a beat pulse, making RX 1712–24 the first discless intermediate polar.
From Buckley *et al.* (1995b).

GK Per *Spin 351 s, orbit 2 days*

GK Per's 2-day orbit means that no X-ray observation has had the frequency resolution to separate the beat
and spin periods. In any case, GK Per's size implies that any disc-overflow would probably fall outside the
magnetosphere. However, during an outburst in 1983 there was a 5000-s quasi-periodic variation which was
possibly caused by disc-overflow (Hellier and Livio 1994). The timescale of the QPO is that of a Keplerian
orbit at the radius of the re-impact point; further, the energy dependence is correct if structure orbiting
at this radius is periodically obscuring the white dwarf and absorbing the X-rays. It is uncertain, though,
whether an overflowing stream could create such structure, so the interpretation remains problematical.
QPOs on the same timescale have also been seen in the 1996 outburst, both in X-rays (Ishida *et al.* 1996)
and optical spectroscopy (Morales-Rueda *et al.* 1996).

TV Col, PQ Gem, XY Ari, YY Dra

These stars have all been observed sufficiently to reveal any obvious X-ray beat pulses, but, to my knowledge, none have been seen. Of course, given the variability in disc-overflow detailed above, this doesn't rule
out such results in the future. For now they are presumed to be pure disc accretors. Note that PQ Gem
has a high field strength comparable to that of RX 1712–24 (~ 10 MG), but still appears to have a disc. In
particular, its polarisation modulation is at the same frequency as the dominant X-ray pulse (e.g. Mason
1997). There are another ~ 10 known intermediate polars (see Patterson 1994, and the updated membership
list in Hellier 1996) but most have not been observed well enough to comment on their accretion mode.

Asynchronous AM Hers

V1500 Cyg, BY Cam and RX 1940–10 are AM Her stars which are asynchronous by ~ 1%. The distinction
between asynchronous AM Hers and intermediate polars is somewhat arbitrary but, so far, clear cut, since
the asynchronism in intermediate polars is at least an order of magnitude larger. Because they are close
to synchronism their beat periods are tens of days and hard to observe. The lightcurves, though, are very
chaotic, and do seem to show the expected pole-flipping on the beat cycle. Further, Mason *et al.* (1995) have
reported seeing the 2ω–Ω pulse predicted by Wynn and King (1992) in optical lightcurves of BY Cam.

UPSHOT

In many of the well studied intermediate polars there is strong evidence, principally the presence of X-ray beat-period pulsations, that disc-overflow accretion occurs. In all cases it is variable, ranging from times when no beat-pulse is seen (and presumably accretion is purely disc-fed) to occasions when the beat and spin pulses have comparable amplitudes. Note that every star (except RX 1712–24) has been observed at least once with no X-ray sideband pulses. This rules out the possibility that the systems are discless, and that the spin pulses are the product of highly asymmetric magnetic fields. In an *ASCA* observation of FO Aqr where the beat pulse had an amplitude of $1/3^{rd}$ of the spin pulse, Mukai *et al.* (1994) used a spectral deconvolution to estimate that 2% of the luminosity originated from disc-overflow, and thus presumably that the overflowing stream carried $\sim 2\%$ of the accretion at that time. The upshot is therefore that a dozen intermediate polars are predominantly disc-fed, with small and variable amounts of disc-overflow, and two stars, RX 1712–24 and AE Aqr, are likely to be stream accretors and may not have discs.

The incidence of disc-overflow amongst cataclysmic variables in general seems to be correlated with a high mass transfer rate. The distorted emission line profiles of the SW Sex group of nova-likes can also be explained by disc-overflow (e.g. Hellier and Robinson 1994) and there is evidence, such as accretion disc winds, that they have a very high mass transfer rate. Further, with the exception of EX Hya in outburst (where the mass transfer rate might be enhanced) no sign of disc-overflow has been reported in any of the lower $\dot{M}$ systems below the period gap.

The amount of accretion carried by the stream relative to the disc has more than incidental importance. Warner and Wickramasinghe (1991) and King and Lasota (1991) have proposed that intermediate polars cycle between disc-fed and discless states over long timescales. King and Lasota (1991) suggest that the tendency for systems to have spin periods near 10% of the orbital cycle is explained if, over timescales of $\sim 10^4$ years, they accrete predominantly from the stream. Known discless systems, though, are a minority, while the stream in the disc-overflow systems appears to be a minor carrier of accretion. Thus, for this explanation to work the disc-fed systems need to be brighter, so that they are over-represented in the known sample. A disc will obviously make a system brighter in the optical, but most known intermediate polars were identified from X-ray surveys. As pointed out by Warner and Wickramasinghe (1991), however, discless accretion is likely to be clumpier than magnetic stripping of an accretion disc. Smooth accretion results in a stand-off shock and the hard bremsstrahlung X-ray emission characteristic of many systems, while clumpy accretion will instead produce most output in the EUV, where interstellar absorption will make it much harder to observe. A snag with this argument, though, is that the known discless system RX 1712–24 has a hard bremsstrahlung X-ray spectrum, rather than the soft black-body spectrum shown by some of the *Rosat* discoveries (Haberl & Motch 1995). In summary, therefore, although the accretion in the known intermediate polars is predominantly disc-fed, the issue for the class overall is not yet resolved.

ACKNOWLEDGMENTS

I am grateful to Koji Mukai, Andrew Beardmore and Andrew Norton for useful comments and for communicating results prior to publication. Koji Mukai also contributed to the arguments in the appendix.

REFERENCES

Armitage, P. J., and M. Livio, *Astrophys. J,* 470, 1024 (1996).

Beardmore, A. P., K. Mukai, A. J. Norton, J. P. Osborne, and P. Taylor, *Proc. Röntgenstrahlung from the Universe, MPE report,* 263, 123 (1996).

Buckley, D. A. H., and A. Schwarzenberg-Czerny, *Annals of the Israeli Physical Society,* 10, 278 (1993).

Buckley, D. A. H., K. Sekiguchi, C. Motch, D. O'Donoghue, A-L. Chen, *et al.,* *MNRAS,* 275, 1028 (1995a).

Buckley, D. A. H., K. Sekiguchi, D. O'Donoghue, A-L. Chen, *et al.* *ASP Conf. Series,* 85, 249 (1995b).

Buckley, D. A. H., and I. R. Tuohy, *Astrophys. J,* 344, 376 (1989).

De Jager, O. C., *ASP Conf. Series,* 85, 373 (1995).

Frank, J., A. R. King and J.-P. Lasota, *Astron. & Astrophys.*, 178, 137 (1987).

Haberl, F., these proceedings, (1996).

Haberl, F., and C. Motch, *Astron. & Astrophys.*, 297, L37 (1995).

Hameury, J.-M., A. R. King, and J.-P. Lasota, *MNRAS*, 218, 695 (1986).

Hellier, C., *MNRAS*, 251, 693 (1991).

Hellier, C., *MNRAS*, 258, 578 (1992).

Hellier, C., *MNRAS*, 265, L35 (1993).

Hellier, C., In: *Proceedings of IAU Colloq. 158*, Kluwer, (1996).

Hellier, C., M. A. Garlick and K. O. Mason *MNRAS*, 260, 299 (1993).

Hellier, C., and M. Livio, *Astrophys. J*, 424, L57 (1994).

Hellier, C., K. O. Mason, and M. S. Cropper, *MNRAS*, 242, 250 (1990).

Hellier, C., K. Mukai, M. Ishida, and R. Fujimoto, *MNRAS*, 280, 877 (1996).

Hellier, C., K. O. Mason, A. P. Smale, R. H. D. Corbet, D. O'Donoghue, *et al. MNRAS*, 238, 1107 (1989).

Hellier, C., and E. L. Robinson, *Astrophys. J*, 431, L107 (1994).

Ishida, M., A. Yamashita, H. Ozawa, F. Nagase, and H. Inoue, *IAU Circ.* 6340 (1996).

King, A. R., and J. P. Lasota, *Astrophys. J*, 378, 674 (1991).

Lubow, S. H., *Astrophys. J*, 340, 1064 (1989).

Lubow, S. H., and F. H. Shu, *Astrophys. J*, 207, L53 (1976).

Morales-Rueda, L., M. D. Still, and P. Roche, *MNRAS*, 283, L58 (1996).

Mason, K. O., *MNRAS*, 285, 493 (1997).

Mason, K. O., S. R. Rosen, and C. Hellier, In: *The physics of compact objects*, *Adv. in Space Res.*, 8, (2)293, Pergamon Press, (1988).

Mason, P. A., I. L. Andronov, S. V. Kolesnikov, and E. P. Pavlenko, *ASP Conf. Series*, 85, 496 (1995).

Mukai, K., M. Ishida, and J. P. Osborne, *Pub. Astron. Soc. Japan*, 46, L87 (1994).

Norton, A. J., A. P. Beardmore, and P. Taylor, *MNRAS*, 280, 937 (1996).

Norton, A. J., I. M. McHardy, H. J. Lehto, and M. G. Watson, *MNRAS*, 258, 697 (1992b).

Norton, A. J., M. G. Watson, A. R. King, H. J. Lehto, and I. M. McHardy, *MNRAS*, 254, 705 (1992a).

Patterson, J., *Pub. Astron. Soc. Pacific*, 106, 209 (1994).

Shafter, A. W., F. V. Hessman and E. H. Zhang, *Astrophys. J*, 327, 248 (1988).

Warner, B., *MNRAS*, 219, 347 (1986).

Warner, B., and D. T. Wickramasinghe, *MNRAS*, 248, 370 (1991).

Wheatley, P. J., *Proceedings of the ASCA 3rd Anniversary Symposium*, in press (1996).

Wynn, G. A., and A. R. King, *MNRAS*, 255, 83 (1992).

Wynn, G. A., A. R. King, and K. Horne, *ASP Conf. Series*, 85, 196 (1995).

APPENDIX

I have relegated the following argument to this appendix because it is complex and not yet secure. In order to explain the Fourier transforms of BG CMi and AO Psc I have proposed that intrinsic power at $\omega-\Omega$ caused by disc-overflow interferes destructively with power at that frequency caused by amplitude modulation of the spin pulse, which requires that the two are out of phase.

The orbital minima are normally caused by material in the stream or disc-impact regions obscuring the white dwarf, and so occur near orbital phase 0.8 (e.g. Hellier *et al.* 1993). At this orbital phase the sidebands caused by amplitude-modulation of ω will be anti-phased with ω (so that the pulse amplitude is reduced).

Now, if, at this orbital phase, the intrinsic $\omega-\Omega$ pulse is in phase with the ω pulse, it will be anti-phased with the amplitude-modulation $\omega-\Omega$ pulse, as required. This will occur if the spin and beat pulses have similar causes — for instance if the spin pulse were at minimum when the upper pole points towards us and the beat pulse were also at minimum when the stream was feeding the upper pole pointing towards us (or if they were both at maximum at these points). While this is plausible we don't yet understand the pulse mechanisms sufficiently to state that it must be the case. However, the argument is open to test simply by deriving the phasings of the $\omega-\Omega$ and $\omega+\Omega$ sidebands.

Adv. Space Res. Vol. 22, No. 7, pp. 981–986, 1998
Published by Elsevier Science Ltd on behalf of COSPAR
Printed in Great Britain
0273-1177/98 $19.00 + 0.00

 Pergamon

PII: S0273–1177(98)00130–6

RADIO PULSARS OBSERVED AT X-RAY ENERGIES BY THE X-RAY TIMING EXPLORER

L.E. Peterson[1], D. Marsden[1], P. Blanco[1], D.E. Gruber[1], W.A. Heindl[1], M.R. Pelling[1], R.E. Rothschild[1], P.L. Hink[2], J. Swank[3], K. Jahoda[3] and A. Rots[3]

[1] *CASS 0424, UC San Diego, 9500 Gilman Dr., La Jolla, CA 92093, USA*
[2] *Washington University, Dept. of Physics, St. Louis, MO 63130, USA*
[3] *NASA Goddard Space Flight Center, Greenbelt, MD 20771, USA*

ABSTRACT

The Rossi X-Ray Timing Explorer observed a number of X-ray emitting radio pulsars soon after launch on 30 Dec. 1995. These included two plerionic systems formed in recent supernova events, the Crab Nebula with its 33 ms pulsar PSR 0531+21 and MSH 15–52 with its 151 ms pulsar PSR 1509-58. Observations of these sources for 20–30 ksec with both the PCA and HEXTE allowed precision phase resolved spectroscopy for the Crab Nebula, confirmation of the light curve and pulsed spectrum for PSR 1509-58, and separation of the Nebular emissions from the pulsed X-rays over the 18–250 keV range for both sources. Within statistical uncertainties, a power law spectrum fits all the observations over this limited energy range. The nebular region for both sources has a photon index of about -2.1, while the pulsed spectra are considerably harder. Significant differences in spectral index are observed for the first pulse, the second pulse, and the interpulse of PSR 0531+21.   Published by Elsevier Science Ltd on behalf of COSPAR.

INTRODUCTION

Only a handful of isolated, rotation-powered radio pulsars have been observed in the X-ray and gamma-ray range. In particular, two of these so far observed are embedded in a remnant produced in a supernova explosion which has occurred within the past several thousand years. These so-called plerionic systems are of considerable interest because the nebular regions are believed powered by the rapidly rotating magnetized neutron star, which also emits the pulsed electromagnetic emissions. In this paper we report results observed with the Rossi X-Ray Timing Explorer (RXTE) on the Crab Nebula and its 33 ms pulsar PSR 0531+21, and on MSH 15–52 and its 151 ms pulsar PSR 1509-58. Results obtained include precision phase resolved spectroscopy of the X-ray pulsars in the 2–250 keV range, of interest to determine the geometry of the emitting region, and to infer its nature in terms of the "polar cap" and "outer gap" emission models. Also results were obtained on the off-pulse, presumably nebular, spectra of both sources over the 18–250 keV range. These spectra are relevant to the injection acceleration and transport of the relativistic electrons in the nebular region. These electrons presumably produce the diffuse X-ray continuum by synchrotron radiation.

OBSERVATIONS

The observations reported here were obtained with instruments on the RXTE during the Initial Orbital Calibration phase (IOC) immediately following the launch on 30 December 1995. Data were obtained with both the 15,000 cm^2 Proportional Counter Array (PCA) which operates over the 2–30 keV range, and with the 1600 cm^2 High Energy X-Ray Timing Experiment (HEXTE) which operates over the 15–250 keV range. Both instruments and the RXTE are described elsewhere (Bradt *et al.* 1991; Gruber *et al.* 1995; Jahoda *et al.* 1996).

Table I indicates intervals the Crab Nebula and MSH 15–52 were observed during January 1996. A total of about 20 ksec were obtained on MSH 15–52 and about 30 ksec on the Crab Nebula. The HEXTE was operated in a rocking on-source, off-source mode so direct background subtraction was obtained. Both the Crab Nebula and MSH 15–52 extend on the order of a few arc minutes, so the entire source region, pulsar + nebula, were in the one degree field of view of the instruments, when observing each source.

Table 1. RXTE Observations of the Crab and PSR 1509-58.

Object	Obs. Date m/d/y	Start (U.T.) hh:mm:ss	End (U.T.) hh:mm:ss	Livetime (s)
Crab	1/10/96	14:46:40	20:17:51	7740
Crab	1/11/96	13:05:20	17:05:04	6855
Crab	1/12/96	13:10:08	15:29:52	4620
Crab	1/15/96	16:33:36	21:57:52	9405
1509-58	1/6/96	00:06:40	03:24:00	7905
1509-58	1/17/96	20:05:51	22:34:40	5865
1509-58	1/26/96	18:48:00	21:08:00	5820

CRAB NEBULA AND PSR 0531+21

Pulsed spectral data on the Crab Nebula and PSR 0531+21 were folded at the radio period, ~ 33.3901 ms. This period was verified with a power spectral analysis, which showed a very strong narrow peak at the expected frequency. Figure 1 shows the light curves of the PCA data, and the HEXTE data in four energy ranges. The light curves confirm previous observations in this X-ray range (Knight 1982; Ulmer *et al.* 1994) which shows a narrow first pulse phase with the radio pulse, a broader second pulse, and considerable emission in the interpulse or bridging region. The region of the lowest emission, phase 0.7–1.0, which we call the off-pulse region, is probably the true flux from the nebula for the HEXTE data, since the detector background is corrected out by the on-source, off-source nature of the measurement. Clearly the second peak has a harder, i.e. less steep, spectrum in this energy range, since the ratio of the fluxes at the peaks reverses with increasing energy. This is consistant with recent observations by OSSE in the 50–550 keV range (Ulmer *et al.* 1994).

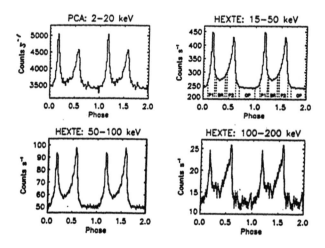

Fig. 1.   The Crab Nebula system folded light curve as seen by RXTE in various energy bands. The dotted lines in the top-right panel show the phase regions used in the spectral analsis.

The precision data with its high statistical significance permits pulse phase spectroscopy to an accuracy unavailable in previous work. Also shown in Figure 1 are the regions the 60 phase bins were accumulated over to allow spectral fitting to the data. These regions are consistant with those identified in previous work (Knight 1982; Ulmer *et al.* 1994). The count rate data, after being accumulated in the defined intervals, were converted to a photon spectrum, using the response matrix in the HEXTE analysis tool XSPEC. Because of the uncertain background correction and response matrix, the PCA data were not included in the spectral analysis. Figure 2 shows the spectral data in the various regions with ± 1 σ statistical errors indicated. The best bit fit power law is indicated as a histogram, computed over the width of the pulse height data channels.

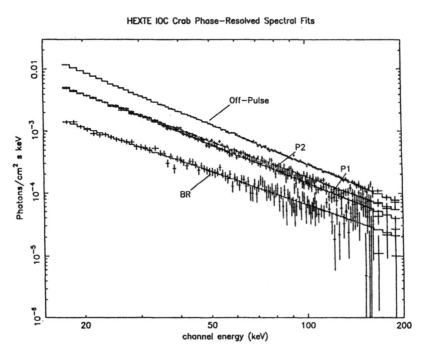

Fig. 2. HEXTE inferred incident photon spectra for the various phase intervals of the Crab Nebula and its pulsar. The best-fit power law component is indicated by a solid line.

The results of the fitting, shown in Table 2, indicate that a power law provides a statistically satisfactory fit to each phase region. This data confirms and quantifies the inference that the leading pulse P1 is significantly softer than the second pulse P2 by 0.15 ± 0.02 of a power law index in photon space. The bridging region, BR, is harder yet in this spectral range, and the off-pulse data are consistent with previous measurements of the nebular flux (Knight 1982; Jung 1989).

Table 2.   Phase-Resolved Spectral Fit Results.

Object	Phase	Energy Range (keV)	Photon Index	χ_ν^2	ν
Crab	P1	18 − 250	2.07 ± 0.01	1.141	230
Crab	BR	18 − 250	1.79 ± 0.02	1.015	230
Crab	P2	18 − 250	1.92 ± 0.01	1.131	230
Crab	Off-Pulse	18 − 250	2.130 ± 0.002	1.217	230
1509-58	P1	18 − 256	1.38 ± 0.12	0.914	237
1509-58	Off-Pulse	18 − 256	2.15 ± 0.18	0.977	237

Pulsed emissions of rotating magnetized neutron stars are generally explained in terms of the polar cap model (Daugherty & Harding 1982) or the outer gap model (Cheng, Ho & Ruderman 1985a,b). For the latter, Ho (1989) has reduced the spectrum descriptor of hard X-rays to a single parameter, which represents the linear fraction of the magnetosphere occupied by the magnetosphere gap. According to the results obtained by OSSE (Ulmer *et al.* 1994) in the 50–550 keV range, the spectrum of the pulsed emission is consistent with the outer gap model if the gap parameter f_G, is 0.46. Our results are in agreement with this supposition.

MSH 15–52 AND PSR 1509–58

The 150.689 ms pulsar in MSH 15–52 was first detected in X-rays by Seward and Harnden (1982), and confirmed in the radio by Manchester *et al.* (1985). The ROSAT image (Greiveldinger 1995) of MSH 15–52 is dominated by a nebular region several arc minutes in size surrounding the pulsar in the center of the image, and a more compact source, associated with an Hα region, $\sim 10'$ minutes to the North. ASCA observations (Tamara *et al.* 1996) have confirmed that the hot Hα is heated by a relativistic jet extending from the region containing the pulsar. The supernova which produced the system is believed to be about 1500 years old (Thorsett 1992; Chin & Huang 1994).

The light curves of PSR 1509–58 obtained by RXTE during January 1996 are shown in Figure 3. Since the total emission of the MSH 15–52 system is only about 60 mCrab at 30 keV, the statistical significance of the data obtained in the 20 ksec observation is appreciably less than on the Crab Nebula and its pulsar. Nevertheless, the observations confirm a single broad hard X-ray pulse, and on "off-pulse" region of about 0.5 phase. Because of the limited exposure time, significant data was not obtained at energies above about 150 keV and phase resolved spectroscopy could be accomplished only over two phase intervals, as shown in Figure 3.

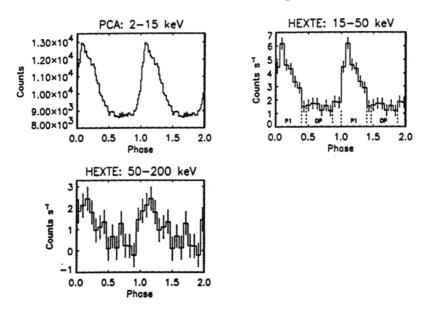

Fig. 3. The folded light curves of the PSR 1509–58/MSH 15–52 system in various energy bands. The dotted lines in the top right hand panel show the phase regions used in the spectral analysis.

The spectra obtained in the "on-pulse" region P1, and the "off-pulse" region OP are also shown in Figure 4. The results of fitting a single power law to each phase interval of the HEXTE data are also shown as a continuous histogram in the Figure. The PCA data was only fit to the main pulse, because of the uncertain background correctic in the off-pulse intervals. The results for the on-pulse intervals in the 18 to ~ 150 keV range are consistent with the measurements reported at lower energies by Seward *et al.* (1984) and are consistent with the less significant spectra index of 1.64±0.4 obtained over the 94–240 keV range from a balloon observation by Gunji *et al.* (1994).

Assuming the off-pulse spectrum shown in Figure 4 is indeed that of the integrated nebula, the HEXTE on RXTE has produced the first measurement of the nebular component spectrum in its energy range. The nebular spectral index, where a statistically significant fit could be obtained is -2.15±0.18. The data do not permit fitting a second power law to determine if a break in the nebular spectrum exists. Changes in the power law shape are indicative of acceleration and transport mechanisms associated with the relativistic electrons presumably injected in the nebular region by the pulsar. The off-pulse, presumably nebular spectrum is remarkably similar in shape to that in the Crab Nebula in this region (Jung 1989). He found that spectrum to be about -2.08 in the 1–100 keV region, with a break at about 150 keV to a slope of $\sim$ 2.5. This data is consistent with more recent observations by OSSE (Ulmer *et al.* 1994) on the Crab Nebula. It is of considerable interest to determine if a similar break occurs in MSH 15–52. Clearly a much longer observation is needed to settle this important question.

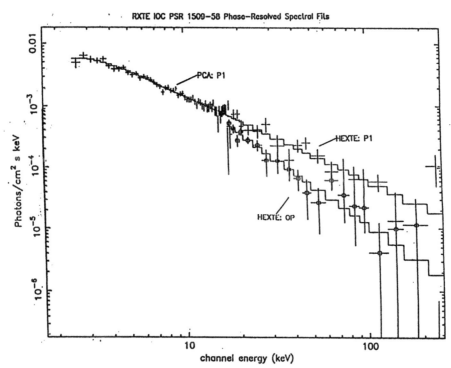

Fig. 4. Inferred incident photon spectra from PSR 1509–58/MSH 15–52, showing the pulsed component (PCA & HEXTE) and the off-pulse component (HEXTE only). The best-fit power law models for the pulsed and off-pulse components are indicated by the solid lines.

CONCLUSION

In this work, we have reported new determinations of pulsed and non-pulsed X-rays for two recently formed plerion systems containing radio and hard X-ray pulsars: the Crab Nebula and its pulsar PSR 0531+21, and MSH 15–52 and its pulsar PSR 1509–558. Phase resolved spectra were obtained for each source, and could be fit with power law spectral shapes. The Crab pulsar is consistent with previous work in that the second pulse is harder than the leading pulse. The off-pulse, presumably nebular, component, is softer than the pulsed component, and is relatively more intense in the 18–250 keV range. The total flux (pulsed and non-pulsed) from MSH 15–52 is about 60 mCrab. The single broad pulse has a power law photon spectrum index of $\sim$ 1.4, harder than the peak emissions of the Crab pulsar. The first measurements of the off-pulse component show the nebula to have a power law index of about 2.1 in the 18–150 keV range. The off-pulse components is softer, and relatively less intense than the Crab Nebula. Longer observations of MSH 15–52 are needed to accomplish phase-resolved spectroscopy in many phase bins, and to search for the possibility of a change in the power law slope in this spectral region.

ACKNOWLEDGEMENTS

This was supported by NASA contract NAS5-30720 at UCSD.

Note Added In Proof: A more complete analysis of the RXTE data on PSR 1509-58/MSH 15-52 has been published by Marsden, *et al.* (1997).

REFERENCES

Bradt, H.V., J.H. Swank, and R.E. Rothschild, The X-Ray Timing Explorer, *Adv. Space Res.*, **11**, No. 8, 243 (1991).

Cheng, K.S., C. Ho, and M. Ruderman, Energetic Radiation from Rapidly Spinning Pulsars. 1. Outer Magnetosphere Gaps, *Ap. J.*, **300**, 500 (1986).

Cheng, K.S., C. Ho, and M. Ruderman, Energetic Radiation from Rapidly Spinning Pulsars. II. Vela and Crab, *Ap. J.*, **300**, 522 (1986).

Chin, Y-N., and Y-L. Huang, Identification of the Guest Star of AD 185 as a Comet Rather than a Supernova, *Nature*, **371**, 398 (1994).

Daugherty, J.K., and Harding, A.K., Electromagnetic Cascades in Pulsars, *Ap. J.*, **252**, 337 (1982).

Greiveldinger, C., S. Caucino, S. Massaglia, H. Ögelman and E. Trussonl, PSR 1509-58 and Its Plerionic Environment, *Ap. J.*, **454**, 855 (1995).

Gruber, D.E., P.R. Blanco, W.A. Heindl, al.., The High Energy X-Ray Timing Experiment on XTE, *Astron. & Astrophy.*, 120, 641 (1996).

Gunji, S., *et al.*, Observation of Pulsed Hard X-Rays/γ-Rays from PSR 1509-58, *Ap. J.*, **428**, 284 (1994).

Ho, C., Spectra of Crab-Like Pulsars, *Ap. J.*, **342**, 396 (1989).

Jahoda, K. *et al.*, In Orbit Performance and Calibration of the Rossi X-Ray Timing Explorer (RXTE) Proportional Counter Array (PCA), *Proc. SPIE*, 2808, 59 (1996).

Jung, J.V., The Hard X-Ray to Low-Energy Gamma-Ray Spectrum of the Crab Nebula, *Ap. J.*, **338**, 972 (1989).

Knight, F.K., Observations and Interpretations of the Pulsed Emission form the Crab Pulsar, *Ap. J.*, **260**, 538 (1982).

Manchester, R.N., J.M. Durdin, and L.M. Newton, A Second Measurement of a Pulsar Braking Index, *Nature*, **313**, 374 (1985).

Marsden, D. *et al.*, The X-Ray Spectrum of the Plerionic System PSR B1509-58/MSH 15-52, *Ap.J. (Letters)*, 491, L39 (1997).

Seward, F.D., and F.R. Harnden, A New, Fast X-Ray Pulsar in the Supernova Remnant MSH 15-52, *Ap. J.*, **256**, L45 (1982).

Seward, F.D., F.R. Harnden, Jr., A. Szymkowiak, and J. Swank, The X-Ray Nebula Around PSR 1509-58, *Ap. J.*, **281**, 650 (1984).

Tamura, K., N. Kawai, A. Yoshida, and W. Brinkmann, Evidence for a Pulsar Jet Producing a Hot Nebula in the Supernova Remnant MSH 15-52, *PASJ*, **48**, L33 (1996).

Thorsett, S.E., *Nature*, **356**, 690 (1992).

Ulmer, M.P., *et al.*, OSSE Observations of the Crab Pulsar, *Ap. J.*, **432**, 228 (1994).

Adv. Space Res. Vol. 22, No. 7, pp. 987–996, 1998

Pergamon

PII: S0273-1177(98)00128-8

CYCLOTRON LINE VARIABILITY

T. Mihara[1], K. Makishima[2], and F. Nagase[3]

[1] *The Institute of Physical and Chemical Research, 2-1 Hirosawa, Wako, Saitama 351-01, Japan*
[2] *Dept. of Physics, School of Science, Univ. of Tokyo, 7-3-1 Hongo, Bunkyo-ku, Tokyo 113, Japan*
[3] *Inst. of Space and Astronautical Science, 3-1-1 Yoshinodai, Sagamihara, Kanagawa 229, Japan*

ABSTRACT

We systematically analyzed the spectra of X-ray binary pulsars observed with *GINGA* (Mihara 1995). A new model NPEX (Negative and Positive power-laws EXponential) was introduced to represent the pulsar continuum. Combining the NPEX continuum with the CYAB factor (cyclotron resonance scattering model), we successfully fit the whole-band spectra of all the pulsars. A possible physical meaning of the NPEX model is the Comptonized spectra.

By using the smooth and concave NPEX model, the cyclotron structures were detected from 12 pulsars, about a half of the 23 sources, including new discoveries from LMC X-4 and GS 1843+00. The magnetic fields were scattered in the range of 3×10^{11} – 5×10^{12} G. The distribution was shown for the first time, which is remarkably similar to that of radio pulsars with a peak at 2×10^{12} G.

The double harmonic cyclotron structures of 4U 0115+63 in 1990 changed to a single structure in 1991. The resonance energy also increased by 40 % as the luminosity decreased to 1/6. If we attribute this change to the height of the scattering region in a dipole magnetic field, the height change is ~ 1.1 km. Such changes of the resonance energies with luminosities are observed from 5 pulsars and can be explained by the accretion column height model.

©1998 COSPAR. Published by Elsevier Science Ltd.

INTRODUCTION

The X-ray binary pulsar is a neutron star in a contact binary system with mostly a high-mass star. The neutron star is highly magnetized ($\sim 10^{12}$G), collimates the accreting matter onto magnetic poles and shows X-ray pulses with the rotation. The electron cyclotron resonance structure in the X-ray spectrum is the only direct method to measure the magnetic fields on the neutron star. The resonance energy is $E_a[\text{keV}] = 11.6B[10^{12}\text{G}]$. The first report of the structure was from Her X-1 (Trümper *et al.* 1978), followed by that from 4U 0115+63 by Wheaton *et al.* (1979). But for the other pulsars, the magnetic fields are estimated by a rather uncertain method using accretion spin up/down theory. Spectra of X-ray binary pulsars look non-thermal and are not explained well. Theories have been proposed (eg. Meszaros 1992), but comparison with the data has not been done much.

GINGA LAC (Turner *et al.* 1989) not only discovered cyclotron structures from many pulsars, but also enabled us to discuss on the continuum spectra. Those two are related with each other. A good representation of the continuum spectra is essential to a precise analysis of the cyclotron structure. *GINGA* observed 23 pulsars including 1 pulsar-candidate with good statistics in its 4 years and 9 months life. First we introduce a new empirical continuum model NPEX, and discuss the meaning. Next we show the magnetic fields distribution. Last we discuss the variability of the cyclotron structure.

NPEX MODEL

It is known that a typical spectrum of a binary X-ray pulsar is a power-law (POWL) below ~20 keV and falls off exponentially at higher energies. The reason of the exponential cutoff (ECUT) was not known. *GINGA* found cyclotron structures at the bottom of the fall-off, which lead to a physical idea that the ECUT is created by the cyclotron resonances of the fundamental, 2nd, 3rd,... harmonics. Thus early studies of pulsars with *GINGA* were done by employing a power-law (POWL) model as a continuum and a cyclotron feature as an absorption (CYAB) (Makishima and Mihara 1992). They employed two resonance, the fundamental and the 2nd harmonics, because those two are within the *GINGA* energy range. This model succeeded to explain the overall spectrum in 8–60 keV of Her X-1 (Mihara *et al.* 1990). It favored an absorption at 34 keV rather than an emission at 50keV which was uncertain in the previous observations.

$$CYAB(E) = e^{-\tau_1}, \qquad \tau_1 = \frac{D_1(\frac{W}{E_a}E)^2}{(E - E_a)^2 + W^2}, \qquad \tau_2 = \frac{D_2(\frac{2W}{2E_a}E)^2}{(E - 2E_a)^2 + (2W)^2},$$

Here τ_1 is optical depth of the fundamental cyclotron scattering in a classical cold plasma. τ_2 is that for the 2nd harmonic. E_a is the resonance energy, W is the width of the resonance, and D_1 and D_2 are the depths of the resonances. Resonance energy and the width of the 2nd harmonic were fixed to the double of those of the fundamental in the fitting, because the 2nd resonance was almost at the end of the energy range and it was difficult to be obtained independently.

The flux of POWL×double CYAB's model goes back to the POWL level far above the resonance. Later the *HEAO-1* A4 spectrum of Her X-1 in 13–180 keV was published by Soong *et al.* (1990), but the data does not show the flux return. Putting the 3rd, 4th and 5th harmonics can reduce the flux, but it is not favorable because it requires larger cross section of the 3rd than the 2nd, larger the 4th than the 3rd. It is possible that optical depth of the 2nd is apparently larger than that of the fundamental because the two-photon decay of the 2nd harmonic may fill up the fundamental, but for higher harmonics than the 2nd, the reverse of the optical depths would not happen.

Another problem is on 4U0115+63, which is the only pulsar with a clear 2nd harmonic observed with *GINGA*. POWL× double CYAB's cannot explain the spectrum. The continuum needs to fall off by itself (Nagase *et al.* 1991).

In order to solve those problems, it is a better and natural idea to assume that the continuum falls off thermally by itself. We tried some continuum models together with a single CYAB to the Her X-1 spectrum which has the best statistics. We started with the simplest $\exp(-E/kT)\times$CYAB, but failed. Next we tried Boltzmann model $E^\alpha \exp(-E/kT)\times$CYAB, which was successful in 13-60 keV. But α became positive ($\alpha = 0.74$) and cannot fit the negative POWL region below 10 keV. Then, by adding negative POWL, we introduce the NPEX (Negative and Positive power-laws

EXponential) model as

$$NPEX(E) = (A_1 E^{-\alpha_1} + A_2 E^{+\alpha_2}) \, \exp\left(-\frac{E}{kT}\right),$$

where kT is a typical temperature of the X-ray emitting plasma, and α_1 and α_2 are the negative and positive POWL indices, respectively.

The NPEX×CYAB model can fit the Her X-1 spectrum very well in the entire 2–60 keV energy band with an iron line included. Moreover the positive index converged to $\alpha_2 = 1.97 \pm 0.26$, which suggests the blackbody ($\alpha_2 = 2$). This model can fit the pulse-phase-resolved spectra, too.

Table 1: The best-fit parameters with NPEX×CYAB model for the pulse averaged spectra. Errors are in 90% confidence level. Positive POWL index α_2 is fixed to 2.0. The units of A_1 and A_2 are [photons/s/keV/4000cm^2] at 10 keV. kT, E_a, W and E_{Fe} are in [keV], N_H is in [cm^{-2}], and I_{Fe} is in [photons/s/4000cm^2].

sources	Negative POWL		Pos. POWL	Exponential	Absorption
	A_1	α_1	A_2	kT	$\log_{10} N_H$
Her X-1	135 ± 8	0.51 ± 0.03	100 ± 23	8.0 ± 0.8	—
4U0115+63 (90)	491 ± 326	0.41 ± 0.48	4960 ± 1220	4.2 ± 0.1	—
4U0115+63 (91)	62 ± 22	0.65 ± 0.29	785 ± 58	4.3 ± 0.1	—
X0331+53	930 ± 63	-0.27 ± 0.05	630 ± 170	6.3 ± 0.5	—
1E2259+586	13 ± 12	1.42 ± 0.47	9 ± 7	2.1 fixed	—
LMC X-4	21 ± 1	0.43 ± 0.06	19 ± 2	7.3 ± 0.3	—
GS1843+00	45 ± 3	0.73 ± 0.08	47 ± 5	8.2 ± 0.2	22.29 ± 0.05
Cep X-4	101 ± 13	0.70 ± 0.05	110 ± 59	6.4 ± 1.5	22.01 ± 0.08
Vela X-1	171 ± 4	0.61 ± 0.05	123 ± 8	6.4 ± 0.1	22.41 ± 0.06
4U1907+09	11 ± 2	1.39 ± 0.29	25 ± 5	6.4 ± 0.7	22.86 ± 0.08
4U1538-52	19 ± 2	1.47 ± 0.20	68 ± 8	4.6 ± 0.2	22.80 ± 0.07
GX301-2	135 ± 71	0.80 ± 0.85	485 ± 184	5.4 ± 0.3	23.37 ± 0.07
	Leaky absorber, Norm ×0.38 ± 0.26				24.44 ± 0.22

sources		Resonance	Width	Depth	Iron Flux	Energy	χ^2_ν
		E_a	W	D	I_{Fe}	E_{Fe}	
Her X-1		33.1 ± 0.3	12.1 ± 1.7	1.53 ± 0.25	30 ± 4	6.65 fix	1.14
4U0115+63 (90)	1st	11.3 ± 0.6	5.9 ± 0.8	0.67 ± 0.08	17 ± 18	6.60 fix	0.69
	2nd	22.1 ± 0.4	5.2 ± 1.0	0.51 ± 0.07			
4U0115+63 (91)		15.6 ± 0.4	9.0 ± 0.6	1.22 ± 0.06	—	—	1.72
X0331+53		27.2 ± 0.3	7.5 ± 0.9	1.62 ± 0.15	38 ± 16	6.59 fix	1.65
1E2259+586		4.2 ± 0.6	2.0 ± 0.9	0.86 ± 0.27	—	—	8.71
LMC X-4		21.4 ± 1.2	5.1 ± 3.8	0.11 ± 0.05	2.7 ± 0.4	6.6 ± 0.1	0.83
GS1843+00		19.8 ± 2.1	9.9 ± 3.6	0.16 ± 0.05	7.6 ± 0.9	6.40 fix	0.49
Cep X-4		28.8 ± 0.4	12.1 ± 3.1	1.67 ± 0.59	7 ± 2	6.5 ± 0.1	0.97
Vela X-1	1st	24.5 ± 0.5	2.2 ± 1.0	0.065 ± 0.015	27 ± 3	6.5 ± 0.1	0.56
	2nd	$2E_{a1}$ fixed	$2W_1$ fixed	0.80 ± 0.26			
4U1907+09		18.9 ± 0.7	7.4 ± 2.1	0.87 ± 0.21	1.7 ± 0.5	6.60 fix	1.18
4U1538-52		20.6 ± 0.2	4.2 ± 0.6	0.83 ± 0.08	2.5 fixed	6.50 fix	1.58
GX301-2		37.6 ± 1.1	16.4 fixed	0.65 ± 0.17	32.1 ± 3.4	6.60 fix	0.98

We applied this model to other pulsars with α_2 fixed to 2. Only NPEX continuum is used to those without a cyclotron structure, NPEX×CYAB is used to those with a single structure, and NPEX×double CYAB's is used to those with two harmonics, which are 4U 0115+63 in 1990 and probably in Vela X-1. A merit of NPEX continuum model is that it is slightly concave as is often seen in the pulsar spectra in 2–10 keV range. With this continuum we discovered cyclotron structures from LMC X-4 and GS1843+00. The fitting parameters are summarized in Table 1.

MEANING OF NPEX MODEL

Let us consider the physical interpretation of the NPEX model. It would be natural to assume that kT is the typical temperature of the X-ray emitting plasma. We normalize the spectra with the energy of kT after correcting the detector efficiency and the absorption by the intervening matter. The flux level is normalized by the flux at $E = kT$ (Figure 1).

In the case of the non-cyclotron sources, the spectra obey a power law in the low energies, but with different indices, and show a round shoulder at around $E = 3kT$. Those are represented by the negative and positive POWL's of the NPEX model, respectively. As the slope of the power-law flattens, the hump at $E = 3kT$ increases, suggesting the existence of ONE hidden parameter which determines the shape of the continuum. The hidden parameter is also suggested from the pulse profiles sliced by some energy bands. Since the pulse shapes do not change below kT and above kT, the two POWL's cannot be independent, but are coupled by a hidden parameter.

In the case of the cyclotron sources, the overall curves are similar to those of the non-cyclotron sources. The difference is that the spectrum shows a steep fall-off at an energy, which is caused by the cyclotron resonance, and in some pulsars reaches a local minimum at the resonance center.

What mechanism creates both the negative POWL and the blackbody ? The multi-blackbody model would be possible, but an artificial distribution of temperature is needed. A better candidate is the Comptonization model. In fact the changes of the spectra in Figure 1 reminds us the Comptonized spectra for different optical depth τ. If a soft photon goes into the hot electron plasma, where scattering is more dominant than absorption, the photon gains energy by the inverse-Comptonization and comes out with a larger energy. When τ is small, the spectrum is a power law, and when τ is large, the Wien peak appears. τ can be the hidden parameter.

An analytic approximate calculation was done by Sunyaev and Titarchuk (1980) for a given soft photon input. The emergent photon spectrum is generally given as $(E^2 + o(E^2))\exp(-E/kT)$, where $o(E^2)$ is the polynomials with lower order than 2. For example, when the input photon has an index of $\alpha = -2$ in the high energy wing, the output photon spectrum $F(x)$ is expressed as

$$F(x) \propto e^{-x}\left(\frac{x^2}{24} + \frac{x}{6} + \frac{1}{2} + \frac{1}{x} + \frac{1}{x^2}\right), \qquad x \equiv E/(kT).$$

x^2 term corresponds to the positive POWL and the rest is combined to a negative POWL.

Figure 1 bottom is another Comptonization model by Lamb and Sanford (1979), (CMPL). It assumes a bremsstrahlung as the input photon. The change of the spectral shape with τ mimics that of the observed spectra. Thus, Comptonization model is a very possible candidate. The CMPL fits to the data ($\chi_\nu^2 = 1 \sim 3$) are not as good as NPEX fits, but it represents the overall shapes well. It would be because the averaged spectrum cannot be represented by an ideal model.

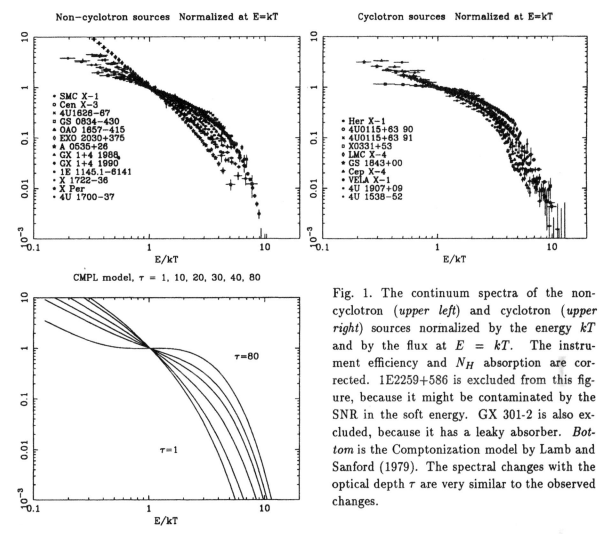

Fig. 1. The continuum spectra of the non-cyclotron (*upper left*) and cyclotron (*upper right*) sources normalized by the energy kT and by the flux at $E = kT$. The instrument efficiency and N_H absorption are corrected. 1E2259+586 is excluded from this figure, because it might be contaminated by the SNR in the soft energy. GX 301-2 is also excluded, because it has a leaky absorber. *Bottom* is the Comptonization model by Lamb and Sanford (1979). The spectral changes with the optical depth τ are very similar to the observed changes.

Then what determines those parameters ? The obtained τ has a negative relation with kT. As τ becomes thick, kT goes down, while kT does not depend on L_X nor spin period. The only parameter which has a possible relation with kT is the resonance energy E_a (Figure 2). In the plasma where scattering is dominant, the energy transfer from an electron to a photon is given by the Kompaneets equation (eg. Rybicki and Lightman 1979), as $\Delta E = E/(mc^2)(4kT - E)$. In an equilibrium, $E = 4kT$. Now the interacting photons are mainly that with the resonance energy because the cross section is extremely large. Consequently the temperature of the electrons is 'adjusted' to satisfy equation $kT = 0.25 E_a$. Monte-Carlo simulations of optically thick media by Lamb *et al.* (1990) find $kT \approx 0.27 E_a$, and it is applied to the γ-ray bursts. The $kT = 0.25 E_a$ relation, shown in Figure 2, is in a rough agreement with the data points.

Let us make sure that the Comptonization is the dominant process in the accretion column. Protons have most of the gravitational energy in the accreting matter. The time scales in which protons give energy to electrons, t_{col}, and electrons lose energy by the Comptonization, t_{comp}, are

$$t_{col} = 5 \times 10^{-5} \, n_{20}^{-1} \left(\frac{kT}{10 \text{ keV}} \right) \text{ [s]}, \qquad t_{comp} = 1 \times 10^{-15} \left(\frac{kT}{10 \text{ keV}} \right)^{-4} \text{ [s]}$$

(Gould 1982, Rybicki and Lightman 1979). Here n_{20} is the density in the unit of 10^{20} cm^{-3}, and Compton cross section is assumed to be $10^4 \sigma_T$ near the cyclotron resonance. Therefore electrons and photons interact much more strongly than protons and electrons, and electrons and photons are in the Comptonization equilibrium.

From the $E_a - kT$ relation one important suggestion can be deduced. Pulsars have kT between 4–14 keV (Table 1), which might indicate the cyclotron resonance energies are fairly constant within 10-60 keV. Moreover, the temperatures of the non-cyclotron sources are relatively higher than those of the cyclotron sources, which might mean that possible resonances are nearly at the high end of the energy range of *GINGA* and they are difficult to detected.

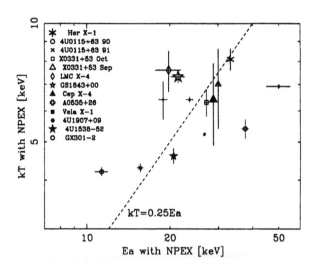

Fig. 2. Correlation between E_a and kT. There is a weak positive relation. If the Comptonization is dominant, $kT \approx 0.25 E_a$ is expected.

What is the source of the input soft photons, then ? The bottom of the accretion column or the neutron star surface are candidates. From the observational view, Her X-1 has a strong soft 0.1keV blackbody component (McCray 1982). Although its origin is said to be the inner accretion disk or the Alfven shell, some of it might come directly from the bottom of the accretion column.

We have used the Comptonization model without magnetic fields. Although the scattering cross section of an electron heavily depends on its energy in the magnetic fields, Meszaros (1992) notices that the continuum spectrum would be similar even in the magnetic fields except for the resonance. An absorption or an emission feature would be formed at around the resonance depending on the geometry and the optical depth of the scattering plasma. Readers might feel as if the Comptonized continuum is absorbed by CYAB, but it is not true. Those two are formed at the same time by the same scattering process.

MAGNETIC FIELDS DISTRIBUTION

We found the cyclotron structures from 11 pulsars among 23 X-ray pulsars including 1 pulsar-candidates. Adding A0535+26 from which *HEXE* discovered the cyclotron line (Kendziorra 1994), the cyclotron structures were detected from 12 pulsars, about a half of the 23 sources. Now we can make a distribution of the magnetic fields (Figure 3). The magnetic fields range between $3 \times 10^{11} - 5 \times 10^{12}$ G, which is similar to the life-corrected distribution of the radio pulsars (*right dotted line*), ranging between $10^{11} - 10^{13}$ G with a peak at 2×10^{12} G. The distribution of X-ray pulsars looks different from that of the *observed* radio pulsars (*right solid line*), which might indicate that the magnetic fields of the X-ray pulsars do not decay within a characteristic time scale of the radio pulsars ($10^6 - 10^7$ y). As the magnetic fields of the radio pulsars are obtained assuming the magnetic dipole radiation, only the dipole component are measured. On the other hand, those measured by the X-ray cyclotron structure are almost on the surface of the neutron

Magnetic fields distributions

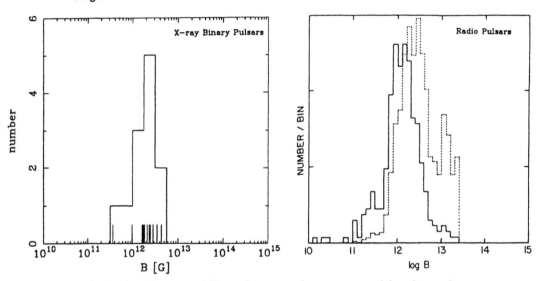

Fig. 3. The magnetic fields distribution of X-ray binary pulsars measured by the cyclotron resonances (*left*). The magnetic fields of the 12 pulsars are indicated with short lines on the horizontal axis. The distribution is similar to the life-corrected distribution of the radio pulsars (*right dotted line*), ranging in $10^{11} - 10^{13}$ G with a peak at 2×10^{12} G. The *right solid line* is that of the observed radio pulsars.

star and contain all multipole components. The agreement of the two indicates that the magnetic fields of the neutron star is dipole, and not multipole.

Let us discuss the selection effect. If there is a pulsar with a cyclotron resonance of less than a few keV, it is expected to show a steep power-law spectrum ($\alpha \sim 3$) in the *GINGA* energy range, as 1E2259+586. But all the other pulsars show a flat power-law in 2–10 keV, which suggests that pulsars with $E_a \lesssim 2$ keV does not exist. Although the detection limit towards the high energy is due to the *GINGA* energy range, kT and the $E_a - kT$ relation predict all the pulsars would have the cyclotron resonances in 4–60 keV. The pulsars with $E_a > 50$ keV are not likely. Since the data in Figure 3 already contain half the sources in the class, eventual inclusion of the others, even if all are at higher and lower energies, will not much change the distribution function of Figure 3. Thus, the magnetic field of X-ray binary pulsars are likely to cluster between $3 \times 10^{11} - 5 \times 10^{12}$ G.

RESONANCE ENERGY CHANGE

As shown in Figure 4 the cyclotron structure of 4U 0115+63 changed between the two observations. In 1990 it had double harmonic structures with the fundamental resonance at $E_a = 11$ keV. In 1991, however, it showed a broad single structure centered at $E_a = 16$ keV. It showed double/single structure throughout the pulse phases in 1990/1991, respectively. The luminosity in 1991 was 1/6 of that in 1990. *GINGA* observations of some sources with different intensities are summarized in Table 2. If we tentatively attribute the change of E_a to the height change of the scattering region ($E_a \propto r^{-3}$) and calculate the height difference assuming r in the weaker state is equal to the radius of the neutron star $R_{NS} = 10$ km, the height change is as much as 1.1 km in 4U 0115+63 as listed in Table 2 Δheight column.

Let us estimate the height of the accretion column employing the model by Burnard *et al.* (1991) to examine whether the change in height cited in Table 2 is reasonable or not. In the case of a pulsar, the accretion stream concentrates on the magnetic poles. Therefore Eddington limit of the emission along the magnetic fields is only $10^{35.7}$ erg/s. However, if the emission is sideward, most of the photon pressure is supported by the magnetic fields without stopping the accreting matter. Then the 'Eddington limit' L_1 becomes

$$L_1 = \theta_c L_{Edd} H_\perp \sim 10^{37.3} \frac{\theta_c}{0.1} \left(\frac{M_{NS}}{1.4 M_\odot}\right) H_\perp \quad \text{ergs/s.}$$

Here θ_c (~ 0.1) is the opening angle of the accretion column, $H_\perp$ (~ 1.3) is the ratio of the Thomson cross section and the Rosseland averaged cross section for the radiation flow across B. L_{Edd} ($= 2.0 \times 10^{38}$ erg/s) is the conventional Eddington Luminosity by the Thomson scattering.

When a pulsar emits as much as L_1, accretion flow yields a mound on the surface, whose height H_s would change in proportion to L_X.

$$H_s \approx \frac{L_X}{L_1} R_{NS} \theta_c = \frac{L_X}{L_{Edd} H_\perp} R_{NS} \tag{1}$$

Detailed calculations by Burnard *et al.* (1991) and Basko and Sunyaev (1976) justify this relation showing that the height H_s is the place where the radiation-dominated shock at Thomson optical depth $\sim 4 - 9$ transforms free fall matter into the subsonically settling on the mound.

We choose kT of the NPEX model as the temperature, assume $R_{NS} = 10$ km and $M_{NS} = 1.4 \, M_\odot$, and calculate $H_\parallel$ and H_s as listed in Table 2. $E_a - H_s$ relations are shown in Figure 5. If we assume a dipole magnetic field ($B \propto r^{-3}$) and the gravitational redshift ($E_a \propto r^{0.34}$ in $r = 10 \sim 11$ km), the predicted H_s agree very well with the observations for X0331+53 and 4U 0115+63 as indicated by the dashed lines. The values of 4U 1538-52 are also consistent, since the low luminosity forms a low mound where the height change is negligibly small when the luminosity changes by 1.3. The measured magnetic field of 4U 1538-52 would be almost that on the surface of the neutron star. These very good agreements of the observations and predictions support the assumption that the change of the resonance energy is caused by the height change of the accretion column by the luminosity and that the magnetic field is dipole. Some questions, however, remain such as why the cyclotron scattering is dominated at the top of the accretion column while the most emission is from the bottom of it.

Her X-1 does not appear to obey the relation, and E_a is changing independently of L_X. However, it has the 35 d intensity cycle and there are many reasons to change the apparent luminosity of Her X-1, such as an increase of the scattering gas, occultation by the accretion disk, change of the X-ray beam and so on. The three points of E_a in 1989 were on one line, and a point in 1990 is off. The circumstances might not have changed much during the same or sequent main-on.

In Cep X-4 we cannot calculate H_s since we do not know the distance to it. But by assuming the relations (eq. 1) and $E_a \propto r^{-2.66}$, we can obtain the distance. Unknown parameters are the distance and the surface magnetic field and we have three data points to be fitted. We obtain the distance to Cep X-4 of 3.2 ± 0.4 kpc. The H_s are calculated to be 210 m, 170 m, and 110 m on 1988/4/3, 8, 14, respectively. The luminosities are $\log_{10} L_X = 36.75$, 36.67, and 36.48, respectively. This can be a new method to estimate the distance to a binary X-ray pulsar.

NOTE: Cep X-4 was optically identified (Bonnet-Bidaud 1997, *IAU Circ.* 6724) using the position by ROSAT from the 1993 outburst (Schulz 1995, *A. & A.*, **295**, 413). The distance is 2.3-2.7 kpc from the reddening assuming the density of 1 H-atom cm^{-3}, or 3 kpc from the strong Na absorption line. Those are roughly consistent with our result.

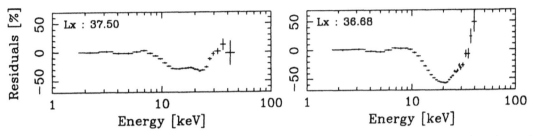

Fig. 4. Residuals from simple NPEX fits of 4U 0115+63 in 1990 (*left*) and 1991 (*right*). The cyclotron structure changed dramatically in the shape and the depth. The NPEX parameters are listed in Table 1.

Table 2: Cyclotron resonance energies with luminosities. $H_\parallel$ is a function of E_a/kT and obtained from Arons *et al.* (1987). H_s is the height of the accretion column calculated by eq. (1) from L_X and $H_\parallel$ assuming $M_{NS} = 1.4\ M_\odot$, $R_{NS} = 10$ km, and $\theta_c = 0.1$.

sources	count rate	E_a	kT	$\log_{10} L_X$	Δheight	$H_\parallel, H_\perp$	H_s
date	[c/s]	[keV]	[keV]	[erg/s]	[m]		[m]
4U 0115+63	3–50 keV						
1990/2/11	4036	11.3 ± 0.6	4.25 ± 0.10	37.50	1100 ± 220	1.23	1280
1991/4/26	661	15.6 ± 0.4	4.34 ± 0.14	36.68	0		203
X0331+53	3–37 keV						
1989/10/1	3586	27.2 ± 0.3	6.3 ± 0.5	37.43	330 ± 70	1.44	930
1989/9/20	2271	30.0 ± 0.5	7.0 ± 1.6	37.29	0		674
Cep X-4	2–37 keV						
1988/4/3	834	28.58 ± 0.5 $(\pm 0.05)^a$	7.5 ± 3.8	36.75^b	82 ± 109 (± 14)	1.38	210^b
1988/4/8	692	28.94 ± 0.4 $(\pm 0.05)^a$	7.1 ± 2.2	36.67^b	40 ± 102 (± 14)		170^b
1988/4/14	450	29.29 ± 0.8 $(\pm 0.11)^a$	6.4 ± 1.7	36.48^b	0		110^b
Her X-1	3–60 keV						
1990/7/27	1154	34.1 ± 0.4	8.1 ± 1.2	37.53	-160 ± 140	1.35	1250
1989/6/3	857	32.5 ± 1.0	10.3 ± 3.2	37.44	0 ± 110		1020
1989/6/6	792	32.5 ± 0.4	8.2 ± 0.9	37.40	0		930
1989/5/3	739	33.9 ± 1.2	9.7 ± 5.2	37.37	-140 ± 123		860
4U 1538-52	3–37 keV						
1988/3/2	184	20.6 ± 0.2	4.7 ± 0.3	36.56	0 ± 50	1.47	120
1990/7/27	130	20.6 ± 0.2	4.6 ± 0.2	36.43	0		91

a: Single parameter error, when other parameters than E_a are fixed to their best-fit values.

b: Estimated in this work assuming that the observed height changes are equal to the H_s changes. They would have very large errors because of the large errors of E_a. The most probable distance to Cep X-4 is 3.2 kpc.

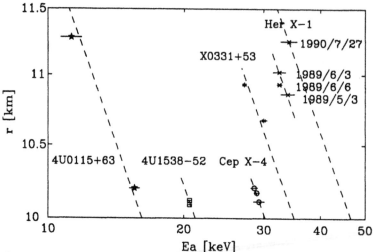

Fig. 5. The observed resonance energies and the heights of the accretion column estimated by a simple theory (eq. 1). r is the height from the center of the neutron star. $R_{NS} = 10$ km and $M_{NS} = 1.4M_\odot$ are assumed. The dashed line indicates $r^{-2.66}$ dependencies of the dipole magnetic field and the gravitational redshift. 4U0115+63, 4U1538-52, and X0331+53 obey this simple law well. Her X-1 does not obey this law, which would have other mechanisms to change the apparent luminosity. Assuming the $r^{-2.66}$ relation, the luminosities of Cep X-4 are calculated, which leads the distance to be 3.2 kpc.

REFERENCES

Arons, J., R. I. Klein, and S. M. Lea, *Astrophys. J.*, **312**, 666 (1987).

Basko, M. M., and R. A. Sunyaev, *Mon. Not. Roy. Astr. Soc.*, **175**, 395 (1976).

Burnard, D. J., J. Arons, and R. I. Klein, *Astrophys. J.*, **367**, 575 (1991).

Gould, R. J., Processes in Relativistic Plasmas, *Astrophys. J.*, **254**, 755 (1982).

Kendziorra, E., P. Kretschmar, H. C. Pan, M. Kuntz, M. Maisack *et al.* , *A. & A.*, **291**, L31 (1994).

Lamb, P. and P. W. Sanford, *Mon. Not. Roy. Astr. Soc.*, **188**, 555 (1979).

Lamb, D. Q., C. L. Wang, and I. M. Wasserman, *Astrophys. J.*, **363**, 670 (1990).

Makishima, K., and T. Mihara, Magnetic Fields of Neutron Stars, *Frontiers of X-ray Astronomy*, p23, ed. Y. Tanaka, and K. Koyama, Universal Academic Press Inc., Tokyo (1992).

McCray, R. A., J. M. Shull, P. E. Boynton, J. E. Deeter, S. S. Holt, *et al.* , EINSTEIN Observatory Pulse-phase Spectroscopy of Hercules X-1, *Astrophys. J.*, **262**, 301 (1982).

Meszaros, P., *High-Energy Radiation from Magnetized Neutron Stars*, University of Chicago Press (1992).

Mihara, T., K. Makishima, T. Ohashi, T. Sakao, M. Tashiro *et al.* , *Nature* , **346**, 250 (1990).

Mihara T., Ph.D. thesis for the physics degree of University of Tokyo (1995).

Nagase, F., Accretion-Powered X-Ray Pulsars, *Publ. Astr. Soc. Japan* , **41**, 1 (1989).

Nagase, F., T. Dotani, Y. Tanaka, K. Makishima, T. Mihara *et al.* , *Astrophys. J.*, **375**, L49 (1991).

Rybicki, G. R. and Lightman, A. P., *Radiative Processes in Astrophysics*, John Wiley & Sons, Inc. (1979).

Soong, Y., D. E. Gruber, L. E. Peterson, and R. E. Rothschild, *Astrophys. J.*, **348**, 641 (1990).

Sunyaev, R. A. and Titarchuk, L. G., *A. & A.*, **86**, 121 (1980).

Trümper, J., W. Pietsch, C. Reppin, W. Voges, R. Staubert *et al.* , *Astrophys. J.*, **219**, L105 (1978).

Turner, M. J. L., H. D. Thomas, B. E. Patchett, D. H. Reading, K. Makishima *et al.* , The Large Area Counter on Ginga, *Publ. Astr. Soc. Japan* , **41**, 345 (1989).

Wheaton, W. A., J. P. Doty, F. A. Primini, B. A. Cooke, C. A. Dobson *et al.* , *Nature* , **282**, 240 (1979).

Pergamon

Adv. Space Res. Vol. 22, No. 7, pp. 997–1001, 1998
© 1998 COSPAR. Published by Elsevier Science Ltd. All rights reserved
Printed in Great Britain
0273-1177/98 $19.00 + 0.00

PII: S0273-1177(98)00135-5

ASCA OBSERVATIONS OF THE BURSTING PULSAR GRO J1744–28

T. DOTANI[1], K. ASAI[1], Y. UEDA[1], F. NAGASE[1], K. MITSUDA[1], H. INOUE[1], Y. MAEDA[2], M. NISHIUCHI[2] and C. KOUVELIOTOU[3]

[1] *Institute of Space and Astronautical Science, Sagamihara, Kanagawa 229-8510, JAPAN*
[2] *Kyoto University, Kitashirakawa-Oiwake-Cho, Sakyo, Kyoto 606-8502, JAPAN*
[3] *Universities Space Research Association, Huntsville, AL 35806, USA*

ABSTRACT

The bursting pulsar GRO J1744–28 was observed with *ASCA* on February 26–27, 1996. The source was detected at a persistent flux level of 0.9 Crab with the X-ray bursts followed by a shallow dip. Clear pulsation with apparent barycentric period of 467.054 msec was also detected. Energy spectrum of GRO J1744–28 shows large photoelectric absorption, which is consistent with the location close to the Galactic center. We detected structures in the energy spectrum of the persistent emission around 7 keV. The structure is most probably due to a partial covering of the X-ray emission region.

©1998 COSPAR. Published by Elsevier Science Ltd.

INTRODUCTION

GRO J1744–28 is a transient bursting pulsar discovered on December 2, 1995 with BATSE in the region near the Galactic center (Fishman *et al.*, 1995; Kouveliotou *et al.*, 1996a). It is a unique X-ray source which shows both X-ray bursts and pulsation (Finger *et al.*, 1996a). The bursts were observed initially at typical intervals of ~ 3 minutes and about 20 sec durations. The burst emission was detected up to ~ 75 keV. The burst rate later decreased to about two per hour. Coherent X-ray pulsations were detected from GRO J1744–28 with a period of 467 msec (Finger *et al.*, 1996c). The pulse profile has a single peak with highly sinusoidal shape, different from that of typical X-ray pulsars. From the pulse timing analysis, the orbital period of the binary system was determined to be 11.8 days. The mass function was found to be very small, 1.36×10^{-4} $M_{\odot}$, and the companion star is most probably a low mass star (Finger *et al.*, 1996b). X-ray flux from GRO J1744–28 gradually decreased after the outburst and the burst activity stopped in early May (Kouveliotou *et al.*, 1996c). However, low level activity was found to continue until about June 96 (Corbet and Jahoda 1996).

Here we report a preliminary analysis of GRO J1744–28 observations obtained with ASCA. Detailed analysis will be presented elsewhere.

GRO J1744−28 light curve

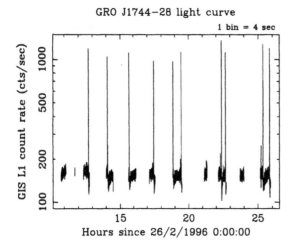

Fig. 1. Light curve of GRO J1744−28 obtained with the ASCA GIS. Spikes in the light curve correspond to the X-ray bursts. Data gaps are due to south atlantic anomaly passages of the satellite and earth occultation of the source.

GRO J1744−28 folded pulse profile

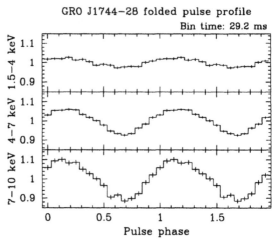

Fig. 2. Folded pulse profile of the persistent emission of GRO J1744−28 in three energy bands.

ANALYSIS AND RESULTS

GRO J1744−28 was observed with the ASCA satellite (Tanaka *et al.*, 1994) from February 26 10:00 (UT) through 27 2:05 (UT) for a net exposure of about 20 ksec. ASCA carries 4 sets of X-ray telescopes, two of which have a gas imaging spectrometer (GIS; Ohashi *et al.*, 1996) at the focal plane; the other two have a solid-state imaging spectrometer (SIS; Burk *et al.*, 1994). ASCA covers an energy range of 0.4–10 keV with moderate spatial (3 arcmin HPD) and high energy (2 % at 6 keV with SIS) resolutions. SIS was operated in the 1 CCD bright mode and GIS was in the PH mode during the observation of GRO J1744−28. Time resolution of GIS was increased to 2 msec in high bit rate and 16 msec in medium bit rate at the sacrifice of the rise time information.

We show the light curve of GRO J1744−28 obtained by GIS in Figure 1. The GIS count rate of the persistent emission was about 130 cts/sec in 1–10 keV (after dead time correction), which corresponds to about 0.9 Crab. We detected 10 large X-ray bursts and a few very small ones, all of which were followed by a dip. Duration of the dips is typically a few hundreds seconds, but can be as short as a few tens of seconds for the dips following the small bursts.

We calculated the pulse period using the epoch folding technique. The apparent barycentric pulse period is found to be 467.054 msec. We also calculated the average pulse profile of the persistent emission (non-bursting and non-dipping parts of the data) in three different energy bands. The results are shown in Figure 2. The pulse profile is highly sinusoidal consistent with the BATSE results obtained in higher energy bands (Finger *et al.*, 1996c). There is no clear phase shift between different energy bands. As seen from the figure, the pulse amplitude ($\equiv (f_{max} - f_{min})/(f_{min} + f_{max})$) is a strong function of the X-ray energy. It is about 2 % at 2 keV and increases up to 10 % around 10 keV. We also found that the pulse amplitude became very large during the bursts. It became as large as 50 %, which is similar to the observations in higher energy bands (Finger *et al.*, 1996a; Kouveliotou *et al.*, 1996b).

We show the energy spectra of GRO J1744−28 obtained by GIS separately for the burst, the persistent and the dip states in Figure 3. At first comparison, the energy spectra of all three states are similar.

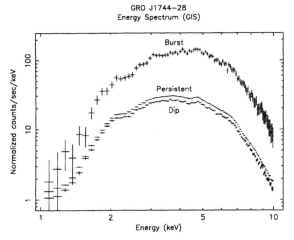

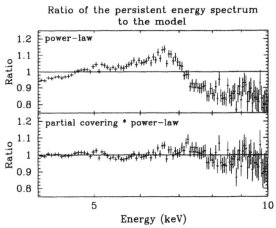

Fig. 3. Energy spectra of GRO J1744–28 obtained with the ASCA GIS. The spectra of burst, dip and persistent emission are similar with one another. The structures seen at 2.2 keV and 4.5 keV are instrumental, but that at 7 keV is intrinsic to the source.

Fig. 4. Ratio of the energy spectrum (SIS) to the best-fit model function. The model assumed is a power-law (upper panel) and a power-law modified by a partial covering absorption (lower panel).

In closer inspection, we notice that the burst energy spectrum tends to be a little harder than the persistent spectrum. The spectra are in general very hard suffering from a large low energy absorption. If we assume a power law for a continuum model, which gives the best fit among various simple models, hydrogen equivalent column density of the persistent emission becomes $(5.1 \pm 0.2) \times 10^{22}$ cm^{-2}, with the best-fit photon index of 1.2.

As seen from Figure 3, some structure is apparent in the energy spectra around 7 keV. To investigate the structure in detail, we calculated energy spectra of the persistent emission using the SIS data. SIS has an energy resolution of about 2 % at 6 keV, which is 4 times better than the GIS and hence is best suited to investigate structures in energy spectra. We show a PHA ratio of the SIS energy spectrum to the best-fitting power law (modified by the low energy absorption) model in figure 4. As seen from the figure, the structure may be interpreted either by an emission line or by an absorption edge. Therefore, we tried two different models to fit the energy spectra.

We first fitted a model consisting of a power law continuum and a gaussian line. This model can reproduce the energy spectra relatively well. The best-fit parameters are listed in Table 1. The line width is found to be vary large, $\sigma = 0.72$ keV. This large line width is probably not due to the co-existence of two lines such as neutral and He-like K$_\alpha$ from iron, because the energy resolution of SIS is high enough to clearly resolve these lines. Another mechanism to produce such a broad emission line may be the Doppler effect in the accretion disk very close to the neutron star, where the relativistic effects become important. This is the so-called disk line. We consider that the disk line is unplausible for GRO J1744–28, because it is likely that the accretion disk does not exist near the neutron star due to its magnetic field (Daumerie et al., 1996). Thus it is not clear how such a broad line is formed. Furthermore, the photon index we obtained for the continuum is significantly smaller than that obtained by RXTE, which is 1.2 at the pulse peak and 1.33 at the pulse minimum (Giles et al., 1996). From these reasons, we consider that a broad line is not plausible as the origin of the structure around 7 keV.

We next fitted a model which includes an absorption edge. However, a simple absorption edge does not fit the data because the model predicts much sharper drop of flux between 6.5–7.5 keV than the observed

Table 1: Best-fit parameters of persistent energy spectra obtained by SIS

Parameters	Unit	Emission line model	Partial covering model
Column density	10^{22} cm^{-2}	4.6 ± 0.2	5.4 ± 0.2
Photon index		0.97 ± 0.02	1.29 ± 0.03
Line center	keV	6.53 ± 0.05	—
Line width (σ)	keV	0.72 ± 0.08	—
Equivalent width	eV	270 ± 30	—
Covering fraction		—	0.45 ± 0.02
Column density[a]	10^{22} cm^{-2}	—	54 ± 4
red-χ^2 (dof)		1.58 (226)	1.79 (228)

Fit was performed in 3–10 keV to avoid the instrumental structures around 2 keV.
Errors quoted are in 90% confidence limit for a single parameter.
[a] Column density responsible for partial covering absorption.

one. Such dull edge structure may be reproduced by the superposition of a highly absorbed component and a non-absorbed component. Thus, we next fitted the following model to the SIS spectra:

$$I(E) = I_0 \exp(-N_H \sigma(E)) \left\{ (1 - f) + f \exp(-N_H^{pc} \sigma(E)) \right\} E^{-\Gamma}, \tag{1}$$

where I_0 is a normalization factor, N_H the hydrogen column density, f the covering fraction, N_H^{pc} the hydrogen column density for partial covering absorption, and Γ the photon index. This is a so-called partial covering model. We found that this model also can fit the SIS energy spectra of persistent emission relatively well; the best-fit parameters are listed in Table 1. The ratio of the energy spectrum to the best-fit partial covering model is also shown in Figure 4.

DISCUSSION

The ASCA observations of GRO J1744–28 showed that the absorption column density to the source is $(5.1 \pm 0.2) \times 10^{22}$ cm^{-2}. This column density is comparable to that to the Galactic center. Unless the column we observe has a significant contribution from circumsteller absorption, GRO J1744–28 is actually located near the Galactic center. Consequently, if we assume a source distance of 8 kpc, the unabsorbed luminosity of the persistent emission of GRO J1744–28 becomes 2.0×10^{38} erg s^{-1} in 2–10 keV. This luminosity corresponds to the Eddington limit of a neutron star. Because the energy spectrum of GRO J1744–28 is very hard, its bolometric luminosity is much larger, and the source becomes super-Eddington. The pulsed emission, which is about 10 % of the persistent emission, is most probably produced at the polar caps, which may subtend only a few percent of the surface area of the neutron star. This means that the local X-ray flux at the polar caps may already exceed the Eddington limit for the persistent emission. At the peak of the bursts, the X-ray flux increased by a factor of 6–7 compared to the persistent emission. Thus the burst peak luminosity is considered to exceed the Eddington limit by at least an order of magnitude.

X-ray bursts from GRO J1744–28 are most probably caused by mass accretion instability (Kouveliotou *et al.*, 1996a; Lewin *et al.*, 1996). If the X-ray emission from the neutron star during the bursts is super-Eddington and isotropic, it prevents mass accretion on to the neutron star and the X-ray bursts cannot be sustained. Thus the X-ray emission from the neutron star might be highly anisotropic. It is not clear at present how the highly anisotropic emission can be produced, and we need further studies to understand the nature of the X-ray burst emission from GRO J1744–28.

The pulse profile of GRO J1744–28 is highly sinusoidal. Its amplitude increases in higher energies, but the profile does not change with energy very much. The sinusoidal profile and its energy independence are very different from those of typical X-ray pulsars, which usually have an energy dependent and complex pulse profile. RXTE observations of GRO J1744–28 showed that there is a cut-off in the energy spectrum around 20 keV (Giles *et al.*, 1996). Thus the surface magnetic field of GRO J1744–28 may be of order 10^{12} G, typical of X-ray pulsars. However, the rate of the spin period change of GRO J1744–28 indicates a dipole magnetic field of $\leq 10^{11}$ G (Finger *et al.*, 1996b; Daumerie *et al.*, 1996). This discrepancy may be reconciled if the magnetic field of GRO J1744–28 has a multi-pole configuration.

We found that the structures in the persistent energy spectrum around 7 keV are reproduced either by a broad emission line or by a partial covering model. However, we prefer a partial covering model because (1) its power law index is consistent with the RXTE results and (2) it is not clear how a broad line is generated in GRO J1744–28. In the case of the partial covering model, about half of the emission is estimated to suffer from absorption by a column of $\sim 5 \times 10^{23}$ cm^{-2}. There are two possibilities for partial covering to work: one is spatial, because the absorber does not cover the entire area of the X-ray emitter, and the other is temporal, because the absorber does not always exist in the line of sight, although it totally covers the X-ray emitter. The former requires a presence of X-ray scattering plasma around the neutron star, and the latter a large number of blobs which frequently cross our line of sight. In the case of transient sources in outburst, it is very likely that the accretion flow onto the neutron star is in highly turbulent motion and inhomogeneity or blobs will be easily formed. Large X-ray flux from the neutron star will completely ionized the accreting matter near the neutron star and it is expected to produce X-ray scattering plasma. Thus, the situation for partial covering to work may be easily realized in the transient sources in outburst. We thus consider that the partial covering is the most plausible model to explain the structure in the persistent energy spectra of GRO J1744–28.

REFERENCES

Burke, B. E., R. W. Mountain, P. J. Daniels, M. J. Cooper, and V. S. Dolat, CCD Soft X-ray Imaging Spectrometer for the ASCA Satellite, IEEE trans. nucl. sci., **41**, 375 (1994).

Corbet, R., and K. Jahoda, IAUC 6399 (1996).

Daumerie, P., V. Kalogera, F. K. Lamb, and D. Psaltis, A Strongly Magnetic Neutron Star in a Nearly Face-on Binary System, Nature **382**, 141 (1996).

Finger, M. H., R. B. Wilson, Harmon, B. A., Hagedon, K., Prince, T. A., IAUC 6285 (1996a).

Finger, M. H., R. B. Wilson, J. van Paradijs, IAUC 6286 (1996b).

Finger, M. H., D. T. Koh, R. W. Nelson, T. A. Prince, B. A. Vaughan et al., Discovery of Hard X-ray Pulsation from the Transient Source GRO J1744-28, Nature **381**, 291 (1996c).

Fishman, G. J., C. Kouveliotou, J. van Paradijs, B. A. Harmon, W. Paciesas et al., IAUC 6272 (1995).

Giles, A. B., J. H. Swank, K. Jahoda, W. Zhang, T. Strohmayer et al., The Main Characteristics of GRO J1744-28 Observed by the Proportional Counter Array Experiment on the Rossi X-Ray Timing Explorer, ApJ, **469**, L25 (1996).

Kouveliotou, C., J. van Paradijs, G. J. Fishman, M. S. Briggs, J. Kommers et al., A New Type of Transient High-Energy Source in the Direction of the Galactic Center, Nature **379**, 799 (1996a).

Kouveliotou, C., J. Kommers, W. H. G. Lewin, J. van Paradijs, G. J. Fishman et al., IAUC 6286 (1996b).

Kouveliotou, C., K. Deal, P. Woods, M. Briggs, B. A. Harmon, G. J. Fishman et al., IAUC 6395 (1996c).

Lewin W. H. G., R. Rutledge, J. M. Kommers, J. van Paradijs, C. Kouveliotou, A Comparison between the Rapid Burster and GRO J1744-28, ApJ **462**, L39 (1996).

Ohashi, T., K. Ebisawa, Y. Fukazawa, K. Hiyoshi, M. Horii et al., The Gas Imaging Spectrometer on Board ASCA, PASJ **48**, 157 (1996).

Tanaka, Y., H. Inoue, and S. S. Holt, The X-ray Astronomy Satellite ASCA, PASJ **46**, L37 (1994).

Adv. Space Res. Vol. 22, No. 7, pp. 1003–1006, 1998
© 1998 COSPAR. Published by Elsevier Science Ltd. All rights reserved
Printed in Great Britain
0273-1177/98 $19.00 + 0.00

Pergamon

PII: S0273-1177(98)00132-X

HARD X-RAY PULSE PROFILE OF GX1+4

A.R. Rao[1], P.C. Agrawal[1], B. Paul[1], R.K. Manchanda[1], K.Y. Singh[2]

[1] *Space Physics Group, Tata Institute of Fundamental Research, Homi Bhabha Road, Colaba, Mumbai 400 005, INDIA*
[2] *Physics Department, Manipur University, Imphal, 795 003, India*

ABSTRACT

We have observed the Galactic center X-ray pulsar GX 1+4 using a balloon-borne hard X-ray telescope. Hard X-ray pulse profiles and spectral parameters were obtained on two occasions; during the high state of December 1993 when the hard X-ray pulse fraction was very low and in March 1995 when there was an indication of the pulse period variation sign reversal. We find significant differences in the pulse shape on these two occasions: the hard X-ray pulse profile changed from a smooth sinusoidal to a double peak structure. The details of the observations are presented and the implications of these results on the accretion process near the compact object are outlined. ©1998 COSPAR. Published by Elsevier Science Ltd.

INTRODUCTION

The hard X-ray pulsar GX 1+4 shows very interesting spin period variation episodes. After the discovery in a balloon observation (Lewin *et al.*, 1971) with a pulse period of 135 seconds, it showed a very fast spin-up rate with a time scale of 40 years. In the early 80's the source made a transition to a low intensity state and attempts using the *EXOSAT* satellite to detect the source failed. The source was observable again in 1987 (Makishima *et al.*, 1988), and by then the spin change had reversed from spin-up to spin-down. Since then, until a recent increase in luminosity, all observations gave a spin down trend with $\dot{P} = 1.5$ s yr^{-1}. We have been carrying out a regular monitoring of GX 1+4 with a balloon-borne large area hard X-ray telescope. Since 1991, four observations were carried out and the source was found in high state during the last two observations and the pulse period was determined accurately.

OBSERVATIONS

The telescope consists of two identical Xenon-filled Multi-layer Proportional Counters (XMPC) with a total effective area of 2400 cm^2 and has a field of view 5° × 5°. For details of the X-ray telescope refer to Rao et al (1991). Observations of the source are carried out by alternately looking at the source and a nearby source-free background region in the tracking mode. The details of the observations are given in Table 1. During the 1991 and 1992 observations, GX1+4 was in a low state and hence detected at significance level of only 5 to 6 σ, and no pulsations were detected.

On 1993 Dec 11 and 1995 Mar 22, it was in a bright state and in the following sections the results of these two observations are discussed.

Table 1. Details of GX1+4 observations

Date of balloon flight	Observation time (UT)	Ceiling altitude (g/cm^2)	Useful source exposure (sec)	GX1+4 count rate in (20–100)keV (count s^{-1})	α	L_o (10^{37} erg s^{-1})
Dec.11,1991	0612-0757	3.2	3100	1.3 ± 0.2	–	–
April 5/6,1992	2118-0118	4.0	6307	1.1 ± 0.2	–	–
Dec 11,1993	0445-0559	3.4	3120	12.4 ± 0.2	0.54±0.18	7.9±0.3
Mar 22,1995	0130-0530	2.5	7980	8.0 ± 0.2	0.67±0.12	2.5±0.3

ANALYSIS AND RESULTS

Spectra from the two detectors were fitted separately for different incident spectra and the parameters for the best fitted spectra were obtained (Rao *et al.*, 1994). To improve on the error bars the spectra from the two detectors were added and a combined fit was also carried out. The observed spectrum was fitted well with a power-law incident spectrum for the 1993 and the 1995 observations. The power law energy index (α) and the corresponding $20 - 100$ keV X-ray luminosity are also shown in Table 1.

Timing Analysis

The counts were binned with 5 sec bin widths and a period search was done in the $50 - 200$ sec range with an FFT algorithm based on the Lomb-Scargle method. In both the observations clear periodograms with single sharp peaks were obtained at a very high significance level. The derived pulse periods are 121.0±0.4 and 121.88±0.09, respectively for the 1993 and 1995 observations. The derived overall spin down rate is 0.72 ± 0.40 sec yr^{-1}, which is somewhat smaller than the average spin down rate of 1.4 sec yr^{-1} since 1986. *BATSE* observations in the intervening period have reported a reversal of the spin change rate, from spin-down to spin-up thereby supporting the smaller rate of change derived from the present observation (Chakrabarty *et al.*, 1994).

Pulse profile of the source was obtained by folding the photon counting rates with the measured periods. The pulse profile in the $20 - 100$ keV energy range is plotted in Figure 1 for two cycles, for the 1995 observations (top panel) and the 1993 observations (bottom panel). The 1995 observations show a structure in the pulse profile which can be interpreted either as a wide pulse with a valley at the center or as two pulses with unequal separation. A double peaked pulse profile similar in structure but narrower in width was seen earlier by Makishima *et al.* (1988) in $2 - 20$ keV range. In the 1993 observation (bottom panel) there is no indication of a double pulse and the detected pulse is also narrower. It is possible that during the recent source brightening there might have been a gradual change in the emission, from a pencil beam to a fan beam, which is more common to a pulsar in its bright state.

Pulse Profile Modeling

Very complex changes in pulse profile with luminosity are seen in many sources and a reproduction of intensity-dependent widely varying pulse profile was observed in the transient pulsar EXO 2030+375 which was modeled with both the fan and the pencil beams of unequal intensity from two offset magnetic poles, the most complex modeling of a X-ray pulsar profile done so far (Parmar *et al.*, 1989). The observed change in the GX 1+4 pulse profile from December 1993 to March 1995 indicates either an activation of the second pole which is possible if the magnetic filed is asymmetric in latitude or a gradual change in the beam pattern, from a pencil beam to a fan beam in spite of a luminosity decrease by a factor of 3 in 20 − 100 keV energy band.

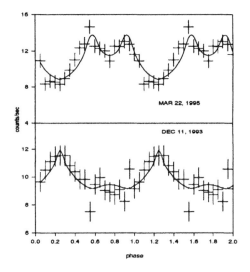

Fig. 1. Pulse profile of GX 1+4 obtained from the XMPC observations.

We have attempted to explain this pulse profile change with simple beam model. We have assumed a simple fan beam pattern with a symmetric magnetic dipole and equal intensity on both sides of the equator with a constant overall emission. The luminosity is maximum towards the magnetic pole and assumed to decay exponentially. The sum of the two angles, θ_m the angle between the magnetic axis and the spin axis and θ_r the angle between the observer line of sight and the spin axis needs to be more than $\frac{\pi}{2}$ so that the line of sight crosses the magnetic equator twice in one period and shows two peaks. Intensity has two minima, corresponding to the phases when the two poles are closest to the viewing axis. Such simple considerations were used successfully to reproduce roughly the pulse profiles of many pulsars by Leahy(1991). A simple geometry as described above gave reasonably good fit and we obtained the following values for the parameters

$\theta_m = 56 \pm 8, \quad \theta_r = 56 \pm 8$ with $\theta_m + \theta_r = 112 \pm 2$

and the exponential intensity decay towards the pole has an angular scale of $\theta_d = 32 \pm 4$.

The model considered here is actually unable to distinguish between θ_m and θ_r because of their interchangeability. However the values we have obtained are the same for both the parameters. The constraint is more on the sum of the two angles which defines the closest position of the second pole with the viewing axis ($180° - \theta_m - \theta_r$) and produces the valley in between the two peaks. The same source geometry also generated the single peaked profile obtained from the 1993 observations, by assuming a pencil beam emission.

DISCUSSION

GX 1+4 had showed both single and double peaked pulse profiles on different occasions (Mony *et al.*, 1991). The parameters obtained from the profile fitting is also consistent with the *GINGA* observations in 1987 and 1988 in 10 − 37 keV range (Dotani *et al.*, 1989). In the first observation a dip with a local maximum was detected and in the second observation about 150° away from the peak there was a dip but without any local maximum. A hollow cylinder of accretion column causing resonance scattering at the energy of the cyclotron line explained the first observation. At a higher intensity level in the second observation, the accretion column was full and the local maximum in the dip was absent. For this to happen the observer has to see just through one of the poles and that is supported by a very close value of θ_m and θ_r we have obtained. The offset of the dip with the peak in the pulse profile as observed in the second *GINGA* observation is also explained with the present value of θ_m and θ_r. In the second observation probably a gradual change from fan beam to pencil beam was taking place with an increase in luminosity, and the peak in the second observation is at the place of the two magnetic equator crossings and the dip is at the phase when one is seeing through the first pole. A larger value of θ_d can produce the wide peak in the second observation and the valley also may become less significant.

ACKNOWLEDGMENTS

It is a pleasure to acknowledge the contribution of Shri M.R. Shah, Electronics Engineer-in-charge of this experiment. We are also thankful to Shri D.K. Dedhia, Shri K. Mukherjee, Shri V.M. Gujar, Shri S.S. Mohite and Shri P.B. Shah for their support in the fabrication of the payload. We thank the Balloon Support Instrumentation Group and the Balloon Flight Group led by Shri M.N. Joshi, under the overall supervision of Prof. S.V. Damle.

REFERENCES

Chakrabarty, D., T.A. Prince, and M.H. Finger, *IAU Circular* no 6105 (1994).

Dotani, T., T. Kii, F. Nagase, K. Makishima, T. Ohashi *et al.*, Peculiar Pulse Profile of GX 1+4 Observed in the Spin-Down Phase, *PASJ* 41, 427 (1989).

Leahy, D. A., Modeling Observed X-ray Pulse Profile, *MNRAS* 251, 203 (1991).

Lewin, W. G. H., G.R. Ricker, and J.E. McClintock, X-rays from a New Variable Source GX 1+4, *ApJ* 169, L17 (1971).

Makishima, K., T. Ohashi, T. Sakao, T. Dotani, H. Inoue *et al.*, Spin-down of the X-ray Pulsar GX 1+4 During an Extended Low State, *Nature*, 333, 746 (1988).

Mony, B., E. Kendziorra, M. Maisack, R. Staubert, J. Englhauser *et al.*, Hard X-ray Observations of GX 1+4, *Astron. Astrophys.*, 247, 405 (1991).

Parmar, A. N., N.E. White, L. Stella, The Transient 42 second X-ray Pulsar EXO 2030+375 II The Luminosity Dependence of the Pulse Profile, *Ap.J.*, 338, 373 (1989).

Rao, A. R., P.C. Agrawal, and R.K. Manchanda, Hard X-ray Observations of 4U 1907+09, *Astron. Astrophys.* 241, 127 (1991).

Rao, A.R., B. Paul, V.R. Chitnis, P.C. Agrawal, and R.K. Manchanda, Detection of a Very Low Hard X-ray Pulse Fraction in the Bright State of GX 1+4, *Astron. Astrophys.*, 289,L43 (1994).

Pergamon

Adv. Space Res. Vol. 22, No. 7, pp. 1007–1016, 1998
© 1998 COSPAR. Published by Elsevier Science Ltd. All rights reserved
Printed in Great Britain
0273-1177/98 $19.00 + 0.00

PII: S0273-1177(98)00136-7

QPO IN TRANSIENT PULSARS

Mark H. Finger[1,2]

[1] *Universities Space Research Association*
[2] *ES 84, Marshall Space Flight Center, Huntsville AL, 35812*

ABSTRACT

QPOs have now been observed from five transient X-ray pulsars. Four of these are in Be/X-ray systems, and one is the newly discovered LMXB transient GRO J1744-28. I review the observations, and discuss their interpretation in terms of the beat or Keplerian frequency models. These models are successful in explaining the observations for two of the Be/X-ray transients. For one of these, A0535+262, the observations constrain the ratio of the radius of the accretion disk inner edge to the Alfvén radius, providing new information on the accretion disk-magnetosphere boundary region. Neither model can explain the QPO in the remaining two Be/X-ray systems, nor that in GRO J1744-28. It is likely therefore that two or more QPO mechanisms are at work. I discuss the potential for gaining new insight into QPOs from analysis tools unique to pulsed systems, and discuss what we can learn from new observations.   ©1998 COSPAR. Published by Elsevier Science Ltd.

INTRODUCTION

The discovery of Quasi Periodic Oscillations (QPOs) in the power spectra of the flux from x-ray binary pulsars (Angelini *et al.* 1989) offers a new probe of accretion physics in these systems. Observations of the transient pulsars EXO 2030+375 and A0535+262 have revealed QPOs at either the Keplerian orbital frequency at the magnetospheric radius, or its beat with the neutron star spin frequency. QPOs can therefore offer direct information on the poorly-understood accretion disk–magnetosphere boundary region. Unlike most unpulsed sources, for many x-ray pulsars the basis system variables of spin period, orbital parameters, and surface magnetic fields have been determined. In addition, pulse timing allows measurements of accretion-induced torques, which can provide a measure of mass-accretion rate unaffected by distance or spectral uncertainties. Study of QPOs in pulsars therefore has great promise for testing QPO models and providing new insights into accretion physics.

I review here QPOs in the transient X-ray pulsar systems, which should be the simplest to interpret. While QPO have been observed in several persistent X-ray binary pulsars, interpretation of these observations is complicated by several factors. Both the beat frequency and the Keplerian frequency QPO models predict a systematic change of QPO frequency with mass-accretion rate, which can be tested by comparing QPO frequency with either X-ray flux or spin-up rate, assuming these are simply related to the mass-accretion rate. However, this assumption may not be valid in all of the persistent systems. For example, in Cen X-3 the X-ray flux, due to variable absorption by circumstellar material,

Table 1: QPO Observations in Transient X-ray Pulsars

Source	ν_{QPO}	ν_{QPO}/ν_{ns}	FWHM	r_{QPO}	r_{cont}	Reference
LMXB Systems:						
GRO J1744-28	40 Hz	18.7	48%	1.5-5.9%	> 30%	1
Be/X-ray Systems:						
V 0332+53	51 mHz	0.22	65%	4.8%	25%	2,3
4U 0115+375	62 mHz	0.22	65%	...	...	4
EXO 2030+375	187–213 mHz	8.0–9.1	24%	3.6%	25%	5
A 0535+262	25–72 mHz	2.6–7.4	50%	9–12%	25%	6

References:

[1] Zhang et al.(1996)   [2] Takeshima et al. (1994)
[3] Makishima et al. (1990), for r_{cont} [4] Soong & Swank (1989)
[5] Angelini, Stella & Parmar (1989)   [6] Finger, Wilson & Harmon (1996a)

is known to be an unreliable indicator of the mass-accretion rate (Van der Klis *et al.* 1980). The disk fed persistent systems may be near equilibrium, with direct accretion of angular momentum being roughly balanced by magnetic braking torques. In this regime the relationship between torque and mass-accretion rate is poorly understood, making comparisons of spin-up rate and QPO frequency more difficult to interpret. In contrast, during transient pulsar outbursts, braking torques are small compared to the direct accretion of angular momentum, resulting in a mass-accretion rate versus torque relation which is nearly a power law. Finally, the range of mass accretion rates explored with the persistent systems is relatively narrow, while the transients experience change by a factor of 10^3 or more.

To date, QPOs have been seen in five transient X-ray pulsars: the transient LMXB pulsar GRO J1744-28, and four pulsars in Be/X-ray systems. Table 1 summarizes the observations, giving the observed QPO frequency, ν_{QPO}; the ratio of the QPO frequency to neutron star spin frequency, ν_{ns}; the full width half maximum width of the QPO relative to the QPO frequency, FWHM; the fractional R.M.S. power of the QPO, r_{QPO}; and of fractional power of the observed continuum, r_{cont}. All of these observations are of broad power spectral features, with several not meeting the strict (FWHM < 50%) definition applied to QPO in LMXB systems. Nevertheless they all represent a substantial concentration of power in a limited frequency range, with fractional RMS's ranging from 1.5 to 12%. In all of the sources the QPO is observed in the presence of a strong red-noise continuum, with RMS of 25% or more. The flux modulation represented by the continuum is often stronger than that of the coherent pulsations, and consistently dominates that due to the QPO.

These five sources are only a small fraction of the 30 known transient X-ray pulsars. However, sensitive wide-band power spectral studies have been performed for only a small fraction of the transient pulsar outbursts that have occurred over the last 20 years. With the combined timing capabilities and sensitivity of RXTE, and the pulsar monitoring capabilities of BATSE/CGRO, I expect in the next few years many new discoveries of QPO in X-ray pulsars. Outbursts from transient LMXB pulsars are clearly rare, but those from Be/X-ray pulsars are fairly common.

THE BEAT AND KEPLERIAN FREQUENCY MODELS

In the beat frequency model (Alpar & Shaham 1985) blobs of matter, in the process of being entrained in the neutron star's magnetic field, orbit the neutron star at approximately the Keplerian orbital frequency of the inner edge of the accretion disk, ν_k, accreting at a rate that is modulated by the magnetic field, which rotates with the neutron star at a frequency ν_{ns}. This produces a QPO at the beat frequency $\nu_{QPO} = \nu_k - \nu_{ns}$. In the Keplerian frequency model (Van der Klis *et al.* 1987) the inner edge of the accretion disk contains structures that persist for a few cycles around the neutron star, and modulate the observed flux by obscuration producing a QPO at the Keplerian frequency, ν_k.

The radius of the disk inner edge, r_0, determines the Keplerian rotation frequency there via Kepler's third law, $2\pi\nu_k = (GM)^{1/2}r_0^{-3/2}$. At the inner disk edge the magnetosphere disrupts the Keplerian rotation of the disk, forcing matter to accrete along magnetic field lines. The dependence of r_0 on the mass-accretion rate may be approximately expressed as (Pringle & Rees 1972)

$$r_0 = K\mu^{4/7}(GM)^{-1/7}\dot{M}^{-2/7} \tag{1}$$

where K is a constant of order 1, μ the neutron star magnetic moment, G Newton's gravitational constant, and M the gravitational mass of the neutron star. Eq. 1 with $K = 0.91$ gives the Alfvén radius for spherical accretion. Ghosh and Lamb (1979) give a value of $K = 0.47$. The accretion torque is

$$N = I2\pi\dot{\nu}_{ns} = \dot{M}(GMr_0)^{1/2} \ . \tag{2}$$

Here I is the moment of inertia of the neutron star. $(GMr_0)^{1/2}$ is the specific angular momentum of matter in a Keplerian orbit at the inner disk radius r_0. Eq. 2, which considers only the transport of orbital angular momentum from the disk to the neutron star, should be valid at high mass-accretion rates. The hard x-ray flux is directly dependent on the mass-accretion rate $\dot{M}$,

$$F = \frac{\alpha\beta}{4\pi d^2}\frac{GM}{R}\dot{M} \tag{3}$$

where α is the fraction of the total luminosity within the observational energy band, β accounts for the effects of beaming, d is the source distance, and R is the neutron star radius.

The predicted dependence of the spin-up rate and flux on the Keplerian frequency are

$$\dot{\nu}_{ns} = \aleph_N\nu_k^2 \quad \text{with} \quad \aleph_N = 1.6K^{7/2}B^2R^6(GM)^{-1}I^{-1} \tag{4}$$
$$\text{and} \quad F = \aleph_F\nu_k^{7/3} \quad \text{with} \quad \aleph_F = 1.4\alpha\beta d^{-2}K^{7/2}B^2R^5(GM)^{-2/3} \ .$$

where $B = 2\mu R^{-3}$ is the polar surface magnetic field.

Mass flow to the neutron star is expected to be stopped at the magnetospheric boundary by centrifugal inhibition of accretion if the Keplerian orbital frequency at the magnetospheric boundary is less than the neutron star spin frequency (Stella *et al.* 1986). This should occur with either wind fed or disk accretion, with the matter that is entrained in the magnetic field at the magnetospheric boundary being flung back by centrifugal force when it is constrained to corotate with the magnetosphere. Therefore QPOs below the neutron star spin frequency cannot be explained by the Keplerian frequency model. In the beat frequency model the mass flow to the neutron star is inhibited as the QPO frequency approaches zero. This occurs at a flux given by

$$F_{CI} = F_{QPO}\left(\frac{\nu_{ns}}{\nu_{ns} + \nu_{QPO}}\right)^{7/3} \tag{5}$$

where F_{QPO} is the flux at the time of the QPO observation. There should be a rapid transition from a low flux level (due to accretion onto the magnetosphere) to F_{CI} at the onset of an outburst, and outbursts should end with a rapid fall in flux once F_{CI} is reached (Corbet 1996).

OBSERVATIONS

GRO J1744-28

The only known transient LMXB pulsar, GRO J1744-28 was discovered in December 1995 due to its unusual bursting behavior (Kouveliotou *et al.* 1996). RXTE observations revealed a 40 Hz QPO, along with smaller, yet significant QPOs at 20 and 60 Hz (Zhang *et al.* 1996). The RMS of the QPO decreased with decreasing flux. An accretion disk was present during the observations as indicated by the high spin-up rate of the pulsar (Finger et al. 1996b). Over the course of the QPO observations the flux changed by a factor of 7.5, with little or no systematic change in the QPO frequency. This eliminates both the beat and Keplerian frequency models for these QPO since these predict a factor of 2.4 change in the QPO frequency for the observed change in flux. Zhang *et al.* (1996) conclude that "there is not a ready model in the literature that can adequately explain in quantitative terms the prominent characteristics of the QPO".

Be/X-ray Binary Outbursts

The remaining four QPO detections are in Be/X-ray transients. They occur fairly regularly, and often are observed in series of outburst repeating once per binary orbit. BATSE has detected a total of 55 outbursts from 11 different Be/X-ray pulsars in the last five years. These outbursts are of two types (Stella *et al.* 1986, Motch *et al.* 1991); "class I" or "normal" outbursts which are of lower flux, are confined to a limited range of orbital phase, and often repeat for several orbits, and "class II" or "giant" outbursts which have higher peak flux, show high rates of spin-up, and can last for several binary orbits. An example of both kinds of behavior is shown in Figure 1, which shows the flux and barycentric frequency history from a series outbursts of 2S 1417-624 observed by BATSE (Finger *et al.* 1996c). While it is clear that accretion disks are present during the giant outbursts, it is unknown whether they are present in the normal outbursts. Since the QPOs observed to date in x-ray pulsars are all associated with accretion disks, one method of investigating this issue would be to search for QPOs during the normal outbursts.

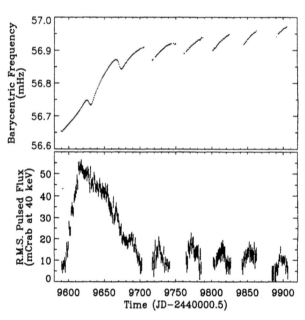

Fig. 1. The barycentric frequency and pulsed flux of 2S 1417-624 observed by BATSE during the August 1994 – July 1995 outbursts.

V0332+53 and 4U 0115+63

For both V0332+53 and 4U 0115+63 there have been single reported observations of QPOs, with these QPOs at a frequencies below the neutron star spin frequencies. A 51 mHz QPO in the flux of V0332+53 was observed with *Ginga* during an outburst of this Be/X-ray pulsar in September and October 1989 (Takeshima *et al.* 1994). This broad feature, which lay below the pulse frequency, was observed on October 1, when the 2.3-37.2 keV flux was 2.9×10^{-8} erg cm^{-2}s^{-1}. An earlier observation on September 19, when the flux was 1.9×10^{-8} erg cm^{-2}s^{-1}, showed no QPO. HEAO observations of 4U 0115+63 during an outburst in January of 1978 revealed a prominent power spectral feature, also below the pulse frequency, centered on 61 mHz (Soong & Swank 1989). During the observation the 2-60 keV flux was 8.7×10^{-9} erg cm^{-2}s^{-1} (Rose *et al.* 1979). Since these QPOs are below the neutron star spin-frequency, they cannot be explained by the Keplerian frequency model.

The peak flux of the September-October 1989 outburst of V0332+53 was simular to that in the giant outburst observed by Vela 5B in 1973, which lasted 100 days (Terrel & Priedhorsky 1984). The centrifugal inhibition threshold calculated from the QPO observation assuming the beat frequency model is $F_{CI} = 1.8 \times 10^{-8}$ erg cm^{-2}s^{-1} (2.3-37.2 keV). A series of three smaller normal outbursts occurring once per orbit were observed from November 1983 – January 1984 (Stella *et al.* 1986). For the first of these, Tenma observed (Makishima *et al.* 1990) a peak flux of 3.2×10^{-9} erg cm^{-2}s^{-1} (2-30 keV). The QPO therefore cannot be explained by the beat frequency model. Rapid X-ray fluctuations with an 0.01-1.0Hz fractional RMS of $\sim 25\%$ were observed during the Tenma observations.

The January 1978 outburst of 4U 0115+634 also appears to be a giant outburst, lasting more than 30 days which is longer than the 24 day orbital period, and having one of the higher peak fluxes of the outbursts that have been observed from this source. The centrifugal inhibition threshold calculated from the QPO observation assuming the beat frequency model is $F_{CI} = 5.5 \times 10^{-9}$ erg cm^{-2}s^{-1} (2-60 keV). Observations made by the Einstein Observatory on January 15 1981 at the end of a small outburst (Kriss *et al.* 1983) detected pulsations and found a 2-10 keV flux of 3.3×10^{-9} erg cm^{-2}s^{-1}. The QPO observed in 4U 0115+634 therefore cannot be explained by the beat frequency model.

EXO 2030+375

QPO was observed with EXOSAT in the flux of the Be/X-ray binary pulsar EXO 2030+375 during the May – August 1985 discovery outburst (Angelini *et al.* 1989). This was clearly a giant outburst, showing high flux and rapid spin-up rates, and lasting several orbits. The QPO ranged in frequency from 187 mHz to 213 mHz, well above the 24 mHz pulse frequency.

As is shown in Figure 2, the QPO frequency is clearly correlated with the observed flux, which is a key prediction of the beat frequency and Keplerian frequency models. The figure shows the trends predicted by the beat and Keplerian frequency model, with the normalizations chosen to fit the data. While the data spans only a small range, the fit for either model is reasonable. Assuming a source distance of 5 kpc, the required normalization of the QPO-flux relations imply a surface (dipole) magnetic field of approximately 5×10^{12} G. The predicted onset of centrifugal inhibition of accretion is at a flux of 2×10^{-10} erg cm^{-2} s^{-1} (1-20 keV), which is nicely bracketed by the last flux measured in the discovery outburst of 4×10^{-10} erg cm^{-2} s^{-1} on August 13 1985, and following the upper limit of 1.3×10^{-11} erg cm^{-2} s^{-1} for the observation of August 25 1985 (Parmar *et al.* 1989).

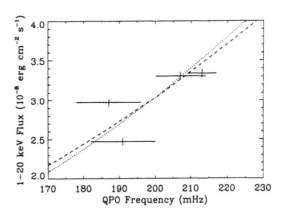

Fig. 2. QPO frequency versus 1-20 keV flux during the May – August 1995 outburst of EXO 2030+375. The data is from Angelini *et al.* (1989). The lines are the trends predicted by the beat frequency (dashed) and the Keplerian frequency (dotted) model.

A0535+262

QPO was observed with BATSE on the Compton Observatory during a giant outburst of
A0535+262 in February-March 1994 (Finger *et al.* 1996a). The giant outburst, which reached a
peak flux of 6 Crab at 40 keV, occurred within
a series of six outbursts spaced roughly once
per 110 day orbit. Figure 3 shows the pulse
amplitude spin-frequency during these observations. An estimate of the binary orbital parameters based on pulsed timing during the first three
normal outbursts (Finger *et al.* 1994) allowed
the spin-up rate $\dot{\nu}_{ns}$ to be determined from pulse
timing measurements during the giant outburst,
in addition to the flux history obtained by occultation measurements.

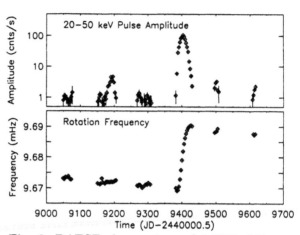

Fig. 3. BATSE observations of A0535+262.

The QPO was observed for 33 days, initially being detected at 27 mHz, rising to 72 mHz at the
peak of the outburst, and falling to 25 mHz in a manner highly correlated with both the flux and
the spin-up rate. The observed relationships between QPO frequency and spin-up rate, and between
QPO frequency and flux are shown in Figures 4 and 5 respectively. In both cases squares are used
for the rise of the outburst, and diamonds for the fall. The rise and fall behavior tracks fairly well.
The plots give the predicted trends for the beat frequency model (dashed) and Keplerian frequency
model (dotted), either of which fit the data reasonably.

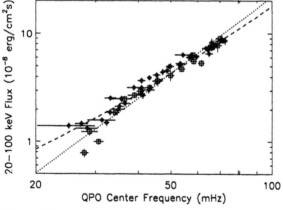

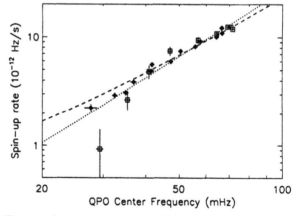

Fig. 4. Comparison of QPO frequency and Flux
for A0535+262.

Fig. 5. Comparison of QPO frequency and spin-
up rate for A0535+262.

Finger *et al.* (1996a) show that the measured normalization constants in the spin-up rate versus
QPO frequency relationship and flux versus QPO frequency relationship (Eq. 4), are consistent
with the 110 keV cyclotron line measured by OSSE during the 1994 giant outburst (Grove *et al.* 1996), estimates of the source distance, and reasonable neutron star parameters, if the constant K
in the magnetospheric radius versus mass accretion rate relationship (Eq. 1) is approximately one.
However, use of the value $K = 0.47$ estimated by Ghosh and Lamb (1979) requires a neutron star
with a very stiff equation of state, or with a mass less than 0.5 $M_\odot$.

BEYOND POWER SPECTRA

The power spectrum is the standard means of characterizing a source's flux noise. If the statistical properties of this noise are unchanged with a translation in time (a stationary noise process) then the mean flux μ_F and the autocovariance function $R_F(\tau)$, or equivalently, the power spectrum $S_F(\nu)$ provide a complete description of the first and second order moments of the flux:

$$< F(t) > \;=\; \mu_F \tag{6}$$
$$< [F(t_1) - \mu_F][F(t_2) - \mu_F] > \;=\; R_F(t_2 - t_1) = \int S_F(\nu) e^{i2\pi\nu(t_2 - t_1)} d\nu \;.$$

In a pulsar the mean flux, of course, depends on the pulse phase $\phi(t)$. In general we expect the statistical properties of the noise to be unchanged with a translation in time by an integer number pulse periods (a cyclostationary noise process). In general the second moments will be phase dependent. A complete description of the first and second order moments of the pulsars flux therefore require a phase dependent mean $\mu_F(\phi)$, and a phase dependent autocovariance $R_F(\tau, \phi)$, both being periodic in phase ϕ:

$$< F(t) > \;=\; \mu_F(\phi(t)) \tag{7}$$
$$< [F(t_1) - \mu_F(\phi(t_1))][F(t_2) - \mu_F(\phi(t_2))] > \;=\; R_F(t_2 - t_1, \phi([t_1 + t_2]/2)) \;.$$

Here I have chosen to specify the phase dependence of the autocovariance using the phase midway between the two times involved. This results in the autocovariance being symmetric in lag for fixed phase. For a pulsar, the power spectrum contains less information than the (phase dependent) autocovariance function.

Using the BATSE 20-50 keV rates during the peak of the A0535+262 giant outburst in 1994, I have estimated the phase dependent autocovariance function. Its phase averaged value versus lag is shown in Figure 6. The bump at 13 s is due to the QPO which was at a frequency of 72 mHz. At a lag of 7 s, QPO peaks align with QPO valleys, causing a dip in the autocovariance. The phase dependence of the autocovariance function is examined in Figure 7 for lags of 7.2 and 13.3 s. The phase dependence is significant in both cases, which show very different signatures.

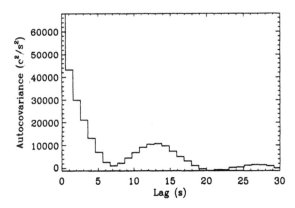

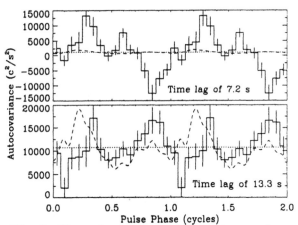

Fig. 6. Phase averaged autocovariance function for A0535+262 on February 16, 1994 using the BATSE 20-50 keV rates. The bump at a lag of 13 s is due to the QPO, which was at a frequency of 72 mHz.

Fig. 7. Phase dependence of the autocovariance function for A0535+262 on February 16, 1994 using the BATSE 20-50 keV rates, at lags of 7.2 and 13.3 s.

We can show that two general classes of models are unable to explain this phase dependent autoco-variance function. The first class of model is additive,

$$F(t) = L(\phi(t)) + N(t) \tag{8}$$

where $L(\phi)$ is periodic, and $N(t)$ is a stationary noise process which contains the QPO and red noise. This could result, for example, from a combination of a pulsed flux component $L(\phi)$, due to magnetically channeled accretion, with an unpulsed flux component $N(t)$, due to matter penetrating the magnetosphere without being channeled. This class of models results in a phase independent autocovariance $R_F(\tau, \phi) = R_N(\tau)$, which is shown by the dotted lines in Figure 7.

The second class of model is multiplicative,

$$F(t) = L(\phi(t)) \times N(t) \tag{9}$$

where $L(\phi)$ is periodic, and $N(t)$ is a stationary noise process. For example, $N(t)$ could represent the mass-accretion rate, and $L(\phi)$ the phase dependent emission profile, or $L(\phi)$ could represent an unabsorbed periodic flux and $N(t)$ absorption as a function of time. In this case the phase dependent autorcovariance function is of the form $R_F(\tau, \phi) = R_N(\tau)L(\phi - \tau\nu_{ns}/2)L(\phi + \tau\nu_{ns}/2)$. This is shown in Figure 7 by the dashed lines.

It is clear from Figure 7 that simple additive or multiplicative models are not capable of explaining the noise behavior in A0535+262. This general conclusion could not have been drawn from the power spectrum alone. The phenomena must be more complex, with accretion, emission, and absorption geometrically interlinked. The phase dependence of the noise can therefore provide us with a new probe of the source geometry.

FUTURE PROSPECTS

In the next few years we should expect to observe a number of transient x-ray pulsars that have QPO in their power spectrum. BATSE will be monitoring the galaxy for the onset of these transients, but can only examine the very brightest for QPO. RXTE, with its excellent timing capabilities and sensitivity, should provide a new wealth of information on their power spectra and associated QPO. At the same time, the spectral capabilities of RXTE should allow the observations of cyclotron lines, establishing the magnetic field of the neutron star.

To date, QPO in Be/X-ray binaries have only been seen during giant outbursts. This may be entirely due to sensitivity considerations, or may be due to an accretion disk only being present in the giant outbursts. By examining normal outbursts for QPO, we could address whether an accretion disk is present during these smaller outbursts, and if so, investigate the QPO in this lower flux regime.

The QPO features observed in 4U 0115+634 and V0332+53 could not be explained by the Keplerian or beat frequency model. Since these were isolated observations, it is difficult to draw any conclusions beyond the fact that more than one QPO mechanism is at work in transient pulsars. To understand these features further we will need to place them in the context of other observations where the dependence of the QPO on flux is determined.

From the example of A0535+262 it is clear that regular monitoring over the course of an outburst is extremely useful. This allows the simultaneous determination of flux, spin-up rate and QPO frequency. With enough range in flux we can demonstrate (or disprove) that a QPO is associated with either the beat for Keplerian frequency. By observing QPO at lower flux we should be able to differentiate between the beat and Keplerian frequency models. Of particular importance is regular monitoring at the tail of an outburst. It is here that counter torques produced by interactions between the accretion disk and the magnetic field should become evident, resulting a spin-down at the lowest fluxes. With an established beat or Keplerian frequency QPO monitored through this transition, a detailed mapping of the torque versus mass-accretion rate relationship would for the first time be established. At the end of an outburst the turn-off due to centrifugal inhibition of accretion might be observable.

Beyond using the QPO behavior to locate the disk-magnetosphere boundary, it should also be studied in its own right. The QPO models that have been developed in any detail are for unpulsed systems. Instabilities at the disk-magnetosphere boundary, magnetic channeling of accretion, and site dependent beaming of radiation all need to be considered in models of the continuum red noise and QPO's in pulsars. These models need to be able to explain the phase dependent autocovariance function as well as the power spectrum of the flux noise. This phase dependence of noise properties in pulsars should provide a level of geometrical information unavailable in unpulsed sources.

ACKNOWLEDGMENTS

I would like to thank Ali Alpar, Fred Lamb, Luigi Stella, and Lars Bildsten for useful discussions. This work have been supported in part by NASA Grant NAG 5-2850.

REFERENCES

Alpar, M.A., and J. Shaham, Is GX 5-1 a millisecond pulsar?, *Nature* **316**, 239 (1985).

Angelini, L., L. Stella, and A. N. Parmar, The Discovery of 0.2 Hz Quasi-Periodic Oscillations in the X-ray Flux of the Transient 42 Second Pulsar EXO 2030+375, *Ap.J.* **346**, 906 (1989).

Corbet, R. H. D., Transient X-ray Sources, Luminosity Gaps, and Neutron Star Densities, *Ap.J.* **457**, L31 (1996).

Finger, M. H., L. R. Cominsky, R. B. Wilson, B. A. Harmon, and G. J. Fishman, Hard X-ray Observations of A 0535+262, in *The Evolution of X-ray Binaries*, editors S.S. Holt and C. S. Day, pp. 459-462, AIP, New York, NY (1994).

Finger, M. H., R. B. Wilson, and B. A. Harmon, Quasi-Periodic Oscillations During a Giant Outburst of A0535+262, *Ap.J.* **459**, 288 (1996a).

Finger, M. H., D. T. Koh, R. W. Nelson, T. A. Prince, B. A. Vaughan, and R. B. Wilson, and D. Chakrabarty, Discovery of hard X-ray pulsations from the transient source GRO J1744-28, *Nature* **318**, 291 (1996b).

Finger, M. H., R. B. Wilson, and D. Chakrabarty, Reappearance of the X-ray binary pulsar 2S 1417-624, *Astron. & Astrophys. Supl.* **204**, 209 (1996c).

Grove, J. E., M. S. Strickman, W. N. Johnson, J. D. Kurfess, R. L. Kinzer, C. H. Starr, G. V. Jung, E. Kendziorra, P. Kretschmar, M. Maisack, and R. Staubert, The Soft Gamma-Ray Spectrum of A0535+26: Detection of an Absorption Feature at 110 keV by OSSE, *Ap.J.* **438**, L24 (1995).

Ghosh, P. and F. K. Lamb, Accretion by Rotating Magnetic Neutron Stars. III Accretion Torques and Period Changes in Pulsation X-ray Sources, *Ap.J.* **234**, 296, (1979).

Kouveliotou, C., J. van Paradijs, G. J. Fishman, M. S. Briggs, J. Kommers, B. A. Harmon, C. A. Meegan, and W. H. Lewin, A new type of transient high-energy source in the direction of the Galactic Centre, *Nature* **379**, 799 (1996).

Kriss, G. A., L. R. Cominsky, R. A. Remillard, G. Williams, and J. R. Thorstensen, The 1980 Outburst of 4U 0115+63 (V635 Cassiopeiae), *Ap.J.* **266**, **806** (1983).

Makishima, K. et al., Observations of the Peculiar Hard X-ray Transient X0331+53 (V0332+53), *P.A.S.J.* 42, **295** (1990).

Motch, C., L. Stella, E. Janot-Pacheco, and M. Mouchet, Accretion Mechanisms in the Be/X-ray Transient System A0535+262, *Ap.J.* **369**, 490 (1991).

Parmar, A. N., N. E. White, L. Stella, C. Izzo, and P. Ferri, The Transient 42 Second X-ray Pulsar EXO 2030+375. I. The Discovery and the Luminosity Dependence of the Pulse Period Variations, *Ap.J.* **338**, 359 (1989).

Pringle J. E. and M. J. Rees, Accretion Disc Models for Compact X-ray Sources, *Astron. & Astrophys.* **21**, 1 (1972).

Rose, L. A., S. H. Pravdo, L. J. Kaluzienski, F. E. Marshall, S. S. Holt, E. A. Boldt, R. E. Rothschild, and P. J. Serlemitsos, Observations of the Transient X-ray Source 4u 0115+63, *Ap.J.* **231**, 919 (1979).

Soong, Y., and J. H. Swank, Timing Analysis of Binary X-ray Pulsars Observed by HEAO 1, in *Proc. 23 ESLAB Symp. on Two-Topics in X-Ray Astronomy*, Vol 1., edited by N. White (Garching:ESA), pp. 617-620 (1989).

Stella, L., N. E. White, J. Davelarr, A. N. Parmar, R. J. Blisett, and M. van der Klis, The Discovery of 4.4 Second X-ray Pulsations from the Rapidly Variable X-ray Transient V0332+53, *Ap.J. Lett.* **288**, L45 (1985).

Stella, L., N. E. White, and R. Rosner, Intermittent Stellar Wind Accretion and the Long-term Activity of Population I Binary Systems Containing an X-ray Pulsar, *Ap.J.* **308**, 669 (1986).

Takeshima, T., T. Dotani, K. Mitsuda, and F. Nagase, Discovery of the Quasi-Periodic Oscillations from V0332+53, *Ap.J.* **436**, 871 (1994).

Terrell, J., and W. C. Priedhorsky, The 1973 X-ray Transient V0332+53, *Ap.J.* **285**, L15 (1984).

Van der Klis, M., J. M. Bonnet-Bidaud, and N. R. Robba, Charactersitics of the Cen X-3 Neutron Star from Correlated Spin-up and X-ray Luminosity Measurements, *Astron. Astrophys.* **88**, 8 (1980).

Van der Klis, M., F. Jansen, J. P. van Paradijs, W. H. G. Lewin, M. Sztajno and J. Trumper, Correlation Between Spectral State and Quasi-Periodic Oscillation Parameters in GX 5-1, *Ap. J.* **313**, L19 (1987).

Zhang, W., E. H. Morgan, K. Jahoda, J. H. Swank, T. E. Strohmayer, G. Jernigan, and R. I. Klein, Quasi-Periodic X-ray Brightness Oscillations of GRO J1744-28, *Ap.J. Lett.* **469**, L29 (1996).

Pergamon

Adv. Space Res. Vol. 22, No. 7, pp. 1017–1024, 1998
© 1998 COSPAR. Published by Elsevier Science Ltd. All rights reserved
Printed in Great Britain
0273-1177/98 $19.00 + 0.00

PII: S0273–1177(98)00137–9

LESSONS FROM QPOS IN X-RAY PULSARS

Pranab Ghosh[1,2,3]

[1] *NASA Goddard Space Flight center, Greenbelt, MD 20771, USA*
[2] *Senior NAS/NRC Resident Research Associate*
[3] *On leave from Tata Institute of Fundamental Research, Bombay 400 005, India*

ABSTRACT

We discuss the diagnostic potential of the studies of quasiperiodic oscillations (QPO) in X-ray pulsars. We show that probing the conditions in the inner accretion disks in these systems, and mapping out the accretion torques on the pulsars are possible by these diagnostics. We suggest that the magnetospheric beat frequency model may not be applicable to all QPOs observed in X-ray pulsars. We suggest possible future lines of investigation with these diagnostics. ©1998 COSPAR. Published by Elsevier Science Ltd.

INTRODUCTION

Quasiperiodic oscillations (QPO) in accretion-powered X-ray pulsars were discovered by Angelini *et al.* (1989, henceforth ASP) in the 42s transient pulsar EXO 2030+375 with a Be-star companion. At the time of this writing, the existence of such QPOs have been either confirmed or strongly suggested in nine X-ray pulsars, the properties of which are summarized in Table 1. In addition, hints of QPO-like features have occasionally been seen in power spectra of other X-ray pulsars, *e.g.*, GX 1+4 (Doty *et al.* 1981). These QPOs have frequencies in the general range 10 mHZ $\lesssim \nu_{QPO} \lesssim$ 400 mHZ, considerably lower than those of both the low ($\sim 6 - 50$ Hz) frequency QPOs and the recently-discovered high ($\sim 300 - 1200$ Hz) frequency QPOs observed in low-mass X-ray binaries (LMXB; for a review of the two classes of QPOs, see van der Klis 1995, 1997 respectively, henceforth K95, K97), although QPOs at both ~ 400 mHz and ~ 40 Hz (Zhang *et al.* 1996) have been reported from the bursting X-ray pulsar GRO J1744-28.

In this paper, we discuss the diagnostic potential of QPOs in X-ray pulsars for probing (a) accretion flows in these binaries, (b) conditions in the inner accretion disks in them, (c) properties of accretion torques exerted on the neutron stars in them, and, (d) the nature of the coupling between stellar magnetic fields and accretion disks. We discuss the extent to which this potential has so far been realized by giving specific instances of the application of diagnostic probes involving QPO observations to X-ray pulsars. Finally, we indicate future possibilities for extending such probes in the era of high-resolution X-ray timing with the *Rossi X-ray Timing Explorer (RXTE)* and upcoming observatories.

Table 1. Properties of QPOs in X-ray Pulsars

Pulsar	ν_s	ν_{QPO}	r_d	ν_K (MBFM[a])	r_0 (MBFM[a])	ω_s (MBFM[a])
	mHz	mHz	cm	mHz	cm	
GRO J1744-28	2100	400	$3.1.10^8$	2500	$9.1.10^7$	0.84
SMC X-1	1410	10	$3.6.10^9$	1420	$1.3.10^8$	0.99
X 0115+63	277	62	$1.1.10^9$	339	$3.5.10^8$	0.82
V 0332+53	229	51	$1.2.10^9$	280	$3.9.10^8$	0.82
Cen X-3	207	35	$1.6.10^9$	242	$4.3.10^8$	0.85
X 1626-67	130	48	$1.3.10^9$	178	$5.3.10^8$	0.73
EXO 2030+375	24	200	$4.9.10^8$	224	$4.6.10^8$	0.11
A 0535+26	9.7	50	$1.2.10^9$	59.7	$1.1.10^9$	0.16
X Per	1.2	54[b]	$1.0.10^9$	55.2	$1.2.10^9$	0.02

[a]These quantities are obtained by applying MBFM: there is currently justification for this in only EXO 2030+375 and A 0535+26 (see text).
[b]Takeshima 1997

DIAGNOSTIC POTENTIAL OF QPOS

The basic diagnostic value of the studies of the X-ray pulsar QPOs derives from the fact that if their timescales are comparable to the dynamical timescales in the gravitational fields of neutron stars, the corresponding radii are $r_d \sim (GM/4\pi^2\nu_{QPO}^2)^{\frac{1}{3}} \sim 10^8-10^9$ cm. Table 1 shows r_d for a neutron-star mass of $1.4M_\odot$; the case of possible ~ 40 Hz QPOs in GRO J1744-28 is discussed later. Thus, QPOs in X-ray pulsars are expected to carry information about processes in the inner accretion disks around the strongly magnetised neutron stars in these pulsars, since these disks are expected to be terminated at radii of the above order by the strong stellar field. In fact, all models proposed so far for both the QPOs shown by X-ray pulsars and the low-frequency QPOs shown by LMXBs on their so-called horizontal branch (*i.e.*, the horizontal-branch oscillations, henceforth HBO) have invoked the dynamical timescale at a characteristic radius in the inner accretion disk as one basic timescale. Of course, this characteristic radius is much smaller for LMXBs than for X-ray pulsars, consistent with our notion that the neutron stars in the former have magnetic fields much smaller than those in the latter. In addition, magnetospheric beat-frequency models (MBFM) have invoked the spin period of the neutron star as a second basic timescale. Extension of these arguments to dynamical timescales $\lesssim$ milliseconds in the close vicinity of neutron stars led to expectations of observing QPOs with $\nu_{QPO} \sim$ kHz around weakly magnetic neutron stars in LMXBs, which were dramatically fulfilled with the discovery of the kilohertz QPOs (kHzQPO) by *RXTE* (K97).

The MBFM, in which the QPO frequency is interpreted as the beat frequency between that of stellar rotation ν_s and that of Keplerian rotation ν_K at the inner edge of the accretion disk (Alpar & Shaham 1985; Shibazaki & Lamb 1987), *i.e.*,

$$\nu_{QPO} = \nu_K - \nu_s,\qquad(1)$$

has been widely applied to HBOs in LMXBs, and also, largely by analogy, to QPOs in X-ray pulsars. As Table 1 shows, if QPOs in all X-ray pulsars are interpreted in this way, then ν_K is in the range $\sim 50 - 2500$ mHz, and the corresponding inner radius of the disk, r_0, obtained from the relation

$$\nu_K = \frac{1}{2\pi}\sqrt{\frac{GM}{r_0^3}},\qquad(2)$$

lies in the range $\sim 10^8 - 10^9$ cm, typically smaller than the lengthscale r_d introduced above by a factor of 3. Table 1 lists the values of r_0 implied by MBFM for a neutron-star mass of $1.4M_\odot$. With this interpretation, the fastness parameter,

$$\omega_s \equiv \nu_s/\nu_K,\qquad(3)$$

for the rotation of the neutron star, which is of central importance in accretion torque theory (Ghosh & Lamb 1979, henceforth GL; Ghosh 1993, henceforth G93) can also be obtained immediately, and is listed in Table 1.

However, a major problem with this interpretation is that it is currently substantiated by observational evidence for only two X-ray pulsars, namely EXO 2030+375 and A 0535+26. For the rest, there is either no information, or contrary evidence, as is clear from the following argument. The radius r_0 scales with the accretion rate $\dot{M}$, the stellar magnetic moment μ, and the stellar mass M as:

$$r_0 \propto \dot{M}^a \mu^b M^c.\qquad(4)$$

As summarised in Table 2, the exponent a is negative for all known models of accretion disks, so that r_0 always decreases as X-ray luminosity increases, which makes it clear from equations (1) and (2) that ν_{QPO} should increase with increasing X-ray luminosity. This signature, which is a well-known characteristic of HBOs in LMXBs (K95), has so far been observed only in EXO 2030+375 (ASP) and A 0535+26 (Finger *et al.* 1996, henceforth FWH). For pulsars such as 4U 1626-67 and Cen X-3, there appears to be a clear contraindication, *i.e.*, that ν_{QPO} does not change significantly between different epochs of widely different luminosity (*e.g.*, in 4U 1626-67, $\nu_{QPO} \approx 48$ mHz stayed essentially the same in the eras of *EXOSAT* and *ASCA*, although the source luminosity decreased considerably, and spinup changed to spindown), and a similar result seems to be valid for the ~ 0.4 Hz QPOs in the bursting pulsar GRO J1744-28. Hence, indiscriminate application of the MBFM to X-ray pulsar QPOs is clearly untenable. Indeed, following the recent detection of significant sidebands at frequencies $n\nu_s \pm \nu_{QPO}$ in the power spectrum of 4U 1626-67 with the lower sidebands stronger than the higher ones in the $4 - 8$ keV range, Kommers *et al.* (1998, henceforth KCL) have argued that MBFM is ruled out for this source, and that the 48 mHz QPOs may be due to a structure orbiting the neutron star at this frequency. Sidebands have also been detected in the power spectrum of Cen X-3 (Finger 1997, private communication to KCL).

Thus, one must establish the applicability of MBFM to each X-ray pulsar individually (by showing, *e.g.*, a positive correlation between ν_{QPO} and X-ray intensity). If this is possible for a given pulsar, then an array of diagnostic probes can be devised from our detailed knowledge of the theoretical underpinnings of the MBFM. We present two specific instances in the next section.

EXAMPLES OF QPO DIAGNOSTICS

Disk Diagnostics

Models of the inner accretion disk leave their characteristic signatures on the inner radius, r_0, of the disk in terms of its scaling with the $\dot{M}$, μ, and the stellar mass M, as determined by the exponents a, b, and c in equation (4) (Ghosh & Lamb 1991, 1992, Ghosh 1996, henceforth G96). The values of these exponents are given in Table 2 for several disk models widely used in the literature. These scalings can be used to construct diagnostic probes of the disk models. Consider, for example, the scaling with the accretion rate. The intensity-dependence of ν_{QPO} can then be used, in principle, to determine the exponent a, and so the disk model. The problem with this scheme is that the actual relation between $\dot{M}$ and the observed X-ray intensity I is not known. While a proportionality between $\dot{M}$ and I is often assumed for simplicity, this need not be correct if there are spectral changes in the source. Indeed, there are counterexamples among the LMXBs: for example, Vrtilek *et al.* (1991) have shown that, for the Z source Sco X-1, $\dot{M}$ increases monotonically as the source moves from the horizontal branch through the normal branch to the flaring branch on the color-color diagram. This would imply that, on the normal branch, $\dot{M}$ is anticorrelated with I.

Table 2. Signatures of Various Disk Models

Disk Model	Value of a	Value of b	Value of c	Value of σ
1T Opt thick GPD (1G)	−0.25	0.58	−0.21	2.33
1T Opt thick RPD (1R)	−0.15	0.51	−0.13	4.11
2T Opt thin GPD Compt brems (2B)	−0.48	0.57	0.05	1.06
2T Opt thin GPD Compt low-energy photon (2S)	−1.70	0.80	0.73	0.059

A diagnostic which avoids using the uncertain relation between $\dot{M}$ and I is that which uses simultaneous observations of ν_{QPO} and the spin-change rate $\dot{\nu}$ of the X-ray pulsar: the latter is given by

$$\dot{\nu} \propto n(\omega_s)\dot{M}\sqrt{GMr_0}\,, \qquad (5)$$

where n is the dimensionless torque, which is a function of the fastness parameter ω_s introduced above (GL). Equations (1), (2), (4) and (5) can be combined to yield

$$n(\omega_s) \propto \frac{\dot{\nu}}{\nu_K^\sigma}\,, \qquad (6)$$

where $\sigma = -\frac{2}{3a} - \frac{1}{3}$. For slow rotators ($\omega_s \ll 1$), the dimensionless torque $n \simeq$ const $\simeq 1$ (GL, G93), as is the case during the outbursts and fast spinup of transient X-ray pulsars like A 0535+262 and EXO 2030+375. In such cases, equation (6) yields

$$\log \dot{\nu} = \sigma \log \nu_K + \text{const.}, \tag{7}$$

showing that the signature of the disk model is contained in the value of the slope σ of the $\log \dot{\nu}$ vs. $\log \nu_K$ plot: values of σ for the disk models are listed in Table 1.

The first application of this diagnostic was made to the data obtained during the large outburst of A 0535+26 in February-March 1994 (FWH). The results are shown in Figure 1 (G96). The data are consistent with model 1G, $i.e.$, the 1-temperature, optically thick, gas-pressure dominated disk (Shakura & Sunyaev 1973), but neither with the radiation pressure dominated model 1R, nor with hot, 2-temperature, optically thin models like 2B or 2S (Shapiro et $al.$ 1976). This identification of the disk model in an X-ray pulsar appears to be the most unambiguous one possible so far.

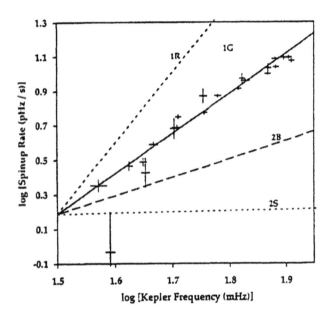

Fig. 1.- The diagnostic $\log \dot{\nu}$ vs. $\log \nu_K$ plot for A 0535+262. Data points from BATSE are superposed on the predictions from disk models 1G (solid line), 1R (short-dashed line), 2B (long-dashed line), and 2S (dots) (see text). Since only the slopes of the lines are relevant here, the lines corresponding to the different disk models have been drawn through a suitable common origin for clarity of display.

Torque Diagnostics

Once the appropriate disk model is identified and σ determined, one can map out the shape of the dimensionless torque function $n(\omega_s)$ for faster rotators (those with $\omega_s \lesssim 1$)

with the aid of equations (3) and (6), if torque-QPO data on such rotators are available. The above data on A 0535+26 does not extend very far into the fast rotator regime, as is obvious from Figure 1, and so is not optimal for this purpose. Nevertheless, it does give us the first map of the torque function in the range $0.1 \lesssim \omega_s \lesssim 0.2$ which is free from the uncertainties in the $\dot{M}$ - I relationship discussed above. Figure 2 shows the observed torque function, superposed on a set of theoretical torque functions taken from Daumerie, Lamb & Ghosh (1998) corresponding to a set of possible values of the maximum magnetic pitch in the outer accretion disk. It is clear that there is a flatness in $n(\omega_s)$, with a considerable scatter, in this range of fastness, but this does not yet constrain the models of magnetic pitch and accretion torque very significantly. Most of the curves in Figure 2 are in reasonable agreement with the data. In particular, the critical fastness ω_c, where the torque changes sign (GL), is not well constrained.

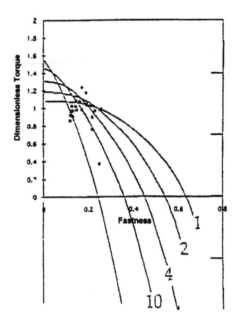

Fig. 2.- Mapping of the dimensionless torque as a function of fastness. Shown are the data points for A 0535+26, superposed on a set of torque curves from the calculations of Daumerie *et al.* (1998). Each curve is labelled by its value of the magnetic pitch in the outer accretion disk

One might naively think that pulsars like SMC X-1, Cen X-3, and 4U 1626-67 would be particularly useful for mapping the torque function for fast rotators, since applications of the MBFM to these indicate values of ω_s which are $\sim 0.75 - 1$, and a similar argument holds for the 0.4 Hz QPOs in the bursting pulsar GRO J1744-28 (Table 1). However, these are precisely the sources for which there is currently either no justification for applying the MBFM or clear evidence against it, as detailed in the previous section. Indeed, for six of the nine pulsars listed in Table 1, ω_s lies in the range $\sim 0.75 - 1$ if one applies the MBFM interpretation to ν_{QPO}, so that all these pulsars should show secular spindown trends according to standard accretion torque theory, but, with the exception of 4U 1626-67, none of these does so. The paradox is particularly striking for SMC X-1 ($\omega_s = 0.99$), for which *all* theoretical torque profiles, including that of Wang (1995), predict strong spindown, contrary to observation. A natural resolution of this paradox is that the QPOs in these X-ray pulsars are not the standard magnetospheric beat-frequency phenomenon: this is consistent with the arguments recently advanced by KCL for 4U 1626-67 (see above).

DISCUSSION

We have indicated above some of the diagnostic capabilities of studies of QPOs in X-ray pulsars through the use of MBFM, when this model is applicable. When it is not, we need a quantitative formulation of the relevant QPO model (*e.g.*, that suggested by KCL for 4U 1626-67) in order to evolve alternative diagnostics. Whereas a radius $\sim r_d$ can be immediately assigned to these oscillations (Table 1) on dynamical grounds, phenomena which might occur selectively at such a radius need to be identified and understood. In general, there appears to be no reason to argue that this is always the inner radius of the disk: a variety of processes in inner parts of accretion disks can lead to such oscillations.

The discovery of kHzQPOs from LMXBs by *RXTE* (K97) has given us exciting diagnostic tools for probing accretion flows in the immediate environs of neutron stars. To date, however, no kHzQPOs have been detected from X-ray pulsars. If we assume that the inner part of an accretion disk is an essential ingredient for the production of kHzQPOs, then this negative result is consistent with our general notion that, whereas the accretion disks in LMXBs can extend to inner radii $r_0 \sim 10^6 - 10^7$ cm, quite close to the surfaces of the weakly magnetized neutron stars in these systems, the disks in X-ray pulsars are terminated at radii $r_0 \sim 10^8 - 10^9$ cm by the strong magnetic fields of the neutron stars in the latter systems. This would further imply, at one remove, that other possible physical phenomena expected to occur at $\sim$ kilohertz frequencies in or near neutron stars, *e.g.*, stellar vibrations, or photon bubble oscillations in accretion columns, are not by themselves capable of generating such X-ray signals as we currently identify as kHzQPOs.

Finally, the nature of the ~ 40 Hz QPO in the bursting X-ray pulsar GRO J1744-28 (Zhang *et al.* 1996) remains to be understood. If we apply MBFM to this QPO, we would reach the conclusion that this 2.1 Hz pulsar is a very slow rotator ($\omega_s \simeq 0.05$), and that the inner radius of the disk is $r_0 \simeq 1.4 \times 10^7$ cm. This would imply a rather weak magnetic field for the neutron star even if we assume persistent super-Eddington luminosities, in contrast to the canonical values of r_0 and the stellar magnetic field implied by the ~ 0.4 Hz QPOs listed in Table 1. The indication that the two QPOs may have been seen in the same time frame (Kommers *et al.* 1997) implies that at most one of these can be the magnetospheric beat frequency. Further, as explained earlier in the paper, there are inconsistencies in the fast-rotator ($\omega_s \simeq 0.84$) interpretation suggested by the MBFM interpretation of the ~ 0.4 Hz QPOs. It is thus possible that the MBFM is applicable to neither of the two QPOs in this source.

REFERENCES

Alpar, M. A., & J. Shaham, *Nature*, 316, 239 (1985).

Angelini, L., L. Stella, & A. N. Parmar, *Ap.J.*, 346, 906 (1989).

Daumerie, P., F. K. Lamb, & P. Ghosh, in preparation (1998).

Doty, J., J. Hoffman, & W. H. G. Lewin, *Ap.J.*, 243, 257 (1981).

Finger, M. H., R. B. Wilson, & B. A. Harmon, *Ap.J.*, 459, 288 (1996) (FWH).

Ghosh, P., in *The Evolution of X-ray Binaries*, ed. S. S. Holt & C. S. Day (New York: American Institute of Physics), 439 (1993).

Ghosh, P., *Ap.J.*, 459, 244 (1996).

Ghosh, P., & F. K. Lamb, *Ap.J.*, 234, 296 (1979).

Ghosh, P., & F. K. Lamb, in *Neutron Stars: Theory & Observation*, ed. J. Ventura & D. Pines (Dordrecht: Reidel), 363 (1991).

Ghosh, P., & F. K. Lamb, in *X-ray Binaries & Recycled Pulsars*, ed. E. P. J. van den Heuvel & S. Rappaport (Dordrecht: xsKluwer), 487 (1992).

Kommers, J. M., D. W. Fox, W. H. G. Lewin, R. E. Rutledge, J. van Paradijs, *et al.*, *Ap.J.*, 469, L29. (1997).

Kommers, J. M., D. Chakrabarty, & W. H. G. Lewin, *Ap.J.*, 497, L33 (1998).

Shakura, N. I., & R. A. Sunyaev, *A&A*, 24, 337 (1973).

Shapiro, S. L., A. P. Lightman & D. M. Eardley, *Ap.J.*, 204, 187 (1976).

Shibazaki, N., & F. K. Lamb *Ap.J.*, 318, 767 (1987).

Takeshima, T., in Proc. XXIII IAU General Assembly, Session 188, *The Hot Universe*, in press (1997)

van der Klis, M., in *X-ray Binaries*, ed. W. H. G. Lewin, J. van Pradijs & E. P. J. van den Heuvel (Cambridge Univ. Press), 252 (1995).

van der Klis, M., in Proc. NATO ASI, *The Many Faces of Neutron Stars*, Lipari, Italy, in press (astro-ph/9710016) (1997).

Vrtilek, S. D., W. Penninx, J. C. Raymond, F. Verbunt, P. Hertz, *et al.*, *Ap.J.*, 376, 278 (1991)

Wang, Y.-M., *Ap.J.*, 449, L153 (1995).

Zhang, W., E. Morgan, K. Jahoda, J. H. Swank, T. E. Strohmayer, *et al.*, *Ap.J.*, 469, L29 (1996).

Adv. Space Res. Vol. 22, No. 7, pp. 1025–1034, 1998
© 1998 COSPAR. Published by Elsevier Science Ltd. All rights reserved
Printed in Great Britain
0273-1177/98 $19.00 + 0.00

Pergamon

PII: S0273-1177(98)00131-8

THE 5–9 SECOND X–RAY PULSARS

L. Stella,[1,*] G. L. Israel,[2,*] S. Mereghetti,[3]

[1] *Osservatorio Astronomico di Roma, V. dell'Osservatorio 2, I-00040 Monteporzio Catone (Roma), Italy,*
stella@coma.mporzio.astro.it
[2] *International School for Advanced Studies (SISSA–ISAS), Via Beirut, 3, I-34014, Trieste, Italy,*
israel@vega.sissa.it
[3] *Istituto di Fisica Cosmica, CNR, Via Bassini 15, I-20133 Milano, Italy,*
sandro@ifctr.mi.cnr.it
Affiliated to I.C.R.A.

ABSTRACT

A new class of X–ray pulsars has been recently identified. Its members are characterised by: (i) similar spin periods in the 5–9 s range; (ii) secular spin down; (iii) very soft X–ray spectra; (iv) relatively low and constant X–ray luminosities of $\sim 10^{35} - 10^{36}$ erg s^{-1}; (v) association with regions of diffuse X–ray and radio emission, perhaps old supernova remnants. These pulsars contain neutron stars that derive their X–ray emission from matter accretion and are probably rotating close to their equilibrium period. The inferred magnetic field strengths are of a few$\times 10^{11}$ G. While at least one member of the class is in a binary system, it is yet unclear whether the other 5–9 s pulsars accrete from a very low mass companion or are isolated and posses a residual accretion disk. It has been proposed that these pulsars are young ($\sim 10^5$ yr) descendants of the evolution of massive binary systems, probably high mass X–ray binaries after common envelope and spiral in evolution. ©1998 COSPAR. Published by Elsevier Science Ltd.

INTRODUCTION

X–ray pulsations with periods ranging from ~ 0.069 to ~ 1450 s are present in a large number (about 50) of X–ray binaries. This signal originates from the beamed radiation which is produced close to the magnetic poles of an accreting neutron star with a surface field of $\sim 10^{12} - 10^{13}$ G. Most of these pulsars (over 40) are found in high mass X–ray binaries (HMXRB), containing an early type (OB) star with a mass of > 5 M$_{\odot}$. Mass transfer usually takes place because part of the intense stellar wind emitted by the OB star is captured by the gravitational field of the collapsed object. In some cases the OB star fills up its Roche lobe and mass transfer towards the neutron star takes place also through the first Lagrangian point, leading to the formation of an accretion disk. Secular spin period changes arise because of the torque exerted on the neutron star magnetosphere by the accreting matter (see e.g. Henrichs 1983; White, Nagase & Parmar 1995). Disk–fed systems are characterised by a pronounced spin–up with a timescale that can be as short as ~ 100 yr. Alternating spin–up and spin–down intervals are instead frequent among wind fed–systems; these likely result from the variable wind characteristics of the early type mass donor (especially if a Be star), which can cause the angular momentum of the captured material to reverse its sign relative to the neutron star.

The spectra of X–ray pulsars in HMXBs are quite hard. These are usually characterised in terms of a power–law extending up to energies of several tens of keV, followed by a steep nearly exponential decay, beyond which cyclotron absorption features are sometimes detected.

The number of X–ray pulsars that are not in HMXBs is very small, but recent observations have drastically enlarged the sample. This has also resulted from the finding that a few optically unidentified sources have X–ray to optical flux ratios incompatible with the presence of massive companions. Several of these systems have low mass donor stars and, therefore, are Low Mass X–ray Binaries (LMXBs). Three of the LMXBs pulsars are peculiar systems: Her X–1 has a comparatively massive A–star companion ($\sim 2\ M_\odot$), while GX 1+4 and the hard transient GRO J1744–28 have evolved low-mass companions (see, e.g., Rappaport & Joss 1983 and Nagase 1989; Finger, Wilson & van Paradijs 1996). GRO J1744–28 in particular displays an unprecedented variety of bursting behaviours (Kouveliotou *et al.* 1996; Strickman *et al.* 1996; Stark *et al.* 1996).

The remaining five X–ray pulsars which are not in HMXBs (4U 0142+61, 1E 1048.1–5937, 4U 1626–67, RX J1838.4–0301 and 1E 2259+589) have very similar spin periods in the 5 to 9 s range (see Figure 1). This narrow period distribution is remarkable, when one considers that the X–ray pulsars in HMXBs have spin periods ranging from 69 ms (A 0538–66, Skinner *et al.* 1982) to 25 min (RX J0146.9+6121, Mereghetti, Stella & De Nile 1993). Based also on several other similarities, we proposed that these 5–9 s X–ray pulsars belong to a homogeneous class of sources and suggest a possible explanation for their narrow spin period distribution (Mereghetti & Stella 1995). Whether or not the recently discovered 13 s X–ray pulsations from HD49798 originate in an accreting neutron star with similar characteristics is unclear at present (Israel *et al.* 1995, 1996).

Spin period distribution of X–ray pulsars

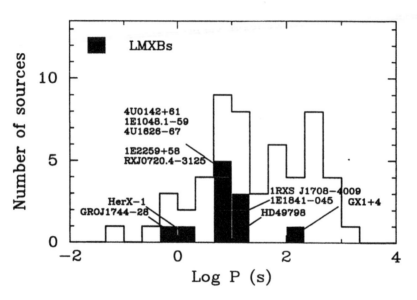

Figure 1: The observed period distribution of X–ray pulsars. The black squares indicate the X–ray pulsars that are not HMXBs.

OBSERVATIONAL CHARACTERISTICS OF THE 5-9 SECOND X-RAY PULSARS

4U 0142+61

The properties of 4U 0142+61 (White *et al.* 1987) remained puzzling for a long time, owing to confusion problems with a nearby pulsating and transient Be/neutron star system (Motch *et al.* 1991, Mereghetti, Stella & De Nile 1993). Despite the small error box (5" radius), no optical counterpart has yet been identified, down to $V > 24$ (Steinle *et al.* 1987), thus excluding the presence of a massive companion. Using data from the EXOSAT archive, Israel, Mereghetti & Stella (1994) discovered periodic pulsations at 8.7 s, which were later confirmed with ROSAT (Hellier 1994). No delays in the pulse arrival times caused by orbital motion were found, with upper limits on $a_x \sin i$ of about ~ 0.37 lt–s for orbital periods between 7 min and 12 hr (Israel, Mereghetti & Stella 1994). The EXOSAT and ROSAT period measurements, obtained in 1984 and 1993, provide a spin–down rate of $\sim 7 \times 10^{-5}$ s yr^{-1}. The 1–10 keV spectrum is extremely soft (power law photon index of ~ 4, White *et al.* 1987) and led to the initial classification of this source as a possible black hole candidate. Recent ASCA spectra provide evidence for an additional ~ 0.4 keV blackbody component contributing some $\sim 40\%$ of the X–ray flux (White *et al.* 1996). The X–ray luminosity of 4U 0142+61 did not show large secular variations around an average value of $\sim 2.5 \times 10^{35}$ erg s^{-1} (assuming a distance of 2 kpc).

<u>1E 1048.1–5937</u>

This source was serendipitously discovered with the Einstein Observatory in 1979, and found to pulsate at 6.44 s (Seward, Charles & Smale 1986). The brightest candidate counterparts in the small error box have V> 20 (Mereghetti, Caraveo & Bignami 1992) indicating that also 1E 1048.1–5937 is a LMXB. 1E 1048.1–5937 was repeatedly observed with ROSAT in 1992 and 1993. While all the previous observations with EXOSAT and GINGA (Corbet & Day 1990) were consistent with a constant spin–down at a rate of $\sim 5 \times 10^{-4}$ s yr^{-1}, the ROSAT data (Mereghetti 1996) indicate a doubling of the spin–down rate (Figure 2) . The power law photon index derived with EXOSAT (2.3, Seward, Charles & Smale 1986) implies also for this source a spectrum somewhat softer than the "canonical" spectrum of X–ray pulsars. The high column density suggests that 1E 1048.1–5937 lies behind the Carina nebula, i.e. at more than 2.8 kpc. The luminosity corresponding to this distance is $\sim 2 \times 10^{34}$ erg s^{-1}.

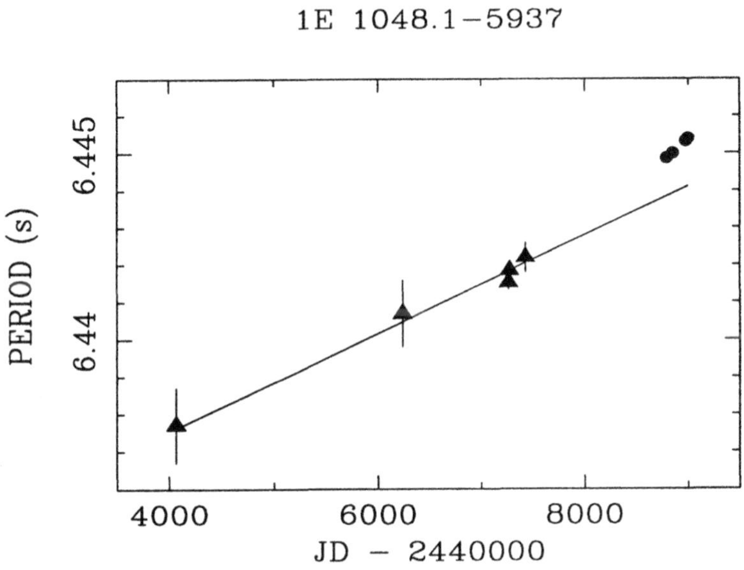

Figure 2: Spin period evolution of 1E 1048.1–5937 (from Mereghetti 1996).

<u>4U 1626–67</u>

This LMXBs is optically identified with the V$\simeq$ 18.5, blue star KZ TrA (Mc Clintock *et al.* 1977). In addition to the X–ray periodicity at 7.7 s (Rappaport *et al.* 1977), a pulsation at a slightly lower frequency is present in the optical band (Middleditch *et al.* 1981). This is probably due to reprocessing of the X–ray pulses occurring near the companion star, and the difference of the two periodicities can be explained with an orbital period of 41.4 min. The high X–ray to optical flux ratio, as well as the very strong limits on the optical mass function ($a_x \sin i < 0.013$ lt–s for 10 min $< P_{orb} < 10$ hr, Levine *et al.* 1988), clearly indicate that 4U 1626–67 is a LMXB. The period measurements obtained before 1990 were consistent with a constant spin–up rate of -1.6×10^{-3} s yr^{-1} (Nagase 1989), but in 1991 the period derivative changed sign (Lutovinov *et al.* 1994, Bildsten *et al.* 1994). On the basis of the luminosity required to explain the spin–up torque ($L_x \sim 2 \times 10^{36}$–$10^{37}$ *erg s*$^{-1}$) Levine *et al.* (1988) estimated a distance of 3 to 6 kpc. With the exception of quasi–periodic flares with a characteristic timescale of ~ 1000 s, little intensity variations are present in 4U 1626–67. The pulse averaged spectrum above a few keV, a flat power law (photon index ~ 0.4) followed by an exponential cut–off at ~ 20 keV (Pravdo *et al.* 1979), is similar to that of HMXB pulsars (White, Swank & Holt 1983). An ASCA observation revealed the presence of a low energy excess (modelled with a 0.6 keV blackbody) and a complex feature around 1 keV, interpreted as emission from hydrogen–like Neon (Angelini *et al.* 1995). The latter implies a large Ne overabundance in 4U 1626–67, testifying that the companion star, now probably a light Ne–rich He white dwarf, once burned He.

RX J1838.4–0301

This pulsar, recently discovered with ROSAT, is embedded in a region of diffuse X–ray and radio emission, which is interpreted as a $\sim$ 32000 yr old supernova remnant at a distance of $\sim$ 4 kpc (Schwentker 1994). The brightest optical object in its 10" radius error box has V=14. The spectrum in the 0.1–2.4 keV band, and therefore the unabsorbed flux, are not well constrained by the data (Schwentker 1994), but there is evidence that also this source is quite soft (best fit power law photon index of $\sim$ 3). Also RX J1838.4–0301 is likely a LMXB.

1E 2259+586

The source 1E 2259+586 was discovered with the Einstein Observatory at the center of the X–ray and radio supernova remnant G109.1–1.0 (Fahlman & Gregory 1981). Extensive searches for optical, IR and radio counterparts were carried out without success (Fahlman *et al.*. 1982; Coe & Jones 1992; Coe, Jones & Letho 1994), but they definitely exclude the presence of a massive companion (Davies & Coe 1991). The spin period of 1E 2259+586 has been increasing at $\sim$ 2 $\times$ 10^{-5} s yr^{-1} until 1992 (Koyama *et al.* 1989, Iwasawa, Koyama & Halpern 1992). Recent ROSAT data revealed the first spin–up episode for this source. A detailed analysis of all the period measurement over the last 15 years (Baykal & Swank 1996) showed that 1E 2259+586 undergoes random angular velocity variations similar to those observed in other accreting binary neutron stars (Figure 3). The upper limit on $a_x \sin i$ is of 0.08 lt–s for

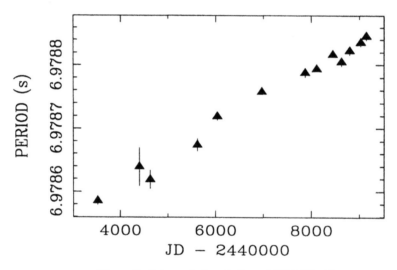

Figure 3: Spin period evolution of 1E 2259+58.

10^3 s $< P_{orb} < 10^4$ s (Koyama *et al.* 1989). 1E 2259+586 has a very soft spectrum, as recently confirmed by BBXRT and ASCA observations (Corbet *et al.* 1995). Also in this source, no long term variability greater than a factor $\sim$2 has been reported. A distance of 3.6$\pm$0.4 kpc for the supernova remnant G109.1–1.0 was estimated by using the classical relation between distance and radio surface brightness (Gregory & Fahlman 1980). Based on improved radio observations, Hughes *et al.* (1984) have subsequently derived a value of 5.6 kpc. For this distance the average X–ray luminosity of 1E 2259+586 is $\sim$2 $\times$10^{35} erg s $^{-1}$.

HD49798

HD49798 is a subdwarf O6 star in a single component spectroscopic binary (with P_{orb} = 1.55 d, $a \sin i$ = 3.60 R$_\odot$, f(m) = 0.263 M$_\odot$; Thackeray 1970; Stickland & Lloyd 1994). Despite extensive studies, the nature of the companion star remained unclear for decades. Being relatively massive ($M \sim 0.7 - 2.1$ M$_\odot$) very hot (T$\sim$ 47000 K) and luminous (Log ($L/L_\odot$) ~ 3.9) the sdO6 star outshines the flux emitted by the companion star (at least in the optical–UV). The level of H depletion together with the strong overabundance of N and underabundance of C, testify that the outer layers of HD49798 were processed through the CNO cycle and the envelope of the progenitor star was lost during a common envelope phase. Bisscheroux *et al.*(1996) argue that the sdO star has a degenerate CO core and is in the phase of shell helium burning. The recent discovery of strong 13 s pulsations in the soft X-ray flux from HD49798, proves that the system hosts an accreting degenerate companion either a white dwarf or, more likely, neutron star (Israel *et al.*1995, 1996). The source X-ray spectrum is extremely soft, with a power law photon index of $\alpha \sim 4$, corresponding to an unabsorbed 0.1–2 keV luminosity of a few $\times 10^{32}$ erg s^{-1} (distance of 650 pc; $\sim$200 pc above the Galactic plane). A luminosity of up to $\sim 10^{35}$ erg s^{-1} might be hidden in the EUV. The relationship of HD49798 to the other 5–9 s X-ray pulsars hinges on the nature of its degenerate star, which remains to be firmly established as a neutron star. Therefore in the next three sections we exclude HD49798 from the sample of "5–9"s pulsars.

A NEW CLASS OF X–RAY PULSARS

The main properties of the 5–9 s X–ray pulsars described above are summarized in Table 1. Their most important similarities are the following. (a) When compared to the other X–ray pulsars, the most striking property of these objects

Table 1: Properties of the VLMXB pulsars

Source	Period (s)	Spin–down (s yr^{-1})	X–ray Flux (erg cm^{-2} s^{-1})	α	L_{eq} (erg s^{-1})	d_{eq} (kpc)	z (pc)
4U 0142+61	8.69	7.2×10^{-5}	5.5×10^{-10}	4	1.4×10^{35}	1.5	-11
1E 1048.1–59	6.44	4.6×10^{-4}	2.2×10^{-11}	2.3	2.8×10^{35}	10.6	+93
4U 1626–67	7.66	1.4×10^{-3}	6.0×10^{-10} ^{a}	0.4	1.9×10^{35}	1.7	-385
RX J1838.4–03	5.45	...	5×10^{-12}–4×10^{-9}	$\sim$3	4.1×10^{35}	27–1	+135
1E 2259+58	6.98	2.3×10^{-5}	5.3×10^{-11}	4–5	2.3×10^{35}	6.2	-108

NOTE — L_{eq} and d_{eq} have been computed assuming $B = 5 \times 10^{11}$ G for all the sources.
^{a} flux during spin–down; when 4U 1626–67 was spinning–up at -1.6 $\times 10^{-3}$ s yr^{-1} its flux was $\sim$4 times higher.

is their narrow spin period distribution (Figure 1). A Kolmogorov–Smirnov test yields a probability of $\sim 4\times 10^{-3}$ that the period distribution of these five sources and that of the HMXB pulsars are drawn from the same parent distribution. (b) With the exception of 4U 1626–67 (which, on the other hand, possesses a low energy excess), their spectra are much softer than those of the other X–ray pulsars (White, Swank & Holt 1983). Figure 4 shows a comparison of the EXOSAT spectra of Vela X–1 and 4U 0142+61. 4U 0142+61 and 1E 2259+586, in particular, have even softer spectra than most LMXBs (White, Stella & Parmar 1988). (c) These sources are probably less luminous than most persistent LMXBs. Indeed, for 1E 2259+586, 1E 1048.1–5937 and 4U 0142+61 firm upper limits of a few$\times 10^{36}$ erg s^{-1} can be derived by requiring that they lie within the Galaxy. The best estimates for the distances of 4U 1626–67 and RX J1838.4–0301 also imply luminosities of this order of magnitude or smaller. (d) Their flux appears to be relatively constant on timescales from months to years. This is unlike most of the other X–ray pulsars (including Her X–1, GX 1+4 and GRO J1744–28) for which large flux variations (encompassing transient activity) have been observed. (e) Two of these sources are likely associated to regions of diffuse X–ray and radio emission, which are likely supernova remnants. Another source, 1E 1048.1–5937, lies in the direction of the Carina Nebula, a complex region of radio, optical and X–ray diffuse emission, clearly associated with recent star formation activity.

Based on these similarities, Mereghetti & Stella (1995) proposed that these sources form a homogeneous subclass of accreting neutron stars, perhaps members of LMXBs, which are characterized by lower luminosities ($10^{35} - 10^{36}$ erg s^{-1}) and higher magnetic fields ($B \sim 10^{11}$ G) than classical LMXBs.

ENERGY PRODUCTION MECHANISMS

Models that have been proposed through the years for some of the X–ray pulsars of the new class (in particular for

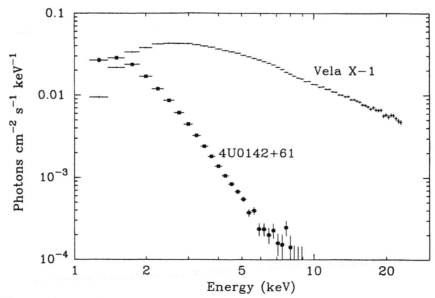

Figure 4: A comparison of the EXOSAT ME spectra of Vela X–1 and 4U 0142+61.

1E2259+586) differ with respect to the energy production mechanism as well as binary versus single nature of the magnetic degenerate star. Among models based on rotational energy dissipation, simple radio pulsar models can be ruled out based on the fact that the measured spin–down rates correspond to a rotational energy loss by a magnetic neutron star which is more than two orders of magnitude lower than the inferred X–ray luminosities. Morini *et al.* (1988) and Paczynski (1990) propose instead a white dwarf equivalent of the standard radio pulsar model. Due to the factor of $\sim 10^5$ larger moment of inertia, the white dwarf rotational energy loss implied by the measured $\dot{P}$ is considerably larger than the measured X–ray luminosity. Within this framework, the spin–down rate changes of 1E 2259+586 observed by Iwasawa, Koyama & Halpern (1992) have been interpreted by Usov (1994) in terms of white dwarf glitches. In a different vein, Thompson & Duncan (1993) suggest that 1E 2559+586 spins down like a standard radio pulsar, while the emitted radiation results from the gradual dissipation of the intense magnetic field ($> 10^{14}$ G) of the neutron star.

All the models outlined above are virtually ruled out by two recent studies. Baykal & Swank (1996) show that the spin period history of 1E 2259+586 consists of short term spin–up episodes superposed on a secular spin–down and that its fluctuation level is similar to that of a number of accreting X–ray pulsars in HMXBs (see figure 3). More crucially Mereghetti (1996) discovered an increased spin–down episode from 1E1048.1–5937 (see Figure 2), which (contrary to the case of 1E 2259+585, Usov 1994) cannot be explained with a glitch, therefore ruling out the unconventional applications of the radio pulsar model described above.

Models based on matter accretion onto a magnetic rotating neutron star are clearly favored. These models, in turn, envisage both possibilities that the neutron star is isolated or in a binary system.

EQUILIBRIUM SPIN PERIODS

The measured spin–down rate of three of the 5–9 s pulsars testifies that the neutron star is accreting close to its equilibrium spin period. We assume that also RX J1838.4–0301, and 4U 1626–67 are (nearly) equilibrium rotators. In the case of the latter source this is supported by the observed reversal from spin–up to spin–down. The equilibrium period is given by

$$L_{eq} \sim 8.6 \times 10^{35} (B/10^{11}G)^2 P^{-7/3} erg\ s^{-1},$$

where L_{eq} is the accretion luminosity at equilibrium and B the surface magnetic field of the neutron star (see, e.g., Henrichs 1983; we use a neutron star mass of 1.4 $M_\odot$ and radius of 10^6 cm). A measurement of the magnetic field strength based on cyclotron features has been obtained only for 1E 2259+586, giving $B \sim 5 \times 10^{11}$ G (Iwasawa, Koyama & Halpern 1992, see, however, Corbet *et al.* 1995 for a different interpretation). The magnetic field of the

EXOSAT ME

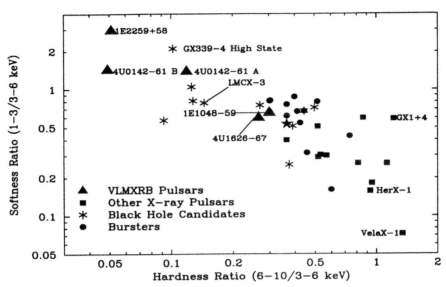

Figure 5: X-ray colour–colour diagram for a sample of X-ray binaries.

other systems can be constrained by using the properties of their X-ray continuum. The spectrum of most X-ray pulsars is characterised by a relatively flat power law (photon index 0.5–1.8), with a cut–off around 10–30 keV above which the spectrum is much steeper (White, Swank & Holt 1983). This high energy cutoff is interpreted in terms of resonant cyclotron absorption in the vicinity of the polar caps. In those X-ray pulsars in which cyclotron line features are observed, the cutoff energy, E_{cut}, was empirically determined to be related to the cyclotron line energy through $E_{cyc} \sim (1.2 - 2.5)E_{cut}$ (Makishima et al. 1990, 1992). Therefore, E_{cut} can be used to approximately estimate the magnetic field strength (see however the case of EXO 2030+375; Reynolds, Parmar & White 1993). With the exception of 4U 1626-67, the photon indeces of the spectra in our sample (measured shortwards of 10 keV) are higher than the range measured in other X-ray pulsars. It is therefore likely that the part of the spectrum above E_{cut} is predominantly observed in the X-ray pulsars in our sample. Due to the combined effects of poor statistics and photoelectric absorption, a cutoff below 2–3 keV would be difficult to detect in the available spectra of 4U 0142+61, 1E 1048.1-5937 and 1E 2259+586 (note that the application of the standard X-ray pulsar spectral model to 4U0142+614 gives $E_{cut} \sim 1.4$ keV, although the fit is quite poor, see White et al. 1996). We conclude that for the 5–9 s pulsars $E_{cut} < 2$–3 keV and therefore $B < 8 \times 10^{11}$ G. These values are substantially lower than those measured (or inferred) from most other X-ray pulsars ($10^{12} - 10^{13}$ G). Assuming $B = 5 \times 10^{11}$ G (based on the analogy with 1E 2259+586), Eq. 1 yields for the sources in our sample luminosities of 1–4×10^{35} erg s^{-1}. The corresponding distances, d_{eq}, are reported in Table 1. These luminosities and distances values are generally in good agreement with the X-ray and optical observational data described in Section 2. The only exception is 4U 1626-67, for which $d_{eq} \sim 1.7$ kpc is outside the range of 3–6 kpc derived by Levine et al. (1988) based on the spin–up rate measured before 1990. This indicates that 4U 1626-67 has a somewhat higher magnetic field (d_{eq} scales linearly with B), as also suggested by its hard spectrum. A similar conclusion is reached if, in addition to Eq. 1, the equation for the accretion torque (cf. eq. 28 in Henrichs 1983) is used together with the period derivative observed during the spin–up phase ($\dot{P}/P \sim -2 \times 10^{-4}$ yr^{-1}) and a luminosity decrease of a factor ~ 4 observed in correspondence to the $\dot{P}$ reversal (Bildsten et al. 1994). This provides an independent rough estimate of $B \sim 1.1 \times 10^{12}$ G and $L_{eq} \sim 9 \times 10^{35}$ erg s^{-1}. In the interpretation above, the narrow spin period distribution of the 5–9 s pulsars should therefore reflect a relatively narrow range of neutron star magnetic field and accretion luminosity.

BINARY VS. ISOLATED ACCRETING NEUTRON STARS

Table 1 gives also the height z above the galactic plane of the 5–9 s X-ray pulsars as estimated from their equilibrium distances (cf. van Paradijs et al. 1995). With the exception of 4U1626-67 and, perhaps, HD49798 (the only two known binary systems of the sample), the 5–9 s pulsars show a narrow z–distribution in the galaxy. This, together with the association (in two cases) with shell–like structures that appear to be $\sim 10^5$ year old supernova remnants, indicates that the 5–9 s pulsars are young descendants (roughly 10^5 yr old) of the evolution of massive stars. Based on their current number and estimated distances Van Paradijs *et al.* (1995) suggest that our galaxy contains some 20–50 X–ray pulsars with similar characterstics, and therefore that their birth rate is in the $\sim 10^{-3}$ yr^{-1} range. However the combination of relatively low luminosities, very soft spectra and absorbing columns in the galactic plane likely introduces a bias against the detection of other 5–9 s X–ray pulsars. Therefore, the numbers quoted above should be regarded as lower limits.

Israel, Mereghetti & Stella (1994) and Corbet *et al.* (1995) suggest that the neutron stars in 4U 0142+614 and 1E 2259+586, respectively, might be accreting matter from a dense region of a molecular cloud. The problem with this scenario is that the highest expected mass capture rates are in the 10^{13}g s^{-1} range, therefore giving rise to an accretion luminosity of $\sim 10^{33}$ erg s^{-1}. This is well below the inferred X–ray luminosities.

Mass accretion from a very low mass companion is perhaps the simplest explanation to account for the observed properties of the 5–9 s pulsars. The faintness of their optical counterpart and the absence of measurable Doppler modulations in the X–ray pulses, indicate that these systems have small orbital periods and/or companions of very low mass (clearly the case of HD49798 could be an important exception). The main contribution to the optical emission is thus expected to come from the X–ray heated accretion disk. An empirical relation between absolute magnitude, X–ray luminosity and orbital period for LMXBs has been derived by van Paradijs & McClintock (1994). For $L_x \sim 1 - 4 \times 10^{35}$ erg s^{-1} it predicts $M_v \sim 5 - 1.5 Log(P_{orb}/1hr)$. Therefore, we would expect to detect the optical counterparts only for the closest and least absorbed of the 5–9 s X–ray pulsars, as indeed is the case (the optical absorption of 4U 1626–67 is only $A_v < 0.3$; van Paradijs *et al.* 1986).

The possibility that such very low mass X–ray binaries, VLMXBs, are formed through the accretion induced collapse of white dwarfs in AM Her–type cataclysmic variables (Mereghetti & Stella 1995) faces serious difficulties with their expected z–distribution ($z_{rms} \simeq 300$ pc for AM Hers) and birth rate ($\sim 10^{-5}$ yr^{-1}; cf. van Paradijs *et al.* 1995).

Van Paradijs *et al.* (1995) propose that the 5–9 s pulsars are the result of the evolution of short orbital HMXRBs, following the expansion of the massive star and the onset of unstable Roche Lobe overflow, before central hydrogen is exhausted (see also Cannon *et al.* 1992). The resulting common envelope phase should lead to the spiral–in of the neutron star and the complete disruption of the companion star after the so–called Thorne–Zytkov stage. Therefore, these sources (with the exception of 4U1626–67 and, possibly, HD49798) should consist of an isolated neutron star accreting matter from a residual disk with a mass in the $10^{-3} - 1$ M$_\odot$ range. A problem with this model, however, is that such a massive disk would be highly unstable to its own gravity. Perhaps the common envelope and spiral in stage could be halted before complete coalescence is reached, even in the absence of a dense helium core, leading to a very low mass, hydrogen depleted companion orbiting the neutron star.

In this context it is interesting to note that independent evidence supports the view that the binary X–ray pulsars 4U1626–67 and HD49798 were formed following a common envelope and spiral–in evolutionary phase (see Angelini *et al.* 1995 and Israel *et al.* 1996 and references therein). The different properties of the companion stars in these two cases may reflect the fact that unstable mass transfer set in at different phases in the nuclear evolution of their progenitors (in turn, reflecting different initial masses and orbital periods; see also Ghosh *et al.* 1996). We note that a post common envelope evolutionary scenario for 4U0142+614 is also suggested by the recent inference that about half of the X–ray absorbing material is close to the source (White *et al.* 1996).

CONCLUSIONS

The most likely interpretation is that the neutron stars in the 5–9 s pulsars derive their X–ray emission from accretion of matter close to the equilibrium rotator regime. Their magnetic fields are likely of the order of a few×10^{11} G, i.e. lower than those of HMXB pulsars, but still much higher than those of classical LMXBs (see Figure 5). That these pulsars are members of binary systems is proven only in one, or perhaps two, cases. Whether the other 5–9 s pulsars are in very low mass X–ray binary systems, or are isolated and accrete from a residual disk remains to be established. There is evidence, however, that the 5–9 s pulsars are young ($\sim 10^5$ yr) and originate in the common envelope evolution that follows the HMXB stage.

Future studies of this new class of X–ray pulsars should concentrate on: (a) obtaining more accurate X–ray pulse timing that could reveal unambiguously the presence of a companion; (b) searching for the optical counterparts with deeper observations; (c) searching for pulsations in low luminosity X–ray sources close to the galactic plane; (d) understanding the processes that lead to the formation of a much softer X–ray spectrum than that of standard X–ray pulsars.

ACKNOWLEDGMENTS

LS acknowledges useful discussions with P. Ghosh, A. Tornambé and N.E. White. This work was partially supported through ASI grants. This research has made use of data obtained through the High Energy Astrophysics Science Archive Research Center Online Service, provided by the NASA–Goddard Space Flight Center. EXOSAT ME data were also extracted from the High Energy Astrophysics Database Service at the Brera Astronomical Observatory.

REFERENCES

Angelini, L., *et al.*, Neon Line Emission in the X–Ray Spectrum of the Pulsar 4U 1626-67, *ApJ* , 449, 41 (1995).

Baykal, A., and J.H. Swank, Pulse Frequency Changes of 1E 2259+586 and the Binary Interpretation, *ApJ* , 460, 470 (1996).

Bildsten, L., *et al.*, Observations of accreting pulsars, *AIP Conf. Proceedings*, 304, 290 (1994).

Bisscheroux, B.C., *et al.*, The nature of the bright subdwarf HD49798 and its X–ray pulsating companion, *A&A* , 317, 815 (1996).

Cannon, R.C., P.P. Eggleton, A.N. Zýtkow, and P. Podsiadlowski, The structure and evolution of Thorne–Zýtkow objects, *ApJ* , 386, 206 (1992).

Coe, M.J., and L.R. Jones, The continuing search for the optical counterpart of the enigmatic X–ray pulsar 1E2259+586, *MNRAS* , 259, 191 (1992).

Coe, M.J., L.R. Jones, and H. Letho, 1994, A deep radio search for the counterpart of the enigmatic X–ray pulsar 1E2259+586, *MNRAS* , 270, 178 (1994).

Corbet, R.H.D., and C.S.R. Day, Ginga observations of the 6–s X–ray pulsar 1E1048.1–5937 *MNRAS* , 243, 553 (1990).

Corbet, R.H.D, A.P. Smale, M. Ozaki, K. Koyama, and K. Iwasawa, The spectrum and pulses of 1E 2259+586 from ASCA and BBXRT observations, *ApJ* , 443, 786 (1995).

Davies, S.R. and M.J. Coe, Optical and IR imaging of the locality of 1E2259+586, *MNRAS* , 249, 313 (1991).

Fahlman, G.G. and P.C. Gregory, An X–ray pulsar in SNR G109.1–1.0, *Nature* , 293, 202 (1981).

Fahlman, G.G., P.C. Gregory, J. Middleditch, P. Hickson, and H.B. Richer, A possible optical counterpart to the X–ray pulsar 1E 2259+586, *ApJ* , 261, L1 (1982).

Finger, M.H., R.B. Wilson, and J. van Paradijs, *IAU Circular N.*6286 (1996).

Ghosh, P., and F.K. Lamb, Accretion by rotating magnetic neutron stars. II – Radial and vertical structure of the transition zone in disk accretion, *ApJ* , 232, 259 (1979).

Ghosh, P., L. Angelini, N.E. White, The Nature of the "6 Second" and Related X–Ray Pulsars: Evolutionary and Dynamical Considerations, *ApJ* , 478, 713 (1996).

Gregory, P.C., and G.G. Fahlman, An extraordinary new celestial X–ray source, *Nature* , 287, 805 (1980).

Hellier, C., A ROSAT observation of the X–ray pulsars X0142+614 and X0146+612, *MNRAS* , 271, L21 (1994).

Henrichs, H.F., in "Accretion Driven Stellar X–ray Sources", ed. W.H.G. Lewin & E.P.J. van den Heuvel (Cambridge University Press), p. 393 (1983).

Hughes, V.A., L.A. Nelson, M.R. Viner, R.H. Harten, and C.H. Costain, Extended radio observations of the SNR CTB 109, *ApJ* , 283, 147 (1984).

Israel, G.L., S. Mereghetti, and L. Stella, The discovery of 8.7 second pulsations from the ultrasoft X–ray source 4U 0142+61, *ApJ* , 433, L25 (1994).

Israel, G.L., L. Stella, L. Angelini, N.E. White, and P. Giommi, *IAU Circular N.*6277 (1995).

Israel, G.L., *et al.*, The Discovery of 13 Second X–Ray Pulsations from the Hydrogen–depleted Subdwarf O6 Star Binary HD 49798, *ApJ* , 474, L53 (1996).

Iwasawa, K., K. Koyama, and J.P. Halpern, Pulse period history and cyclotron resonance feature of the X–ray pulsar 1E 2259+586, *PASJ* , 44, 9 (1992).

Kouveliotou, C., *et al.*, A new type of transient high–energy source in the direction of the Galactic Centre, *Nature* , 379, 799 (1996).

Koyama, K., *et al.*, Ginga observation of the X–ray pulsar 1E 2259+586 in the supernova remnant G109.1–1.0, *PASJ* 41, 461 (1989).

Leisawitz, D., F.N. Bash, and P. Thaddeus, A CO survey of regions around 34 open clusters, *ApJ Supp.* , 70, 731 (1989).

Levine, A., *et al.*, 4U 1626–67 – The binary with the smallest known mass function, *ApJ*, 327, 732 (1988).

Lutovinov, A.A., S.A. Grebenev, R.A. Syunyaev, and M.N. Pavlinskii, Timing of X-ray pulsars using the data from the ART-P telescope of the GRANAT observation in 1990–1992, *Astronomy Letters*, 20, 631 (1994).

Makishima, K., *et al.*, Observations of the peculiar hard X-ray transient X0331+53 (V0332+53), *PASJ*, 42, 295 (1990).

Makishima, K., *et al.*, in "Frontiers of X-ray Astronomy", ed. Y. Tanaka & K. Koyama, (Univ. Academy Press), p.23 (1992).

McClintock, J.E., *et al.*, Optical candidates for two X-ray bursters and an X-ray pulsar, *Nature*, 270, 320 (1977).

Mereghetti, S., A Spin–down Variation in the 6 Second X-Ray Pulsar 1E1048.1–5937, *ApJ*, 455, 598 (1996).

Mereghetti, S., and L. Stella, The very low mass X-ray binary pulsars: A new class of sources ?, *ApJ*, 442, L17 (1995).

Mereghetti, S., P. Caraveo, and G.F. Bignami, CCD imaging and spectroscopy in the field of the X-ray pulsar 1E1048.1–5937, *A&A*, 263, 172 (1992).

Mereghetti, S., L. Stella, and F. De Nile, On the nature of the 25–min periodicity from 4U 0142+614: A nearby, slowly spinning neutron star/Be system?, *A&A*, 278, L23 (1993).

Middleditch, J., K.O. Mason, J.E. Nelson, and N.E. White, 4U 1626–67 – A prograde spinning X-ray pulsar in a 2500 s binary system, *ApJ*, 244, 1001 (1981).

Morini, M., N.R. Robba, A. Smith, and M. van der Klis, Exosat observations of the supernova remnant G109.1–1.0 and the X-ray pulsar 1E 2259+586, *ApJ*, 333, 777 (1988).

Motch, C., *et al.*, A Rosat glance at the galactic plane, *A&A*, 246, L24 (1991).

Nagase, F., Accretion-powered X-ray pulsars, *PASJ*, 41, 1 (1989).

Paczynski, B., X-ray pulsar 1E 2259+586 – A merged white dwarf with a 7 second rotation period?, *ApJ*, 365, L9 (1990).

Pravdo, S.H., *et al.*, HEAO 1 observations of the X-ray pulsar 4U 1626–67, *ApJ*, 231, 912 (1979).

Rappaport, S.A. and Joss, P.C., in "Accretion Driven Stellar X-ray Sources", ed. W.H.G.Lewin & E.P.J. van den Heuvel (Cambridge University Press), p.1 (1983).

Rappaport, S. A., *et al.*, Discovery of a 7.68 second X-ray periodicity in 3U 1626–67, *ApJ*, 217, L29 (1977).

Reynolds, A.P., A.N. Parmar, and N.E. White, The luminosity dependence of the X-ray spectrum of the transient 42 second pulsar EXO 2030+375, *ApJ*, 414, 302 (1993).

Schwentker, O., Evidence for a low–luminosity X-ray pulsar associated with a supernova remnant, *A&A*, 286, L47 (1994).

Seward, F., P.A. Charles, and A.P. Smale, A 6 second periodic X-ray source in Carina, *ApJ*, 305, 814 (1986).

Skinner, G.K., *et al.*, Discovery of 69 ms periodic X-ray pulsations in A0538–66, *Nature*, 297, 568 (1982).

Stark, M.J., A. Baykal, T. Strohmayer, and J.H. Swank, Pulse Arrival Time Glitches in GRO J1744–28, *ApJ*, 470 L109 (1996).

Steinle, H., W. Pietsch, M. Gottwald, and U. Graser, CCD observations of X-ray binaries, *A&A*, 131, 687 (1987).

Stickland, D.F., and C. Lloyd, Spectroscopic binary orbits from ultraviolet radial velocities – XIV – HD49798, *The Observatory*, 114, 41 (1994).

Strickman, M.S., *et al.*, Hard X-Ray Spectroscopy and Pulsar Phase Analysis of the Bursting X-Ray Pulsar GRO J1744–28 with OSSE, *ApJ*, 464, L131 (1996).

Thackeray, A.D., The spectroscopic orbit of the O–type subdwarf HD49798, *MNRAS*, 150, 215 (1970).

Thompson, C., and R.C. Duncan, Neutron star dynamos and the origins of pulsar magnetism, *ApJ*, 408, 194 (1993).

Usov, V.V., Glitches in the X-ray pulsar 1E 2259+586, *ApJ*, 427, 984 (1994).

van Paradijs, J., and J.E. McClintock, Absolute visual magnitudes of low–mass X-ray binaries, *A&A*, 290, 133 (1994).

van Paradijs J., R.E. Taam, and E.P.J. van den Heuvel, On the nature of the "anomalous" 6–s X-ray pulsars, *A&A*, 299, L41 (1995).

van Paradijs, J., S. van Amerongen, E. Damen, and H. van der Woerd, Five–colour photometry of early–type stars in the direction of galactic X-ray sources, *A&A Supp.*, 63, 71 (1986).

White, N.E. *et al.*, A 25 min modulation from the vicinity of the unusually soft X-ray source X0142+614, *MNRAS*, 226, 645 (1987).

White, N.E., J.H. Swank, and S.S. Holt, Accretion powered X-ray pulsars, *ApJ*, 270, 711 (1983).

White, N.E., L. Stella, and A.N. Parmar, The X-ray spectral properties of accretion discs in X-ray binaries, *ApJ*, 324, 363 (1988).

White, N.E., F. Nagase, and A.N. Parmar, in "X-ray Binaries", eds. W.H.G. Lewin, J. van Paradijs & E.P.J. van den Heuvel (Cambridge: Cambridge Univ. Press), p.1 (1995).

White, N.E., L. Angelini, K. Ebisawa, Y. Tanaka, and P. Ghosh, The Spectrum of the 8.7s X-Ray Pulsar 4U0142+61, *ApJ*, 463, L83 (1996).

Pergamon

Adv. Space Res. Vol. 22, No. 7, pp. 1035–1038, 1998
© 1998 COSPAR. Published by Elsevier Science Ltd. All rights reserved
Printed in Great Britain
0273-1177/98 $19.00 + 0.00

PII: S0273-1177(98)00139-2

THE DISCOVERY OF 13 s X-RAY PULSATIONS FROM THE HYDROGEN DEPLETED SUBDWARF O6 STAR BINARY HD49798

G. L. Israel[1,*], L. Stella[2,*], L. Angelini[3,4], N.E. White[3], T. R. Kallman[3], P. Giommi[5] and A. Treves[1]

[1] *International School for Advanced Studies (SISSA-ISAS). Via Beirut 2-4, I-34014 Trieste, Italy,* israel@vega.sissa.it treves@astmiu.mi.astro.it
[2] *Osservatorio Astronomico di Roma, V. dell'Osservatorio 2, I-00040 Monteporzio Catone (Roma), Italy,* stella@coma.mporzio.astro.it
[3] *Laboratory for High Energy Astrophysics, Code 662, NASA Goddard Space Flight Center, Greenbelt, MD 20771, USA,* white@adhoc.gsfc.nasa.gov, angelini@lheavx.gsfc.nasa.gov, tim@xstar.gsfc.nasa.gov
[4] *University Space Research Association*
[5] *SAX Science Data Center. ASI, Viale Regina Margherita 202. I-00198 Roma, Italy.* giommi@sax.sdc.asi.it
*Affiliated to I.C.R.A.

ABSTRACT

We discovered strong 13 s pulsations in the X-ray flux of HD49798 (Israel *et al.* 1995), a 1.55 d single-component spectroscopic binary containing a hydrogen depleted subdwarf O6 star. The source X-ray spectrum is extremely soft, corresponding to an unabsorbed 0.1–2 keV luminosity of a few $\times 10^{32}$ erg s^{-1} (distance of 650 pc). A higher luminosity might be hidden in the EUV. Our results imply that the unseen companion is an accreting degenerate star, either a white dwarf or a neutron star. In either case HD49798 corresponds to a previously unobserved evolutionary stage of a massive binary system, after common envelope and spiral-in.

©1998 COSPAR. Published by Elsevier Science Ltd.

INTRODUCTION

Subdwarf O stars form a fairly inhomogeneous group, spreading over a large range of temperatures, surface gravity and chemical compositions. Different evolutionary and nuclear histories probably contribute to the group of sdO stars. The case of HD49798 is especially intriguing because this 8th magnitude sdO6 star, is also a spectroscopic binary (with $P_{orb} = 1.55$ d, $a \sin i = 3.60$ R$_\odot$, f(m) = 0.263 M$_\odot$; Thackeray 1970; Stickland & Lloyd 1994). Despite extensive studies, the nature of the companion star remained unclear for decades. Being very hot (T~ 47000 K) and luminous (Log $(L/L_\odot) = 3.90$) the sdO6 star outshines the flux emitted by the companion star (at least in the optical–UV). The level of H depletion together with the underabundance of C and strong overabundance of N, testify that the outer layers of HD49798 were processed through the CNO cycle and the envelope of the progenitor star was lost during a common envelope phase. Bisscheroux *et al.* (1996) argue that the sdO star has a degenerate CO core and is in the phase of shell helium burning. Attempts at detecting the optical continuum of the companion provided conflicting conclusions: Thackeray (1970) and Kudritzki & Simon (1978) suggested a F4–K0 main sequence companion; Goy (1978) favored instead a compact star.

X-RAY OBSERVATIONS AND DATA ANALYSIS

HD49798 was observed on 1992 Nov 11 (exposure time of 5453 s) with the Position Sensitive Proportional Counter (PSPC, 0.1–2.0 keV energy range). Two X-ray sources were clearly detected within 1 arcmin from the center of the ROSAT pointing. The ROSAT error circle of the brightest of the two sources (1WGA J0648.0–4418 center of RA = 06 48 04.6, DEC = –44 18 54.4 and radius of ~ 10 arcsec, equinox 2000) includes the optical position of HD49798 (~ 5 arcsec offset). The

ROSAT light curve and spectrum of HD49798 contained ∼ 1000 photons extracted from a circle of ∼ 32 arcsec radius around the X–ray position. The arrival times of the 0.1–2.0 keV photons from HD49798 were corrected to the barycenter of the solar system and a power spectrum calculated over the entire observation duration (3.7 hr). The peaks around 0.0049 and 0.0125 Hz are seen in a number of ROSAT sources and arise from the wobble in the pointing direction (see Figure 1). The pronounced peak at a frequency of 0.076 Hz is instead unique to HD49798 and has a significance of ∼ 15σ over the sample of ROSAT light curves analyzed (∼ 25000). The period was determined to be 13.1789±0.0007 s by using a phase fitting technique.

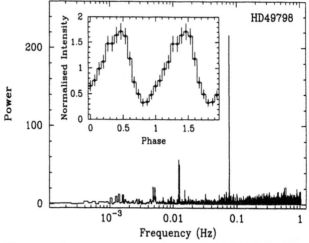

Fig. 1.   Power spectrum of the 0.1–2.0 keV ROSAT PSPC light curve of HD49798. The highest peak, centered around 0.076 Hz, corresponds to the 13.2 s pulsations. The folded lightcurve is shown in the inner panel.

An upper limit to the period derivative of $|\dot{P}| < -2.3 \times 10^{-7}$ s s^{-1} was derived. The folded light curve folded is shown as an insert in Figure 1. The pulsed fraction (semiamplitude of modulation divided the mean source count rate) of the nearly sinusoidal modulation is ∼ 60%. The ROSAT PSPC spectrum of HD49798 is extremely soft. Among simple one–component models, a power law produced by far the best fit (χ^2/dof of 22.1/14) for a photon index of $\alpha = 4.7 \pm 0.5$ and an interstellar column density of $N_H = (2.1^{+0.7}_{-0.4}) \times 10^{20}$ cm^{-2}. The corresponding X–ray flux at the earth is $F \sim 8 \times 10^{-13}$ erg cm^{-2} s^{-1}. For the estimated distance of 650 pc (Kudritzki & Simon 1978) this converts to an unabsorbed luminosity of ∼ 4×10^{32} erg s^{-1}. A large contribution to the χ^2 derives from a deficit of photons in the 0.4–0.6 keV range. The peculiar chemical composition and temperature of the sdO star atmosphere, should play a dominant role in determining the photoelectric optical depth in the wind for photon energies in the PSPC band. Including an edge at 0.46 keV (NIV) in the power law model significantly improves the fit (χ^2/dof of 10.0/13); we derive an optical depth of $\tau_N = 3.4^{+3.0}_{-1.9}$. Note that the ROSAT PSPC spectrum of HD49798 can also be interpreted in terms of the sum of a black body spectrum plus a high energy excess, described by a power law (or a thermal bremsstrahlung spectrum) (Israel *et al.* 1995; Bisscheroux *et al.* 1996). For this model, the best fit is obtained for $kT_{bb} \simeq 18.4$ eV, $\alpha \simeq 3$ and $N_H \simeq 5 \times 10^{20}$ cm^{-2}. Only a lower limit of $N_H > 10^{20}$ cm^{-2} and an upper limit of $kT_{bb} < 50$ eV can be deduced from the ROSAT spectrum. A five decade range of bolometric black body luminosity is allowed, starting from 2×10^{32} erg/s. This is due to the fact that, for decreasing temperatures, a larger and larger fraction of the black body flux would be hidden in the extreme ultraviolet.

DISCUSSION

The discovery of 13 s pulsations in the X–ray flux of HD49798 proves that the long–sought companion is a degenerate star, either a white dwarf or a neutron star. The inferred radius of the sdO star is $R_* \sim 10^{11}$ cm, a factor of ∼ 2 smaller than its Roche Lobe. Therefore, the mass transfer towards the companion must be driven by the sdO star wind. The estimated mass loss rate $\dot{M}_*$ ranges between 5×10^{-10} and 10^{-8} M$_\odot$ yr^{-1}. Bisscheroux *et al.* (1996) argue that $\dot{M}_*$ can be as high as ∼ 3×10^{-8} M$_\odot$ yr^{-1} and the bulk of the wind matter (mainly H and He) as slow as $v_w \sim 800$ km s^{-1} (see also Springmann & Pauldrach 1992). By using the whole range of estimated v_w and $\dot{M}_*$, and the standard theory of wind accretion in binary systems we obtain a mass capture rate of $\dot{M}_x \sim 10^{11} - 6 \times 10^{14}$ g s^{-1} by the degenerate companion (for more details see Israel *et al.* 1997). By using $\dot{M}_x$ given above, an accretion luminosity of $2 \times 10^{28} - 2 \times 10^{32}$ erg s^{-1} is predicted in the case of a white dwarf and of $1 \times 10^{31} - 1 \times 10^{35}$ erg s^{-1} in the case of a neutron star. While the former range is only marginally consistent with the inferred unabsorbed luminosity in the ROSAT PSPC energy range, the second overlaps comfortably and allows also for a larger luminosity in the EUV as suggested by the steep spectrum. By using the maximum angular momentum of the accretion flow relative to the companion, a circularization radius of $\leq 10^7$ cm is derived. This is insufficient to form an accretion disk in the case of a white dwarf. A disk could form outside a neutron star magnetosphere only for magnetic fields of $B \leq 3 \times 10^8 (\dot{M}_x/10^{15} \text{g s}^{-1})^{1/2}$ G. The galactic HI column in the direction of HD49798 is $\sim 6 \times 10^{20}$ cm^{-2} (Stark et al. 1992). Moreover a fraction of the column might be beyond HD49798, despite its height above the galactic plane (~ 210 kpc). We conclude that the estimate above is not in contradiction with the range of $\sim 1 - 3 \times 10^{20}$ cm^{-2} derived from the ROSAT PSPC spectrum. Given the abundances in the sd0 star atmosphere, He and N should dominate the wind photoelectric absorption in the ROSAT PSPC energy range (H is 100% ionized); due to the UV flux from the sdO star the most populated stages are probably HeII and NIV–NVI. In the white dwarf interpretation, the required $\dot{M}$ implies a very large optical depth at the HeII edge. For the spectrum not to display a larger low energy cutoff than observed, a large fraction of fully ionized helium must be present in the wind, possibly as a result of additional photoionization from the X-ray flux. In any case, a conspicuous N K–edge should be present around energies of $0.46 - 0.55$ keV (ionization stages above He-like N would not be populated), favoring the power law plus N K–edge model (see Israel et al. 1997). If the accreting object is a neutron star, the observed 0.1-2 keV luminosity can be produced for smaller mass inflow rates which do not require appreciable wind absorption from HeII and NIV-VI. On the other hand, accretion luminosities of up to $\sim 10^{35}$ erg s^{-1}, as suggested by the extrapolation of the steep power law–like PSPC spectrum to EUV, would still be compatible with the wind parameters. In this case the EUV luminosity could photoionize the He to the required level, while a NIV–VI K–edge optical depth of a few could still be present. Moreover for bolometric luminosities in the $10^{32} - 10^{35}$ erg s^{-1} range, the blackbody plus power law model implies $R_{bb} \sim 20 - 2000$ km a difficult range to reconcile with a neutron star. The orbital dependence of N–edge optical depth should be tested with future phase resolved soft X–ray measurements (variations of a factor of 3 or larger are expected for the allowed inclination).

White Dwarf Accretor

If the degenerate star in HD49798 is a white dwarf, it would be the first example of a white dwarf accreting from an early type (though very unusual !) companion. The 13 s X–ray pulsations might originate from magnetic polar cap accretion, as in the case of intermediate polars (IPs). The 13 s rotation period would then be the shortest known. By requiring that the magnetospheric radius is smaller than the corotation radius ($r_m < r_c$), such that the "centrifugal barrier" is open and accretion can take place, we derive $B \leq 5 \times 10^3 (\dot{M}_x/10^{15} \text{g s}^{-1})^{1/2}$ G, which is substantially lower than the range inferred for IPs ($\sim 10^5 - 10^7$ G). While most IPs display hard X–ray spectra, a small subgroup with spin periods of 5-15 min is characterized by an additional very soft spectral component, likely originating from the reprocessing at the white dwarf surface of the primary hard X–ray radiation emitted at the end of the accretion column(s) (Duck et al. 1994; Haberl & Motch 1995). The very soft X–ray spectrum of HD49798, however, must have a different origin. Indeed in this case, $r_c \sim 8 \times 10^8$ cm sets an upper limit of ~ 3 white dwarf radii, r_{wd}, to r_m. Accretion is therefore expected to take place over a large fraction of the white dwarf surface, $f \sim r_{wd}/r_m \sim 1/3$. This is far larger than the R_{bb} derived from blackbody fits for luminosities of $\sim 10^{32}$ erg s^{-1}. The limit on the coherence Q of the 13 s signal in the ROSAT light curve of HD49798 ($Q \geq 10^6$) suggests a possible analogy with the $\sim 10 - 30$ s quasi–coherent oscillations, QCOs, ($Q \sim 10^4 - 10^6$) that are seen at soft X–ray and EUV energies during the outbursts of dwarf novae, such as SS Cyg and U Gem (Córdova et al. 1980; Jones & Watson 1992; Mauche 1996). Note that the outburst X–ray spectra of these systems are also very soft (kT_{bb} of $\sim 25 - 30$ eV).

Neutron Star Accretor

If HD49798 hosts an accreting magnetic neutron star, then the 13 s pulsations arise from its rotation. The condition that the centrifugal barrier is open requires $B \leq 6 \times 10^{11} (\dot{M}_x/10^{15} \text{g s}^{-1})^{1/2}$ G. An accretion disk cannot form unless B is three orders of magnitude lower. The X–ray spectra of most X-ray pulsars are hard ($\alpha \sim 0 - 2$) and extend to energies of several tens of keV. However a group of 4-5 accreting pulsars with 5-9 s periods and very soft X-ray spectra ($\alpha \sim 2 - 5$) has been recently identified (Mereghetti & Stella 1995). This group is also characterized by X-ray luminosities of $\sim 10^{35} - 10^{36}$ erg s^{-1} and secular spin-down indicating that the neutron stars are close to their equilibrium period and have of relatively low B of $\sim 10^{11.5} - 10^{12}$ G. On the other hand, these X-ray pulsars either have a very low mass companion, or are isolated and accrete from a residual disk (van Paradijs *et al.* 1995). Moreover their distribution is highly concentrated in the galactic plane. In the accreting neutron star interpretation, HD49798 would be the first X-ray binary with an early type H-depleted mass donor. Such system would likely be the remnant of a high mass X-ray binary after common envelope and spiral-in, an evolutionary phase that has never been seen before. If the sdO star is not massive enough to undergo gravitational collapse, the system will evolve in $\sim 10^6$ yr into a non-interacting binary consisting of a massive white dwarf and a neutron star.

CONCLUSIONS

Our results prove that HD49798 hosts either a white dwarf or, more likely, a magnetic neutron star. The value and sign of any changes in the 13 s X-ray period should clarify the nature of the accreting degenerate star. The dwarf nova oscillation scenario requires quasi–coherent pulsations. If instead the 13 s pulsations arise directly from the rotation of the degenerate star, they would be coherent enough to make HD49798 a "double spectroscopic" binary; in this case the system would hold a great potential for accurate mass measurements of a previously unobserved evolutionary stage.

REFERENCES

Bisscheroux, B.C., *et al.* , The nature of the bright subdwarf HD49798 and its X-ray pulsating companion, *A&A*, 317, 815 (1996).

Córdova, F.A., T.J. Chester, I.R. Tuohy, and G.P. Garmire, Soft X-ray pulsations from SS Cygni, *ApJ*, 235, 163 (1980).

Duck, S.R., *et al.* , ROSAT observations of a soft X-ray emission component in the intermediate polar RE 0751+14, *MNRAS*, 271, 372 (1994).

Goy, G., HD49798 – A strange composite sdO star, *A&A*, 64, 445 (1978).

Haberl, F. and C. Motch, New intermediate polars discovered in the ROSAT survey: two spectrally distinct classes, *A&A*, 297, L37 (1995).

Israel, G.L., L. Stella, L. Angelini, N.E. White, and P. Giommi, HD49798 and 2E 0050.1–7247, *et al.* , IAUC 6177 (1995).

Israel, G.L., *et al.* , The Discovery of 13 Second X-Ray Pulsations from the Hydrogen–depleted Subdwarf O6 Star Binary HD49798, *ApJ*, 474, L53 (1997).

Jones, M. H., and M.G. Watson, The Exosat observations of SS Cygni, *MNRAS*, 257, 633 (1992).

Kudritzki, R.P., and K.P. Simon, Non–LTE analysis of subluminous O-stars the hydrogen–deficient subdwarf O-binary HD49798, *A&A*, 70, 653 (1978).

Mereghetti, S., and L. Stella, The very low mass X-ray binary pulsars: A new class of sources?, *ApJ*, 442, L17 (1995).

Springmann, U.W.E., and A.W.A. Pauldrach, Radiation–driven winds of hot luminous stars. XI – Frictional heating in a multicomponent stellar wind plasma and decoupling of radiatively acceler-ated ions, *A&A*, 262, 515 (1992).

Stark, A.A., *et al.* , The Bell Laboratories H I survey, *A&A Supp.* , 79, 77 (1992).

Stickland, D.F., and C. Lloyd, Spectroscopic binary orbits from ultraviolet radial velocities – XIV – HD49798, *The Observatory*, 114, 41 (1994).

Thackeray, A. D., The spectroscopic orbit of the O-type subdwarf HD49798, *MNRAS*, 150, 215 (1970).

van Paradijs, J., R.E. Taam, and E.P.J. van den Heuvel, On the nature of the "anomalous" 6–s X-ray pulsars, *A&A*, 299, L41 (1995).

Adv. Space Res. Vol. 22, No. 7, pp. 1039–1042, 1998
© 1998 COSPAR. Published by Elsevier Science Ltd. All rights reserved
Printed in Great Britain
0273-1177/98 $19.00 + 0.00

Pergamon

PII: S0273-1177(98)00145-8

THERMAL X-RAY EMISSION OF THE EXTENDED LOBE NEAR 1E 2259+586 WITH CTB 109

J.-H. Rho[1,2], R. Petre[2], and J. Ballet[1]

[1] *Service d'Astrophysique, Centre d'Etudes de Saclay, F-91191 Gif-sur-Yvette CEDEX, FRANCE*
[2] *Laboratory for High Energy Astrophysics, Code 662, NASA Goddard Space Flight Center, Greenbelt, MD 20771, USA*

ABSTRACT

A "jet-like" lobe runs northeastward from the pulsar 1E2259+586 to the shell within the supernova remnant CTB 109, and it was previously suspected to be an "X-ray jet" in a binary system analogous to SS433/W50 (Gregory & Fahlman, 1983). We report X-ray spectra of the lobe relative to elsewhere in the remnant using ASCA and ROSAT, and comparison with other X-ray jets. X-ray emission from the "jet-like" lobe is thermal with dominant lines of Ne IX, Ne X, Mg XI, and Si XIII, and has softer emission than elsewhere in the remnant. The spectral fit for the lobe indicates that the plasma is in ionization non-equilibrium and its best fit is $T_s \sim 10^{6.5}$ K and $n_o t \sim 48,000$ cm^{-3}yr.

This lobe contrasts with the double lobes in W50 where non-thermal X-ray emission is found (Yamauchi *et al.*, 1994). If the lobe in CTB109 is a jet, it would be similar to the Vela jet because it was recently discovered to have thermal X-rays (Markwardt & Ögelman, 1995). The Vela jet is connected with the pulsar, providing the loss mechanism for most of the pulsar's rotational energy, and it has high pressure of the "cocoon" of shocked jet material. On the contrary, the "jet-like" lobe in CTB 109 is not connected with 1E 2259+586 in the ROSAT image (Rho & Petre, 1997) and does not have higher pressure than elsewhere in the remnant. Therefore, this lobe may not be a supersonic beam; it is neither powered by nor related to the central pulsar. We suggest this lobe was produced by Rayleigh-Taylor instability due to a reflected shock arising from the blast wave-molecular cloud interaction.   ©1998 COSPAR. Published by Elsevier Science Ltd.

INTRODUCTION

The supernova remnant (SNR) CTB 109 is a composite SNR, with a central pulsar 1E2259+586 and shell emission. The remnant is semicircular in shape, missing its western part possibly due to interaction with a large molecular cloud (Tatematsu et al., 1985), and an X-ray bright jet-like structure (or lobe) runs northeastward from the pulsar and the shell. CTB109/1E2259+586 has been considered analogous to W50/SS433: both have lobe structures suggested to originate from the binaries. Gregory and Fahlman (1980, 1983) originally proposed a binary system and precessing jet model of CTB 109. According to this model, the jet-like feature represents synchrotron radiation of the precessing jet. The main arguments against the binary model for 1E2259+586 are that no optical counterpart and no Doppler shifts have been found.

In this paper, we report high resolution of X-ray spectra using ASCA of the lobe and shell emission. The X-ray emission from the jet-like lobe is thermal, showing Ne IX, Ne X, Mg XI (1.35keV), Si XIII (1.86 keV), and S XV (2.46 keV) lines, which is contrary to the precessing jet model and observations of W50 where the lobes are non-thermal (Yamauchi et al., 1994). We discuss whether the X-ray thermal and/or nonthermal emission could be the pulsar's influence, and compare to Vela and MSH15-52. We also apply non-equilibrium ionization (NEI) model to the spectra, and examine the conditions of plasma across the remnant. The analysis of 1E2259+586 itself has been reported in Corbet et al., (1995).

OBSERVATION

CTB 109 was observed using ASCA May 30-31, 1993 during PV phase of ASCA, for approximately 1 day of exposure time. The pointing center was R.A. $23^h01^m18.7^s$ and Dec. $58° 54' 18.7''$ (J2000). ASCA has two detector pairs: Gas Imaging Spectrometers (GIS - GIS 2, GIS 3), and Solid-state Imaging Spectrometers (SIS - SIS 0, SIS1). While the field of view of the GIS is circular with a diameter of 40', SIS has a 22' square field of view. The energy resolution of the SIS is 2% at 5.9 keV, and the GIS has 8% resolution. In the range of 1.5-5 keV the effective areas are similar. Below 1.5 keV the SIS is more sensitive and above 5 keV, the GIS is. For GIS 2, the elevation angle to cutoff the data, 10 degree has been used to result in effective observation time 40 ks. The exposure time used for SIS is 30 ks. The count rates for the lobe structure are 0.29±0.003 and 0.32±0.003 cts s^{-1} for GIS2 and GIS3, and 0.666±0.005 and 0.797±0.005 cts s^{-1} for SIS0 and SIS1, respectively. The details of instrumentation of ASCA can be found in Tanake et al., (1994) and Serlemitsos et al., (1994) and for GIS in Ohasi et al., (1991) and Burke et al., (1991).

IMAGES

We have constructed GIS and SIS images in five energy bands: 0.5-10 keV, 0.5-1 keV, 1-2 keV, 2-3 keV, and 3-10 keV. The GIS and SIS images are largely similar to the ROSAT PSPC: the jet-like structure and northern shell appeared bright. The connection between the pulsar and jet-like structure is unclear in the ASCA images: however, the ROSAT/PSPC image had shown no connection between them (Rho & Petre, 1997). X-ray emission from the western half of the remnant is not detected: if the lack of X-rays is only due to the absorption, ASCA would have revealed hard X-ray emission. The exposure map has been corrected, and for the GIS image has been restored using the Lucy-Richardson algorithm and an energy dependent point spread function. The channel maps of 0.5-1 keV, 1-2 keV, 2-3 keV, and 3-10 keV, are shown in Figure 1A. The pulsar appeared in all channel maps. The jet-like lobe and northern shell are prominent in the 0.5-1 keV and 1-2 keV maps, but they are not in the 2-3 keV map. The count rate above 3 keV was not sufficient to reveal the diffuse structures. A GIS spectral hardness map was made by the ratio between channels (~1.4-9 keV) and channels (~0.5-1.4 keV), showing the pulsar is the hardest and the diffuse emission north part of the pulsar seems to be harder than elsewhere in the remnant, where a molecular cloud is in the line of sight. The jet has no harder emission that the diffuse emission.

SPECTRAL OBSERVATIONS

We have extracted the spectra from the jet-like lobe and the remainder diffuse emission using the GIS and SIS. The GIS2 and SIS0 spectra of the jet-like lobe are shown in Figure 1B. The "jet-like" obe has thermal X-ray emission with dominant lines of Ne IX (0.9 keV), Ne X (1.0 keV), Mg XI 1.35 keV), Si XIII (1.86 keV), and S XV (2.46 keV). The spectra of GIS2, GIS3, SIS0 and SIS1 vere simultaneously fit with ROSAT/PSPC using a Raymond model and ionization non-equilibrium NEI) models based on Hamilton et al. (1983). Two-temperature thermal model or one-temperature ion-equilibrium model yielded acceptable fits while one-temperature thermal or power-law models vere rejected. The two temperature Raymond model gives $N_H= 7.5\times10^{21}$ cm^{-2}, kT_{low}=0.3 keV, ind kT_{high}=0.7 keV. The plasma of the lobe has not completely reached ionization equilibrium, and :he NEI best fit ($\chi^2_\nu \sim 1.1$) gives $T_s\sim10^{6.5}$ K and an ionization age of $t\sim$48,000 n_o^{-1} yr, where n_o .s the pre-shock density. This temperature is similar to that in the southern shell using BBXRT lata (Rho & Petre, 1997), and the ionization age is larger than $t\sim$17,000 n_o^{-1} yr in the southern shell. This can be understood if the jet has a higher density than the southern part of the remnant, which is consistent with emission measurement from the PSPC spectra. The fit of the remainder of liffuse emission did not require a second thermal component, and one temperature model yielded an acceptable fit with $N_H= 7.0\times10^{21}$ cm^{-2}, $kT = 0.57$ keV and nearly solar abundances.

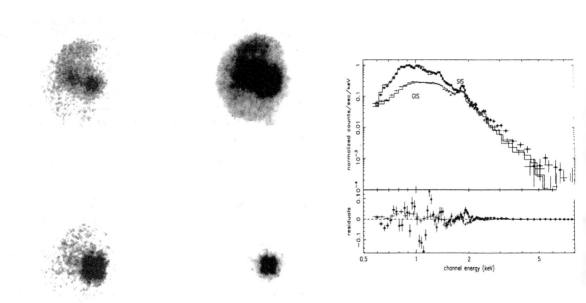

Fig. 1 (A:left:images) The GIS channel maps with (a:top left) 0.5-1 keV, (b:top right) 1-2 keV, (c:bottom left) 2-3 keV, and (d: bottom right) 3-10 keV. (B:right:spectra) The best fit of GIS and SIS spectra using a non-equilibrium model is superposed on the data, and the residuals are also shown.

DISCUSSION

A large fraction of rotational energy loss of a pulsar is believed to be released in the form of relativistic charged particles, which produce non-thermal X-rays due to synchrotron radiation. For the Crab nebula, the emission is powered by a wind of particles accelerated in the equatorial plane of the rotating neutron star (Brinkmann et al., 1985). The lobe in CTB109 is clearly different from the X-ray emission of plerions, and also contrasts with the double lobes in W50 where synchrotron radiation of X-ray emission is found (Yamauchi et al., 1994). W50 is known to have double lobes, which are 50 pc away from the central binary SS433 to the east and west. It is often considered analogous to CTB 109/1E2259+586. X-rays from the lobes of W50 are produced by the interaction with the jet with the surrounding material, probably the SNR shell. It indicates when the jet is interacting with the surrounding material, X-rays are also non-thermal due to synchrotron radiation.

However, an X-ray emitting jet from the Vela pulsar which begins at the pulsar and extends roughly 7 pc to the south-southwest and along the pulsar spin axis, is interpreted as thermal gas produced by the pulsar jet (Markwardt & Ögelman, 1995). If the lobe in CTB109 is a jet, it would be similar to the jets in Vela or MSH 15-52 due to the X-ray characteristic of thermal emission. The Vela jet is connected with the pulsar, providing the loss mechanism for most of the pulsar's rotational energy, and it has high pressure of the "cocoon" of shocked jet material. As another example, the jet of MSH 15-52 is a stream of relativistic electrons generated by the pulsar, and the thermal nebula is created by collision of the jet with the interstellar material (Tamura et al., 1996).

Contrary to the situation for Vela, the jet-like lobe in CTB 109 is not connected with 1E 2259+586 in the ROSAT images (Rho & Petre, 1997; Fesen & Hurford, 1997) and does not have higher pressure than elsewhere in the remnant. Also the high energy map (3-10 keV) where emission can be interpreted as synchrotron radiation, does not extend toward the jet-like lobe, while MSH 15-52 has shown this structure. The pulsar 1E2259+586 exhibits a near steady spin-down trend, and the loss of rotational energy is 3 orders of magnitude lower than the X-ray luminosity of $\sim 10^{35}$ erg s^{-1}. Therefore, the jet-like lobe is not a supersonic beam, and neither it is powered by nor related to the central pulsar. The high resolution ASCA analysis shows the X-ray spectra are not primarily different from the diffuse emission elsewhere in the remnant and supports that this lobe was produced by Rayleigh-Taylor instability due to a reflected shock arising from the blast wave-molecular cloud interaction (Rho & Petre, 1997; Tenorio-Tagle et al., 1985).

REFERENCES

Brinkmann, W., B. Aschenbach,, A. Langmeier, Nature, 313, 662 (1985)

Burke B.E., R.W. Mountain R.W., D.C. Harrison., M.W. Bautz, J.P. Doty, IEEE Trans. ED-38, 1069 (1991)

Corbet, R.H.D, A.P. Smale, M. Ozaki, K. Koyama, and K. Iwasawa, ApJ, 443, 786 (1995)

Fesen, R.A. and A. P. Hurford, AJ, 110, 747 (1995)

Gregory, P.C. and G. G. Fahlman, 1980, Nature, 287, 805 (1980)

Gregory, P.C. and G. G. Fahlman, in IAU symposium, "Supernova Remnants and their X-ray Emission", p429 (1983)

Hamilton, A.J., C. L. Sarazin, and R.A. Cevalier, ApJS, 51, 115 (1983)

Markwardt, C. B., and H. Ögelman, Nature, 375, 40 (1995)

Ohashi T., K. Makishima, M. Ishida, T. Tsuru, M. Tashiro et al., SPIE 1549, 9 (1991)

Rho, J.-H., and R. Petre ApJ, 484, 828 (1997)

Serlemitsos, P.J., et al., "Frontiers of X-ray Astronomy", ed. Y. Tanaka & K. Koyama (Tokyo: Universal Academy), p221 (1992)

Tanaka Y., S.S. Holt, H. Inoue, PASJ, 46, L37 (1994)

Tatematsu, K., M. Nakano, S. Yoshida, S. D. Wiramihardja, and T. Kogure, PASJ, 37, 345 (1985)

Tenorio-Tagle, G., P. Bodenheimer,and H. Yorke, A&A, 145, 70 (1985)

Tamura, K., N. Kawai, A. Yoshida, and W. Brinkmann, PASJ, preprint (1996)

Yamauchi, S., N. Kawai, and T. Aoki, PASJ, 46, L109 (1994)

Adv. Space Res. Vol. 22, No. 7, pp. 1043–1047, 1998
© 1998 COSPAR. Published by Elsevier Science Ltd. All rights reserved
Printed in Great Britain
0273-1177/98 $19.00 + 0.00

 Pergamon

PII: S0273-1177(98)00146-X

PRODUCTION OF X-RAYS BY COMET HYAKUTAKE

R. BINGHAM[1], J. M. DAWSON[2], V. D. SHAPIRO[3], D. A. MENDIS[3], and B. J. KELLETT[1]

[1] *Rutherford Appleton Laboratory, Chilton, Didcot, Oxon, OX11 0QX*
[2] *Department of Physics, University of Physics, University of California, Los Angeles, CA 90024*
[3] *Department of Physics, University of California, San Diego, La Jolla, CA 92091*

ABSTRACT

An explanation of recent observations of a strong X-ray emission from comet Hyakutake (Science 72, 194, 1996) is proposed. It is based on an idea that the interaction of the solar wind and photoionized cometary plasma produces strong plasma turbulence in the lower hybrid frequency range which is responsible for the acceleration of suprathermal electrons with energies in the range $100eV$ up to several keV. A nonlinear theory of the instability responsible for the generation of lower-hybrid electric field, typical energies and flux of suprathermal electrons are obtained. Two possible mechanisms of X-ray production by non-thermal energetic keV electrons are analyized; bremsstrahlung and cometary gas (mostly oxygen) K-shell line radiation. It is shown that for conditions under investigation line radiation is dominant resulting in total X-ray luminosity of $2.5 \times 10^{15} erg/sec$ in excellent agreement with ROSAT observed emission of $3.0 \times 10^{15} erg/sec$.

©1998 COSPAR. Published by Elsevier Science Ltd.

INTRODUCTION

Observations of X-ray emission from comet Hyakutake (Ganz 1996) appear to have been a surprise to most comet researchers. This was the first time that an X-ray image of a comet has been obtained. However, plasma data already exists which suggests that this observation is not as surprising as one would suspect. This data was obtained from the encounter with comet Halley by an armada of spacecraft. The data shows the presence of strong plasma wave turbulence (Klimov et al., 1986) together with energetic electrons (Gringauz et al., 1986) in the region of comet Halley which corresponds closely to the region where the X-rays were observed from comet Hyakutake. The plasma wave turbulence observed (Klimov et al., 1986) is close to the frequency range of lower-hybrid waves which are very effective in energising electrons and it is therefore not surprising that an energetic electron population up to several keV was also observed (Gringauz et al., 1986) in the same region as the lower-hybrid wave activity. Physical conditions at comet Hyakutake are quite close to those found at comet Halley with a gas production rate only several times smaller ($2 \times 10^{29} mol/s$ at $1AU$ for Hyakutake and $7 \times 10^{29} mol/s$ for Halley comet) so there are strong reasons to believe in the existence of a similar highly energetic electron population at Hyakutake with the energies required for X-ray emission.

A previous attempt to take an X-ray image of a comet (Bradfield 1979) was made by Hudson et al (1981) using the Einstein X-ray satellite. This yielded a negative result putting an upper limit $\sim 10^{14} erg\ s^{-1}$ for the total X-ray power at the comet. These authors assumed that the X-rays resulted from the precipitation of 'auroral' type keV electrons following the cometary analogy of a terrestrial substorm. Contrary to that we propose here a mechanism wherein the suprathermal electrons are produced by a mechanism based on lower-hybrid turbulence (Bingham et al., 1997, Dawson et al., 1997, Shapiro et al., 1998).

In this article we will demonstrate that the interaction of the cometary plasma and the solar wind produce waves in the lower-hybrid frequency range which are responsible for the production of suprathermal electrons with energies in the range 100eV up to several keV, which are necessary for bremsstrahlung and cometary gas k-shell radiation of X-rays. Lower-hybrid waves play a special role in plasma astrophysics due to the fact that these waves, which propagate almost at right angles to the magnetic field, have high phase velocities along the magnetic field direction and slow phase velocities perpendicular to the magnetic field. The waves provide the intermediary step in transferring energy between the ions and electrons.

Close to the sun every cometary nucleus is surrounded by an expanding gas cloud, which consist mostly of water with some carbon dioxide, and is ionised largely by photo-ionisation. These ions, created in the solar wind, immediately see the $\underline{v}_{SW} \times \underline{B}_{SW}$ electric field which accelerates them to high energies. ($\underline{v}_{SW}, \underline{B}_{SW}$ are the solar wind velocity and magnetic field respectively). These ions, called pick-up ions, form in the solar wind frame an ion beam gyrating in the solar winds magnetic field. The energy source for this process is the relative motion between the solar wind and newly created cometary ions which results in the modified two stream instability (McBride et al., 1972) of an ion beam. The excitation of waves in the lower-hybrid frequency range by this instability and subsequent absorption of the wave energy by electrons is the main mechanism for converting energy from the solar wind flow into plasma electrons, accelerating them parallel to the magnetic field forming a suprathermal electron component.

Other astrophysical areas where this problem is of interest is in the study of planetary ion exospheres (Hartle et al., 1973) interacting with the solar wind, in particular the pick-up of planetary ions in the mantle region of unmagnetised planets such as Mars and Venus, (Shapiro et al., 1996) the pick-up of helium and oxygen ions of interstellar origin (Wu et al., 1973) and in acceleration and heating of electrons and ions in supernovae remnants (Bingham 1998). In fact experiments were carried out in the 1980's by the Active Magnetospheric Particle Tracer Explorers (Valenzuela et al., 1986) (AMPTE) satellite mission to study the coupling of the solar wind to neutral gas. In the AMPTE experiments neutral lithium and barium atoms were released in the solar wind. These neutral atoms rapidly ionised by photoionisation and were picked up by the solar wind. Measurements of suprathermal electrons up to keV energies were obtained (Hall et al., 1986). In these experiments both the solar wind and released plasma are collisionless as is the case in the cometary interaction and any coupling between them must depend on wave-particle interactions. Lower-hybrid waves were identified as being the most likely candidate for the suprathermal electron production (Hall et al., 1986).

Alternative mechanisms have been proposed for X-ray emission from Comets. Charge exchange between heavy solar wind minority ions with cometary neutrals (Cravens 1997). For this process the emission is a line spectrum and origionates from a very large region, the linear size of $\sim 10^6 km$ used by Cravens (1997) is probably too large for significant charge exchange to be effective. Another mechanism is scattering of solar wind X-rays with cometary dust (Owens et al., 1997, Krasnopolsky et al., 1997). The spectrum in this case would be similar to the solar X-ray spectrum which it is not, it is a broad continium and agrees more with the predictions of the model proposed by Bingham et al., 1997, Dawson et al., 1997, and Shapiro et al., 1998.

X-RAY PRODUCTION MECHANISM

The cometary ion flux F_i at a distance r from the surface of the comet can be obtained by equating the photo-ionised part of the cometary gas outflow to the flow of the cometary ions picked-up by the solar wind and is given by

$$F_i = 4\pi r^2 n_{ci} u = Q_s - Q_s e^{-\frac{r}{v_g \tau}} = \frac{Q_s r}{v_g \tau} \tag{1}$$

where Q_s is the initial flux of gas molecules at the comet surface ($= 2 \times 10^{29}$ molecules/s) τ^{-1} is the rate of photoionisation ($\tau = 10^6 s$ at 1 A.U.), v_g is the initial gas velocity ($v_g = 10^5 cm/s$) and n_{ci} is the cometary ion density. A more accurate calculation (Wallis and Ong, 1975) based on the analysis of the solar wind dynamics, mass-loaded by the picked-up cometary ions lead to the same formula for the ion density.

From Eq.1 we can estimate the density of the cometary ions, n_{ci} to be $\approx 10cm^{-3}$ for $u = 3.10^6 cm/s$ corresponding to the downstream shocked solar wind velocity, at a distance $r = 50000km$ which is close to the position where X-ray emission is being generated.

These cometary photo-ions picked up by the solar wind excite the lower-hybrid waves which in turn are absorbed by the suprathermal electrons through Cherenkov resonance with the waves.

A theoretical analysis for this process, given by Bingham et al., (1997), predicts an average energy of the suprathermal electrons to be

$$\epsilon_e \sim 100eV$$

and their density

$$n_{Te} \simeq n_{ci} \left(\alpha^2 \frac{m_e}{m_i} \right)^{\frac{1}{5}} \simeq 10cm^{-3}$$

This is a powerful source of suprathermal electrons with an energy flux of $10^{20} erg/sec$ through the region of X-ray emission. The electron energy flux is about two orders of magnitude less than the total power available from the solar wind indicating a reasonably high efficiency of energy transformation. This region also corresponds closely to the region where intense lower-hybrid waves were observed by the Vega satellite (Klimov et al., 1986) during its encounter with Comet Halley. The presence of intense lower-hybrid waves can result in field aligned electrons accelerated up to energies in the keV range exceeding significantly the average energy derived above (Bingham et al., 1997).

The equation for the electron energy ϵ_e is (Bingham et al., 1997)

$$\epsilon_e \simeq \left[\frac{e^2}{2\pi m_e^{1/2}} \ell < E_f^2 > \right]^{2/3} \tag{2}$$

where ℓ is the interaction length $\sim 10^9 cm$. Using for the wave energy spectral density values obtained from observations at Halley comet by the Vega spacecraft $\left(< E_f^2 > \sim 1. \frac{mV^2}{m^2 Hz} \right)$ it is possible to estimate from Eq.2 the maximum energy of accelerated electrons as

$$\epsilon_e \sim 5keV \tag{3}$$

We now make an estimate of the luminosity of the X-rays assuming that they are a combination of bremsstrahlung emission and inner shell radiation from collisionally excited oxygen atoms. For bremsstrahlung emission the power radiated by one electron in $1cm^{-3}$ is given by

$$g = \int \hbar \omega n_o v_e d\sigma(\omega) \tag{4}$$

where n_o is the number density of water molecules $\left(= \frac{Q_s}{2\pi r^2 v_g} \right)$ at distance r, v_e is the electron velocity and $d\sigma(\omega)$ is the differential cross-section for bremsstrahlung given by

$$d\sigma(\omega) = \frac{16}{3}\frac{Z^2 e^2 c}{\hbar v^2} v_o^2 \ell n \left\{ \frac{b_{max}}{b_{min}} \right\} \frac{d\omega}{\omega} \tag{5}$$

r_o is the classical electron radius $\left(= \frac{e^2}{m_e c^2} \right)$, Ze is the nuclear charge of the water molecule, ω is the radiation frequency and, $b_{max}/b_{min} = \frac{mv^2}{\hbar\omega}$ note that $\ell n \left\{ \frac{mv^2}{\hbar\omega} \right\}$ is of order 1.

The power radiated in $1 cm^3$ is $\int_{v_{min}}^{v_{max}} g f_e dv_e$ where f_e is the electron distribution function of suprathermal electrons. If the distribution function is constant in velocity up to a maximum value v_{max} then $f_e = \frac{n_e}{v_{max}}$ and the total luminosity (in erg/s) from a shell of thickness Δr surrounding the comet nucleus is given by

$$L = 2\pi \int g f_e dv_e r^2 \Delta r \simeq 3 \times 10^{-25} \frac{Q}{v_g} \Delta r n_e Z^2 \sqrt{\epsilon_{emax}(eV)} \tag{6}$$

In addition to bremsstrahlung radiation there is also radiation from partially ionized oxygen and other heavy cometary ions whose bound electrons are excited by collisions with the suprathermal electrons. The bound electrons quickly de-excite and radiate any energy they receive from this source. Line radiation produced by excitation of bound electrons in inelastic collision with keV plasma electrons is found to be more effective than bremsstrahlung. Detailed calculations of the intensity of this radiation known in fusion research as impurity radiation (Post et al., 1977) have shown that this radiation is a powerful source of energy loss in fusion reactor. The basic physics of this radiation can be explained using the simple picture proposed in (Dawson 1981) where it is assumed that the cross-section of the process is determined by inelastic Coulomb electron-electron collisions with small impact parameter thereby leading to relatively large angle scattering of plasma electrons. The Coulomb logarithm, which usually increases the cross-section for electron-electron collisions in a plasma by a factor of $10 \sim 20$ due to the input from distant collisions, for impurity radiation is a strong function of the Z factor and is of the order ≤ 1. The total luminosity due to recombination or impurity radiation from the cometary shell, described earlier is given by (Post et al., 1977)

$$L_I \approx 5 \times 10^{-18} \Lambda n_e \frac{Q\Delta r}{v_g} \cdot \frac{1}{\sqrt{\epsilon_e(er)}} \tag{7}$$

The numerical factor Λ for oxygen atoms has been estimated in (Dawson 1981) as 0.1 which is close to the value obtained by detailed calculations (Post et al., 1977).

From Eq's 6 and 7 we estimate the total luminosity of X-ray radiation for the parameters we have used above as $L_{TOTAL} \simeq 2.5 \times 10^{15} ergs/sec$ for photon energies in the keV energy range, in excellent agreement with the ROSAT measurement of $3. \times 10^{15} erg/sec$.

In conclusion we have shown that the instability of newly picked-up cometary ions generates lower-hybrid waves which are capable of accelerating electrons to the keV range of energies. These electrons are responsible for the X-rays from comet Hyakutake seen by the ROSAT satellite. These observations may lead the way to a better understanding of cometary make-up, and in particular, spectral resolution of k shell lines should lead to the determination of comet composition.

ACKNOWLEDGMENTS

This work was supported in part by NSF grant PH-9319198;003 and NASA NAGW-1502.

REFERENCES

Bingham, R, J.M. Dawson, V.D. Shapiro and B.J. Kellett, submitted to *ApJ*, 1998.

Bingham, R, J.M. Dawson, V.D. Shapiro and B.J. Kellett, *Science*, 275, 49, 1997.

Cravens, T.E., *Geophys. Res. Lett*, 24, 105, 1997.

Dawson, J M, Fusion Vol 1 part B page 465, edited by E Teller, Academic Press, New York (1981).

Dawson, J M, R. Bingham, and V.D. Shapiro, *Plasma Physics and Controlled Fusion* 39, 105, 1997.

Ganz, J, *Science*, 272, 194, (1996).

Gringauz, K I, A.P. Remizov, M.I. Verigin, A.K.Richter, M. Tatrallyay *et al.*, *Nature*, 321, 282, (1986).

Hall, D.S., D.A. Bryant, C.P. Chaloner, R. Bingham, and D.R. Lepine, *J. Geophys.Res.*, 91, 1320, (1986).

Hartle, T E, K.W. Ogilvie, and C.S. Wu, *Planet Space Science* 21, 2181, (1973).

Hudson H S, W-H. Ip, and D.A. Mendis, *Planet. Space Sci.*, 29, 1373, (1981).

Klimov, S, S Savin, Yu. Aleksevich, G Avanesova, V. Balebanov *et al.*, *Nature*, 321, 292 (1986),

Krasnopolsky.V.A., M.J.Mumma, M. Abbott, B.C. Flynn, K.J. Meech *et al.*, *Science*, 277, 1488, 1997.

McBride, J B., E. Ott, P.B. Jay and J.H. Orens, *Phys.Fluids*, 157, 2367 (1972).

Owens A., T. Oosterbroek, A. Orr, A.N. Parmar, L.A. Antonelli *et al.*, *ApJ*, in press 1998.

Post, D F, R.V. Jensen, C.B. Tarter, W.H. Grasberger, and W.A. Lokke, Princeton Plasma Physics Laboratory Report PPPL-1352 (1977).

Shapiro, V.D., R. Bingham, J.M. Dawson, Z. Dobe, B.J. Kellett and D.A.Mendis *Physica Scripta*, in press 1998.

Shapiro, V D., K. Szego, S.K. Ride, A.F. Nagy, and V.I. Shevchenko, *JGR*, 100, 21289, (1996).

Valenzuela, A,, G. Haerendel, H. Föppl, F. Melzner, N.F. Ness, E. Reiger, J. Stocker, O. Bauer, H. Höfner, and J. Loidl, *Nature*, 320, 700, (1986).

Wallis, M K, and R.S.B. Ong, *Planet and Space Science*, 23, 713, (1975).

Wu, C S, T.E. Hartle, and K.W. Ogilvie, *J. Geophys. Res.*, 78, 306, (1973).

Pergamon

Adv. Space Res. Vol. 22, No. 7, pp. 1049–1052, 1998
© 1998 COSPAR. Published by Elsevier Science Ltd. All rights reserved
Printed in Great Britain
0273-1177/98 $19.00 + 0.00

PII: S0273–1177(98)00140–9

REAL TIME SCIENCE DISPLAYS FOR THE PCA EXPERIMENT ON THE ROSSI X-RAY TIMING EXPLORER

A. B. Giles [1]

Code 662, NASA Goddard Space Flight Center, Greenbelt, MD 20771, USA
[1] also Universities Space Research Association

ABSTRACT

The Rossi X-ray Timing Explorer (RXTE) spacecraft was launched on 30th December 1995 and contains a large Proportional Counter Array (PCA) experiment developed at GSFC. Telemetry from RXTE is returned via the NASA Tracking and Data Relay Satellite System (TDRSS) which, apart from specific gaps in coverage, provides a steady stream of nearly continuous real time data packets. The PCA has an area of 7000 cm^2 and produces high count rates for many x-ray sources. This provides the opportunity for some serious interpretation and decision making in real time. The display programs developed by the PCA team fall into 4 classes: Housekeeping, General performance (Instrumental and calibration bias), Spectral (Science bias) and Temporal (Science bias). These displays are used by the duty scientist and experiment controllers to monitor the observation in progress to try and ensure that the observation is proceeding as planned and that modifications to the observing modes are not required. Guest Observers (GO's) can be present in the Science Operations Facility (SOF) for their observations. The PCA team can also monitor their experiment using these programs. The SOF are planning to make some or all of these display tools available to GO's at their home institutions for the specific duration of their observations. This paper briefly describes the available display options. ©1998 COSPAR. Published by Elsevier Science Ltd.

INTRODUCTION

During 1992 it was decided to use C++ for all the software associated with the RXTE Science Operations Center (SOC) but this decision was revised in mid 1993. At this time the Guest Observer Facility (GOF) elected to stay in line with the well established practices of the Office of Guest Investigator Programs (OGIP) and High Energy Astrophysics Science Archive Research Center (HEASARC) activities within NASA at GSFC. The SOF was committed to using C++ and continued with this approach. The SOF Build deliveries were driven by the inexorable requirement to keep up with the spacecraft's agressive schedule and the need to support various tests and Mission Simulations. The GOF development was under less pre-launch pressure. Since almost all GOF code is written in FORTRAN using FITS & FTOOLS there could be no commonality with the SOF object oriented C++ environment. This fundamental decision meant that the original requirement for an integrated system with extensive analysis capability in the SOF was no longer feasible. Resources did not allow the duplication of the functionality of the analysis tools that the GOF was required to produce.

The RXTE software effort was widely distributed and the Instrument Teams were contracted to deliver much supporting code for their hardware experiment. The principal PCA contributions to the GOF were in the following areas: field of view response maps, energy response matrices, dead time corrections and orbit background model. The PCA C++ code produced for the SOF was developed using Object Center[tm] from CenterLine Software Inc. The GUI used is TAE+ which is a commercially available NASA product. The delivered code falls mainly in the following SOF subsystems:

- Command Generation (CG)
- Mission Monitoring (MM)
- Health and Safety (HS)
- Science Monitoring (SM)

In this paper we discuss only the PCA contributions to the SOF for the SM subsystem. The various software displays are all modular in design and are selected from a main PCA menu. Most modules are fully developed though a few are still evolving and the need for some additions emerged in the first few weeks after launch. New options are tested on the PCA monitoring and development system at GSFC and are not immediately available in the configuration controlled SOF environment. Future SM options may be added as needs arise if resources are available. The PCA experiment contains 5 similar detectors, each of which produces identical housekeeping data packets for the HS display software. These packets contain many detector parameters and some X-ray rates but no spectral information. Most science data packets are generated by the Electronic Data System (EDS) which is provided by the Massachusetts Institute of Technology (MIT). This sophisticated data selection and compression system allows many pre-programmed data modes to be run simultaneously in the six Event Analyzers (EA's) devoted to the PCA. Extensive details on the PCA/EDS combination can be found in the 1995 RXTE NRA. Further details of the RXTE spacecraft and mission can be found in Swank *et al.* (1994) and Giles *et al.* (1995).

RXTE was a fixed price program that was completed within budget and on time with respect to goals set ~4 years prior to the planned launch date of 31st August 1995. Given the practicalities following the GOF/SOF split the PCA team chose to emphasize the following aspects in SOF displays:

- Real Time or near real time graphical displays
- Visual impact and clarity
- Multiple options within a pre-defined (limited) set of constraints
- Limited analysis capability but export of data to external tools such as IDL
- Only support a few of the many EDS modes. Concentrate on the always present Standard Modes 1 & 2
- Real Time display or access to SOF short term Data Base

If GO's are present at the SOF during their observations and need to do higher level tasks such as background subtraction or spectral fitting in near real time, the GOF's XTE Fits Formatter (XFF) system can be used. A SOF copy of XFF can be run to produce temporary FITS files on an estimated 15 - 30 minute time frame. The precise time scale depends on the data flow from the NASA PACOR system into GSFC and "fast" FITS files made in this way may be very incomplete. Normally a 24 hour period elapses before all late packets are assumed to have arrived. Once the FITS files are created they can be studied using the GOF FTOOLS. This paper is intended to give an impression of the PCA displays available. There are too many features and options for a "User Guide" (Rhee, 1995) type detailed discussion (also see Design Guide, Giles, 1995).

SCIENCE MONITORING REQUIREMENTS

The intention of the science monitoring displays can be summarized as follows, in approximate order of increasing scientific interest. Remember that "detailed analysis" is not intended with this real time system:

- Do spectra (maybe just background for a weak source) look typical. Are they free of spikes, gaps and noise
- Are the spectra from all xenon layers in all detectors similar (total of 30 spectra)
- Are all the internal calibration spectra normal

- Has the source been detected
- How does the source intensity vary with time
- What sort of spectral shape does the X-ray source have
- How does the spectrum vary with time
- How do the detected source intensity & spectrum translate into telemetry loading
- What is the source hardness ratio and how does it vary with time
- What sort of "slow" time scale periodic or aperiodic features are visible
- What sort of "faster" time scale features are visible in the on-board EDS FFT & Delta Time Binned modes

The set of display programs developed to address these questions is detailed below in Table 1. These can all be run individually or more normally selected from a main GUI interface. The GUI allows multiple instances of programs to be started e.g. it is common to run 4 light curve options at the same time - 0.125, 1, 8 and 16 seconds temporal resolution. These displays then span 128 seconds, 1024 seconds, 2.28 hours and 18.2 hours respectively.

Table 1.   PCA Science Monitoring Displays

Main Group	Sub Group	Description
STD Mode 1	Calibration Spectra	Raw 256 channel spectra every 128 seconds
	Light Curves (LC)	Eight 1024 point light curves every 128 seconds 0.125 second resolution
Temporal	Power Spectra (PS)	Power spectrum of STD Mode 1 light curves
Science	PS History	2 D plot, frequency vs. time (power color coded)
	Layer LC's	LC's (select keV energy range) from STD Mode 2
	Energy LC's	LC's (4 adjustable energy ranges) from STD Mode 2 data
STD Mode 2	Spectra	Xenon & Propane spectra every 16 second updates
	Rates	29 (x 5 PCU's) diagnostic rates monitored every 16 seconds
Spectral	Base Summation	Derived from STD Mode 2, all xenon signals can be added raw or with gain & offset corrections applied
Science	Recommend	Base summation split into the 6 energy bands defined for the GOF RECOMMD program
	Colors vs. Time	3 definable keV bands on Base Mode, plots of color vs. time
	Spectral History	2-D plot, spectra (vertical axis) vs. time (horizontal axis) with signal being color coded
Others	EDS FFT	Selected on-board EDS FFT spectra mode
	EDS Delta Time	Selected on-board delta time binned mode
	Event Selection	A tool to make selections from PCU
	Slew Detection	A derivative of Event Data selection optimized for detection of weak sources during slews

Each program contains extensive menu selections and a typical display appearance is shown in Figure 1 (the first item from the list in Table 1). This can be shown in refresh mode (new display every 128 seconds) or integrated up over time as actually shown in Figure 1. This figure is a direct screen snapshot but the colors are lost in this black and white reproduction. The spectra cover the energy range 0-60 keV and show the Am^{241} calibration source peaks in each of the 6 anode chains within a single detector. The energy resolution and gain uniformity are both excellent for a large proportional counter. The broad low energy peaks in the top two plots are from an astrophysical source whose photons are mostly absorbed in the front layer of the detector.

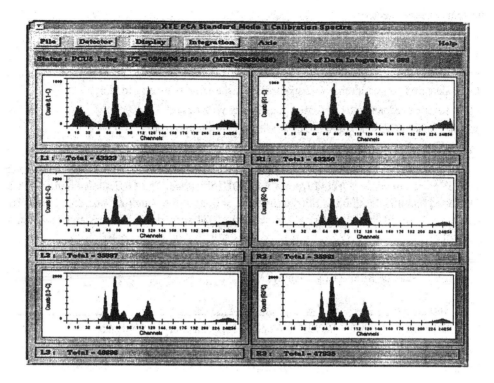

Fig. 1. Typical PCA screen display - integrated calibration spectra for one detector

ACKNOWLEDGMENTS

The design of these software systems has benefited from much comment and input from my fellow scientists in the PCA Instrument Team. The actual subsystem coding for HS has been written by Ramesh Ponneganti and Vikram Savkoor, for CG & MM by Vikram Savkoor and for SM by Hwa-Ja Rhee. All these programmers are with the Hughes STX Corporation working under contract at GSFC. David Hon, also with STX but located with the main SOF programming team, has provided the real time client interface to the SOF system. The entire EDS system is the responsibility of MIT and they have provided us with code to unpack data packets for a few of their many Event Analyzer modes.

REFERENCES

1st RXTE NASA Research Announcement, January (1995).

X-ray Timing Explorer - Instrument Operations User's Guide, Version 1.0, May (1995).

Giles, A.B., PCA Science Monitoring, Design Concepts, Screen Functions, SOC User Interface, Version 6.1, December (1995).

Giles, A.B., K. Jahoda, J.H. Swank, and W. Zhang, Publ. Astron. Soc, Aust, 12, 219 (1995).

Rhee, H., PCA Science Monitoring User's Guide, Version 4.1.1, August (1995).

Swank, J.H., *et al. The Lives of the Neutron Stars*, eds. M.A.Alpar et al., NATO ASI Series, pp. 525-542, Kluwer Academic Publishers, Dordrecht, The Netherlands (1994).

Pergamon

Adv. Space Res. Vol. 22, No. 7, pp. 1053–1056, 1998
© 1998 COSPAR. Published by Elsevier Science Ltd. All rights reserved
Printed in Great Britain
0273-1177/98 $19.00 + 0.00

PII: S0273-1177(98)00142-2

X-RAY TIMING IN THE ASTROPHYSICAL EXPERIMENT ON-BOARD "MIR-SPECTR" ORBITAL COMPLEX

M.I. Kudryavtsev[1], S.I. Svertilov[2], V.V. Bogomolov[2], and A.V. Bogomolov[2]

[1]*Space Research Institute, Russian Academy of Science, Profsoyznaya st. 84/32, Moscow 117810, Russia*
[2]*Skobeltsyn Institute of Nuclear Physics, Moscow State University, Vorob'evy Gory, Moscow 119899, Russia*

ABSTRACT

A multipurpose experiment with a number of instruments is being conducted on-board the *"Mir-Spectr"* orbital complex. Observations began on October 5, 1995. The main astrophysical instrument is a wide-field oriented scintillator spectrometer (~300 cm^2 effective area, 10-300 keV energy range). It consists of 7 identical detector units with the axes turned 5° to each other, which allows us to determine the direction on the source of the registered photons. The significant peculiarity of the experiment is the possibility of the continuous monitoring of gamma-quanta and relativistic electrons fluxes, which are responsible for those background components in the X-ray spectrometer, which can not be removed by instrumental methods. Due to the long exposure time and accurate background patrol, the X-ray timing of different galactic sources (including *Sco X-1, Cyg X-1, Crab*) as well as the registration of hard X-ray bursts (especially from the Soft Gamma-ray Repeaters) and transients are quite realistic in the experiment (the fluxes of weakest detectable bursts from SGR is $\leq 10^{-8}$ erg/cm^2).

©1998 COSPAR. Published by Elsevier Science Ltd.

INTRODUCTION

A multipurpose experiment *"Grif-1"* (Pankov *et al.*, 1990) was conducted on-board the *"Spectr"* module of *"Mir"* orbital station since October 5, 1995. The experimental data were obtained up to June 23, 1997.

The observational equipment included four instruments:
- the wide-field oriented spectrometer of hard X- and gamma-rays *PX-2* (energy range ΔE_{γ} = 10-300 keV, effective area S_{γ} ~ 300 cm^2, field of view (FOV) ~ 1 sr);
- the spectrometer of gamma-quanta and neutrons *NEGA-1* (ΔE_{γ} = 0.05-50 MeV, $\Delta E_n \geq 20$ MeV, S_{γ} ~ 250 cm^2, S_n ~ 20 cm^2, FOV ~ 4π sr);
- the spectrometer of electrons, protons and nuclei with large effective area *FON-1* (ΔE_e = 40-500 keV, ΔE_p = 1-3 MeV, $\Delta E_{nuclei} \geq 3$ MeV/nucleon, $S_{e,p}$ ~ 80 cm^2, FOV ~ 1 sr); the instrument is intended for monitoring of low-intensive charged particles fluxes beyond the Earth radiation belts;
- the spectrometer of electrons, protons with small effective area *FON-2* (ΔE_e = 0.04-1.5 MeV, ΔE_p = 2-200 MeV, $S_{e,p}$ ~ 1 cm^2, FOV ~ 0.5 sr); the instrument is intended for measurements of charged particles fluxes in the zones of trapped radiation.

With the exception of *NEGA-1*, the detector units of other instruments are placed outside the *"Spectr"* module - beyond its air tight compartment.

During the experiment all instruments are operating continuously. There are two modes of instruments output to data storage in the on-board memory. They have the time resolutions ~5.1 s and ~0.64 s.

INSTRUMENTAL APPROACH AND EXPERIMENT'S BACKGROUND CONDITIONS

PX-2 Instrument Description

The spectrometer *PX-2* is the main instrument for astrophysical observations. It consists of 7 identical phoswich (*CsI(Na)* with 8.0 cm diameter, 0.3 cm thickness - plastic scintillator) detectors with crossed FOVs: the axes of these detector units are turned 5° respectively to each other. This allows us to observe almost the same area of the sky with all detectors simultaneously, and on the other hand, in the case of burst-like event registration to determine the direction on the source by the output data from each detector. The instrument provides flux measurements in energy ranges: 10-50, 25-50, 50-100, 100-200 and 200-300 keV. Also there is an opportunity to improve the time resolution for some output parameters - up to ~0.4 s (by continuous observations).

Background Conditions

The obtained background measurements allow to estimate the experiment's capabilities in the X-ray timing of some galactic sources as well as its sensitivity to the gamma-ray bursts registration. The typical background variations of *PX-2* output parameters are illustrated by time dependence of 25-50 keV count rates N_x^{25-50}, presented in Figure 1. The N_x^{25-50} values are strongly correlated with 0.15-0.5 MeV count rates $N_\gamma^{0.15-0.5}$, measured by *NEGA-1* instrument. There is also some correlation between N_x^{25-50} values and *FON-1* output data (N_e^{80-160}) in the case of significant increasing of electrons count - more than 2 orders. These variations correspond two main background components caused by the local gamma-quanta and Magnetospheric electron bremsstrahlung. Due to the small thickness of *CsI(Na)* crystal the activation background component was not sufficient in *RX-2* instrument outputs.

Local Gamma-Rays. It is necessary to note, that the *NEGA-1* instrument registers mainly local gamma-quanta, produced as the result of galactic cosmic rays (GCR) interactions in the station's material. Thus the temporal dependencies of *NEGA-1* parameters correspond to the latitude variations of GCR intensity. The "latitude" effect in *PX-2* parameters can be connected with the registration of the part of local gamma-quanta, scattered in the shield materials due to the Compton process. In order to exclude all other variations except those caused by the "latitude" effect, the time intervals corresponding to the *PX-2* orientation toward the Earth were chosen. Because in such orientation the instrument FOV is significantly shadowed by the Earth's Atmosphere, it guarantees that X-ray fluxes, registered in FOV will be constant. The coefficient of linear regression α between N_x^{25-50} and $N_\gamma^{0.15-0.5}$ values for the data corresponding to this time intervals was determined: $\alpha \sim 0.01$. Then the time set of $N_x^{25-50} - \alpha N_\gamma^{0.15-0.5}$ values pure from "latitude" effect were obtained for different intervals of observations (a similar procedure can be made for any other PX-2 parameter).

Magnetospheric Electron Bremsstrahlung. The registration of electron bremsstrahlung, produced in the thin (~0.03 cm) layer of Al, which cover detectors as the light and other damages protector, can be the reason of some correlation between N_x^{25-50} and 80-160 keV electrons count rates N_e^{80-160}, measured by *FON-1* instrument (see Figure 1). For the measurements in the trapped radiation zones, where the electron fluxes can be very high, the *PX-2* output data was distorted (the corresponding level of electron count rate is marked in Figure 1). But for the near space areas the influence of electron fluxes variations on the *PX-2* output parameters can be taken into account. For this purpose the coefficient of linear regression β between $N_x^{25-50} - \alpha N_\gamma^{0.15-0.5}$ and N_e^{80-160} values was determined (also for the data corresponding to time intervals of full *PX-2* FOV shadowing by the Earth's Atmosphere). The obtained value of $\beta \sim 0.001$ corresponds to the assumption about bremsstrahlung mechanism of those additional count in *PX-2* parameters, which is connected with electron flux variations

The time dependence of $N_x'^{25-50}$ values pure as from "latitude" effect as from the influence of electron fluxes variations are also shown in Figure 1. The character of $N_x'^{25-50}$ time dependence is quite different from primary one of N_x^{25-50} values. Without predominance of "latitude" variations, the quasimonotonous trend caused by the registration of cosmic X-ray diffuse background and X-rays from the sources in the instrument FOV opening by its going out from the Earth's shadow (the time interval when the FOV was shadowed by the Earth Atmosphere is marked in Figure 1), become visible. Some count rate variations, which can be connected with the real X-ray fluxes (from the sources in the instrument FOV) registration, should be also noted. In that interval of observations the central axis of PX-2 instrument was constantly oriented toward the point on the sky with coordinates: $\alpha = 15^h59^m$, $\delta = -26°$, that is near the bright galactic X-ray source *Sco X-1* ($\alpha = 16^h17^m$, $\delta = -15.5°$).

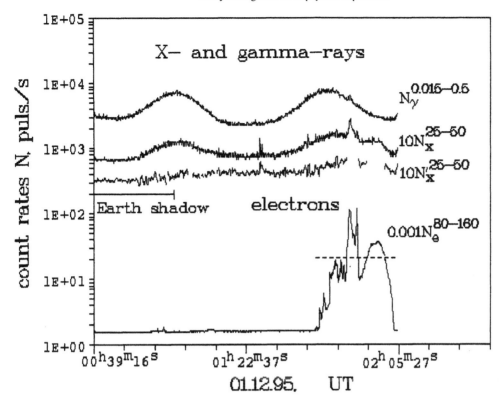

Fig. 1. The time profiles of primary (N_γ, N_x, N_e) and pure from instrumental background variations (N'_x) count rates obtained for 1 orbital cycle. The dotted line shows the electron count rate level, corresponding to the boundaries of the space areas (near the Brazil anomaly), where *PX-2* instrument becomes unable to give correct output data.

The simultaneous registration of different types of space radiation during the experiment allows to obtain the accurate determination of the background variations caused by cosmic rays and radiation belt particles and hence to reduce effectively the influence of space environment on the *PX-2* instrument's background. It also makes possible to identify the imitations of burst-like events by the precipitating electrons fluxes short increases.

ASTROPHYSICAL OBSERVATIONS

Along with large effective area and wide FOV of *PX-2-* instrument this provides high sensitive monitor observations of large parts of the sky. It is proposed that experiment will be carried out during about several years, that gives an opportunity to realize practically observations of the all sky. Thus astrophysical observations during *"Grif-1"* experiment (Kudryavtsev *et al.*, 1995) correspond the goals of all-sky monitor in hard X-rays:

- research of temporal phenomena such as GRBs, SGRs and hard X-ray transients:
- X-ray timing of some galactic X-ray binaries and pulsars.

The GRBs Observations

Due to the complex shield of detector elements in *PX-2* instrument, the part of the locally produced X- and gamma-rays in instrument's background is rather small in the energy range 10-100 keV. For the low-latitude observations background in the 25-50 range is all almost caused by the cosmic X-ray diffuse radiation. The 5σ level of burst-like event registration in the 25-50 keV range in the case of typical burst parameters (duration about 10 s, thermal bremsstrahlung spectrum with $kT \sim 25$ keV) corresponds to a fluence of $\sim 10^{-8}$ erg/cm^2.

The example of one of the registered GRB-candidates is presented in Figure 2 (the main criterion of the reality of burst-like X-ray flux increasing besides the absence of accompanying electron fluxes increasing is the event registration of at least two different detector units - N1 and N2 for the presented case).

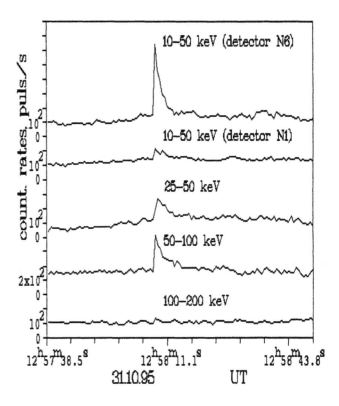

The rather high sensitivity of *PX-2* instrument in the energy range 10-100 keV makes the most attractive its use for detailed measurements of sufficiently soft bursts and transients including SGR bursts. As it supposed, the SGR-like objects belong to a population separate from the common GRB sources (Kouveliotov *et al.*, 1994). Although the recent *CGRO BATSE* results are in accordance with cosmological models (Meegan *et al.*, 1992), GRBs still remains a not quite understandable phenomenon. In particular, Lingenfelter and Higdon (1992) supposed that GRBs statistical properties can not be accurate due to registration of SGR bursts. Thus the supplement of GRBs and SGRs statistics in *"Grif-1"* experiment can be useful both for solving of GRB problem and for understanding the nature of SGR-like objects.

Fig. 2. The time profiles of *PX-2* instrument parameters for GRB-candidate.

The X-ray Timing

The experiment's possibilities on the X-ray sources timing can be based on the *PX-2* 25-50 keV output data pure from the variations caused by the charged particles influence. Such data were obtained for the cases of different instrument orientations: for the sky areas containing bright galactic sources - *Sco X-1* (see Fig. 1), *4U1700-37* and for the regions placed rather far from the Galactic plane. As the result of these observations, it can be concluded that for the output data averaged over 20-min time interval (that corresponds to the duration of sky observation pure from the FOV shadowing by the Earth Atmosphere) on the one orbital cycle, it is possible to reveal the X-ray fluxes variations on the level of ~400-500 mCrab with characteristic times ~1 min. Thus the timing of the sources, characterized by the fluxes about hundreds mCrab by the energies in dozens keV (such as *Sco X-1*, *Crab*, *Cyg X-1* and some other) can be realized in *"Grif-1"* experiment.

REFERENCES

Kouveliotov, C., G.J. Fishman, C.A. Meegan, W.S. Paciesas, J. van Paradijs, *et al.*, The Rarity of Soft γ-ray Repeaters deduced from Reactivation of SGR1806-20, *Nature*, 368, 125 (1994).

Kudryavtsev, M.I., V.M. Pankov, A.V. Bogomolov, V.V. Bogomolov, Yu.I. Denisov, *et al.*, The "Mir-Spectr" Gamma-Astronomy Experiment, *Proc 24 ICRC*, 3, 567 (1995).

Lingenfelter, R.E. and J.C. Higdon, Are Repeaters responsible for Gamma-ray burst <V/V$_{max}$> <0.5?, *Astrophys. J.*, 397, 576 (1992).

Meegan, C.A., G.J. Fishman, R.B. Wilson, W.S. Paciesas, G.N. Pendleton, *et al.*, Spatial Distribution of γ-ray Bursts observed by BATSE, *Nature*, 355, 143 (1992).

Pankov, V.M., M.I. Kudryavtsev, Yu.G. Shkurkin, A.V. Bogomolov, V.V. Bogomolov, *et al.*, Instrumentation for Background Monitoring in Astrophysical and Ecological Programs, *Proceedings of the IVth International Seminar "Manufacturing of Scientific Space Instrumentation"*, 3, 42 (1990).

Adv. Space Res. Vol. 22, No. 7, pp. 1057–1060, 1998
© 1998 COSPAR. Published by Elsevier Science Ltd. All rights reserved
Printed in Great Britain
0273-1177/98 $19.00 + 0.00

Pergamon

PII: S0273-1177(98)00143-4

THE PROJECT OF ALL-SKY HARD X-RAY AND SOFT GAMMA-RAY MONITORING ON-BOARD INTERNATIONAL SPACE STATION

M.I. Kudryavtsev[1], S.I. Svertilov[2], V.V. Bogomolov[2], and A.V. Bogomolov[2]

[1]*Space Research Institute, Russian Academy of Science, Profsoyznaya st. 84/32, Moscow 117810, Russia*
[2]*Skobeltsyn Institute of Nuclear Physics, Moscow State University, Vorob'evy Gory, Moscow 119899, Russia*

ABSTRACT

Further progress in hard X-ray and soft gamma-ray observations can be achieved in the experiments with wide-field code-aperture instruments, which allow, due to long exposure times, to achieve the timing of different astrophysical objects (possible even in the near-Earth missions). The deep all-sky survey at the level of registration ~100 new sources (mainly Active Galactic Nuclei) in the energy range ~0.3-1 MeV is possible in the observations made during ~10 years with the instrument with ~2π FOV, ~1000-3000 cm^2 effective area and ~3°-5° angular resolution. The gamma-telescope *"Gammascope"* with such characteristics is now being elaborated. It is proposed for the experiment on the International Orbital Station *"Alpha"*. The instrument is based on a combination of the quasi-spherical coding-mask and *NaI-CsI* position-sensitive detector (PSD).

©1998 COSPAR. Published by Elsevier Science Ltd.

ALL-SKY MONITOR FOR LOW ENERGY GAMMA-RAY ASTRONOMY.

Since the first X-ray- and gamma-ray astronomical experiments on-board the spacecrafts (*Uhuru, SAS-2* and others), in which the discrete sources, more or less constantly existing in the sky, were studied, the properties of the astrophysical objects constantly emitted gamma-rays in the energy range 0.1-1 MeV remain at least not enough known. The number of studied objects in the nearby energy ranges is equal to some tens (>1 MeV) and hundreds (10-100 keV), but only few (<10) objects were detected in the 0.1-1 MeV range (without gamma ray bursts). Among these objects are *Crab* (diffuse region and pulsar), *Cyg X-1*, the sources near the Galactic Center, the active galaxies nuclei (AGN): *Cen A, NGC4151, 3C273*.

Survey Sensitivity of Modern Experiments

The curves in Figure 1 illustrate the break in the experiment's sensitivity - about an order between the >1 Mev and <1 MeV energy ranges. At these energies the change of gamma-astronomical methods takes place. The measurements in the <1 MeV range were made mainly by the scintillation spectrometers with the narrow field of view (FOV) - collimated as *HEAO1 A4* (Matteson et. al, 1978), *CGRO OSSE* (Kurfess et al., 1991), or coded aperture - *GRANAT SIGMA*). In the >1 MeV range Compton telescopes are used (*CGRO COMPTEL* (Schonfelder, 1991). But really, only the *HEAO1 A4* experiment remains the survey experiment allowed to study the all sky. The significant values (not upper limits) of the fluxes were obtained in this experiment for ~10 sources near the low border of the energy range of our interest (0.1-0.2 MeV). Regular inspection of the areas near the Galactic Center region made during 5 years in *GRANAT SIGMA*, allowed us to detect several (including new) sources and to give the information about the temporal properties of these objects (also only near the lower border of the energy range 0.1-1 MeV).

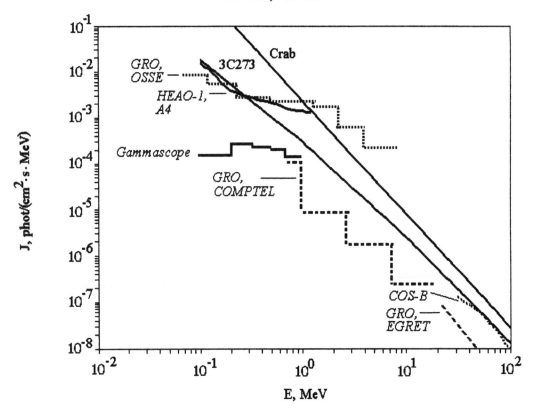

Fig. 1. The curves, characterizing the survey sensitivity of some experiments, with spectra of known sources Crab and *3C 273*.

To overcome the existing break in the survey sensitivity, the one-order lowering of the minimal detectable fluxes level in the survey monitoring experiment should be achieved. This is vital for solving a number of important problems of low-energy gamma ray astronomy, such as:
- What is the role of nuclear reactions in the gamma-ray producing in some objects?
- Why are many gamma-ray sources with the falling spectra, detected in E>100 MeV range, not seen in the X-ray range?
- What are the spectral and temporal peculiarities of the sources - the black hole candidates?

The Problem of Gamma-Ray Bursts

Although in *CGRO BATSE* experiment rather rich gamma-ray burst (GRBs) statistics was obtained, their sources still remains unknown. Certain progress in the GRBs nature understanding can be connected with the study of their statistical characteristics, thus it is necessary to combine in GRB search experiments the wide instrument's FOV with the possibility of accurate GRB sources localisation on the sky - even for weak events ($\leq 10^{-7}$ erg/cm^2).

THE WIDE-FIELD GAMMA-RAY TELESCOPE *GAMMASCOPE*

The situation in the gamma-ray astronomy in the energy range 0.1-1 MeV at the present time is similar to the situation in the X-ray astronomy in the epoch before *Uhuru* lunching. Thus the long-term and expensive experiment should guarantee the break in the soft gamma-ray range, similar to that *Uhuru* break in 2-20 keV range. This goal can be achieved with the use of the instruments with maximal FOV and at the same time with narrow enough angular resolution. It may be a coded aperture telescope based on a combination of coding mask and the position-sensitive detector (PSD), which was used in astronomical imaging systems (Caroli *et al.*, 1987).

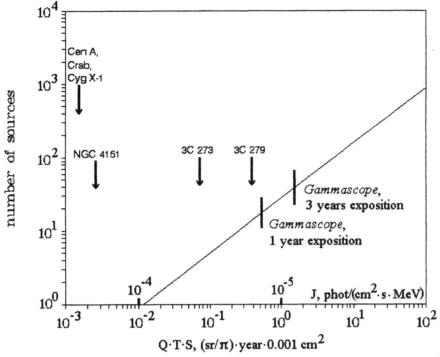

Fig. 2. The number of detectable sources versus $\Omega \times T \times S$ (Ω·in π sr, T in years, S in 1000 cm²).

The Survey Sensitivity and FOV

The widest instrument's FOV allows us to achieve the longest exposure time of given source and the richest GRBs statistics. But it is also necessary for maximal sensitivity by the all-sky survey. For the energies 0.1-1 MeV the background is mainly determined by the natural and induced radioactivity of the instrument's and spacecraft's materials, and it is practically independent on the FOV value, thus its maximisation will not lead to background increase (Dean et al., 1991). Thus the maximal FOV gives a combination of maximal sensitivity in the all-sky survey with the maximal exposure time of the given sky area (the exposure time $T = (\Omega/4\pi)T_{total}$, where T_{total} is the total time of the experiment). But the FOV of the telescope, intended for observations on-board the near-Earth space should not exceed 2π (because a half of the sky will be always shadowed by the Earth).

The Survey Sensitivity and the Angular Resolution

To detect the weakest sources, the maximal number of the gamma-ray bursts etc., the instrument's effective area and exposure time must be maximal. Most of modern results were obtained by the instruments with the areas ~1000 cm² and the experiment's time about several years. The angular resolution in the survey experiment should be enough to locate the observed object with accuracy, sufficient for its identification. As the universal characteristic of the experiment's survey sensitivity, the value, equal to ΩTS., (i.e. the multiplication of instrument's FOV solid angle Ω, total time of experiment T and effective area S) can be chosen. The minimal detectable fluxes J_{min}, corresponding the ΩTS values can be calculated for a given background level (that may be the same, which was observed on-board the *Mir* orbital station).The number of observable objects (mainly the AGN-type) will grow up with the experiment sensitivity increase. In the case of their uniform space distribution, the number of sources N with the fluxes, exceeding J_{min}, $\sim J_{min}^{-3/2}$. If we assume, that all AGN-like sources with the fluxes, exceeding that of 3C273 ($\sim 3 \cdot 10^{-2}$ phot./cm²/s/MeV at $E_\gamma \sim 0.5$ MeV) are known and its number (4-6) corresponds to achieved sensitivity level (i.e. ΩTS value), then the expected dependence $N(\Omega TS)$ can be obtained (see Figure 2). In the experiment characterizing by the ΩTS value ~10 (for example, $S \sim 1000$ cm², $\Omega \sim \pi$ sr, $T \sim 10$ years), about 100 AGN-like sources on the all sky can be discovered. To detect separately the objects with such sky density, the angular resolution $\sim 5^o \times 5^o$ is quite sufficient.

The *Gammascope* instrument's Configuration

The optimal telescope's configuration, that can gave the FOV $\sim 2\pi$, can be chosen between two variants:

- a plane circle detector and hemispheric mask with the diameter, some times (at least 2) larger, than the detector's diameter;
- the hemispherical detector and hemispherical mask with the size, closed to the detector's size.

Both variants have their own advantages. In the first configuration all the detector's area can be used, but the instrument's size and mass will be too large, because the main part to full instrument's mass gives the coding mask, which should cover PSD. In the second configuration mask's mass can be made ~ 4 times smaller, but the detector's surface will be used 2 times less effectively. Because the transparent part of coding mask area practically can not be more $\sim 30\%$ (Caroli, 1987), PSDs geometrical area should be at least 3 times larger, than necessary effective area (i. e. ~ 3000 cm^2). Due to background behaviour in the energy range 0.1- 1 MeV there is an opportunity to change the coding mask onto "antimask" without any losses of telescope's imaging properties. By given instrument area the coding mask and PSD masses are determined really by their thickness. The detector's thickness (including the shield from the background gamma-rays) must be >30 g/ cm^2 (it may be for example phoswich - the ~ 3 cm layer of *NaI(Tl)* actively shielded by ~ 4 cm *CsI(Tl)*). The similar thickness, but of the matter with larger Z (lead, tungsten) for the effective gamma-ray absorption is necessary for the coding mask. Thus in the case of the plane circle PSD detector with the geometric area S ~ 3000 cm^2, the full instrument mass will be ~ 500 kg, in the case of hemispherical PSD the necessary detector area (in view of less effectively use of detector surface) - ~ 400 kg.

THE *GAMMASCOPE* EXPERIMENT'S PERSPECTIVES FOR GAMMA-RAY ASTRONOMY

If the experiment will be carried out about 10 years, then up to 100 new extragalactic sources in the range 0.1-1.0 MeV can be detected. Also a new information about the sources known from the measurements in the nearby energy ranges can be obtained. Such information about the objects of AGN-type observed in hard gamma-rays >30 MeV will allow to obtain the continuous spectra of this objects from soft X-rays to high energy gamma-rays. The experiment also allows to obtain new data about galactic X-ray binaries of *Cyg X-1* type as well as X-ray novas, which seem to be the most favourite candidates for black holes. The preliminary estimates allow to conclude, that the *Gammascope* instrument allows to discover such bright gamma-ray source as Cyg X-1 for the time of continuous exposure ~ 1 min, and in the same temporal scale (or some greater for weak sources) to study their variability practically during the all time of the experiment. In the case of AGNs observations (as 3C273) the minimal exposure needed for their detection will be ~ 10 days. The weakest GRBs, which source can be located with given accuracy ($5^0 \times 5^0$) are characterized by a fluence of $\sim 10^{-8}$ erg/cm^2. This allows to hope that more precise GRBs statistical characteristics of cosmic gamma-ray bursts, in particular the possible discovery of their sources sky distribution angular anisotropy, will be obtained.

ACKNOWLEDGEMENTS

This work was supported by the NASA and the Russian Space Agency in scope of contract "Nauka-NASA" (NAS15-10110).

REFERENCES

Caroli, E., J.B. Stephen, G. Di Cocco, L. Natalucci, and A. Spizzichino, Coded Aperture Imaging in X- and Gamma-ray Astronomy, *Space Sci. Rev*, 45, 350 (1987).

Dean, A.J., F. Lei, and P.J. Knight, Background is Space-Borne Low-Energy γ-ray Telescopes, *Space Sci. Rev*, 57, 109 (1991).

Kurfess, J.D., W.N. Johnson, R.L. Kinzer, R.A. Kroeger, M.D. Leising *et al.*, The Oriented Scintillation Spectrometer Experiment on GRO, *Adv. Space Res.*, 11, 323 (1991).

Matteson, J.L., HEAO1 A4 Hard X-Ray Sources Catalogue, *Proceedings AIAA 16th Aerospace Sci. Meeting*, paper 78-35, Huntsville, Alabama (1978).

Schonfelder, V., The Imaging Gamma-ray Telescope Comptel aboard GRO, *Adv. Space Res.*, 11, 313 (1991).

COSMIC GAMMA RAY BURSTS

Proceedings of the E1.2 Symposium of COSPAR Scientific Commission E which was held during the Thirty-first COSPAR Scientific Assembly, Birmingham, U.K., 14–21 July 1996

Edited by

C. KOUVELIOTOU

USRA, ES-84, NASA/MSFC, Huntsville, AL 35812, U.S.A.

and

K. HURLEY

UC Berkeley, Space Sciences Laboratory, Berkeley, CA 94720-7450, U.S.A.

 Pergamon

Adv. Space Res. Vol. 22, No. 7, p. 1063, 1998
© 1998 COSPAR. Published by Elsevier Science Ltd. All rights reserved
Printed in Great Britain
0273-1177/98 $19.00 + 0.00

PII: S0273-1177(98)00196-3

PREFACE

A breakthrough in our understanding of cosmic gamma-ray bursts took place in 1997, when it was at last determined that some bursts are definitely at cosmological distances. The 1996 COSPAR session on cosmic gamma-ray bursts was held on the eve of this breakthrough. The thirteen papers in these proceedings, although written in most cases prior to this discovery, still represent accurate summaries of the data, descriptions of relevant instruments, and theoretical investigations of radiation mechanisms.

The papers include a review of the data from the Burst and Transient Source Experiment aboard the Compton Gamma-Ray Observatory, and a preview of one of the instruments aboard the Italian-Dutch BeppoSAX satellite, which was to play a crucial role in the solution of the burst mystery. The energy spectra of bursts are covered in three papers, and burst time histories are covered in another. Several presentations cover models for burster spatial distributions, energy release, and radiation mechanisms. Observations of the soft gamma repeaters are also treated. The triangulation technique is described, and a method for verifying its accuracy is presented also.

The organizers would like to thank the organizing committee, consisting of C. Meegan, C. Winkler, M. Rees, E. Fenimore, and R. Sunyaev, for their assistance in putting this meeting together. We are grateful to our referees for their work.

C. Kouveliotou
K. Hurley

Pergamon

Adv. Space Res. Vol. 22, No. 7, pp. 1065–1075, 1998
Published by Elsevier Science Ltd on behalf of COSPAR
Printed in Great Britain
0273-1177/98 $19.00 + 0.00

PII: S0273-1177(98)00197-5

GAMMA-RAY BURST OVERVIEW

C. A. Meegan

NASA/Marshall Space Flight Center, Huntsville, AL 35812, USA

ABSTRACT

Over 25 years since their discovery, gamma-ray bursts (GRBs) remain an unsolved mystery. They occur about once a day, last from about 0.1 to 1000 seconds, and emit detectable radiation only at X-ray and gamma-ray energies. The spatial distribution of the sources is isotropic but inhomogeneous, inconsistent with any known galactic component. Two very different hypotheses are currently debated: either the bursts arise from an extended galactic halo, possibly populated by high velocity neutron stars, or they lie at cosmological distances and redshift effects are responsible for the apparant inhomogeneity. The current observational and theoretical status of the GRB problem will be discussed with emphasis on the central question of the distance to the sources.   Published by Elsevier Science Ltd on behalf of COSPAR.

INTRODUCTION

Gamma-ray bursts, discovered by Klebesadel, Strong, and Olsen (1973), are brief flashes of gamma-rays from celestial sources. About 2000 GRBs have now been recorded. The full sky rate, as determined by the most sensitive instruments, is $\sim 10^3$ per year. Occurrence times and directions to the sources appear to be random, and emission at wavelengths shorter than X-rays has not been detected. The rapid variability and high energy of the emitting photons pointed to compact objects as the sources. Until recently, the data favored a class of models in which the burst sources were relatively nearby (< 1 kpc) galactic neutron stars. A wide variety of energy sources had been proposed. The main theoretical difficulty was the problem of efficient conversion of energy into the gamma-ray spectral band, without also producing significant radiation at other wavelengths. Recent observations by the Burst and Transient Source Experiment (BATSE) on NASA's Compton Gamma Ray Observatory (CGRO), however, have led to the abandonment of the original galactic neutron star scenario. The intensity and angular distributions of the bursts (described in later sections) are not consistent with a disk geometry. Only two possibilities for the distance scale currently have significant support. Either the burst sources are at cosmological distances (z~1), or they form an extended Galactic halo with radius of 100-300 kpc. This review will describe the observational properties of the bursts and the evidence bearing on these two possibilities.

TEMPORAL STRUCTURE

The time histories of GRBs reveal a wide range of behaviors. Figure 1 shows some examples from BATSE observations. Durations range from tens of milliseconds to ~1000 seconds. Profiles can be smooth, as in Figure 1b, or may consist of numerous, apparently uncorrelated, short spikes, as in Figure 1c. Figure 1e shows an example of a fairly common feature, a pulse showing a fast rise and exponential decay (FRED), with wide ranges of decay time constants. Bursts often contain several FRED pulses, or may consist of a single FRED. In general, bursts and pulses within bursts are asymmetric, with shorter rise times than fall times (Link, Epstein, & Priedhorsky 1993; Nemiroff et al. 1994).

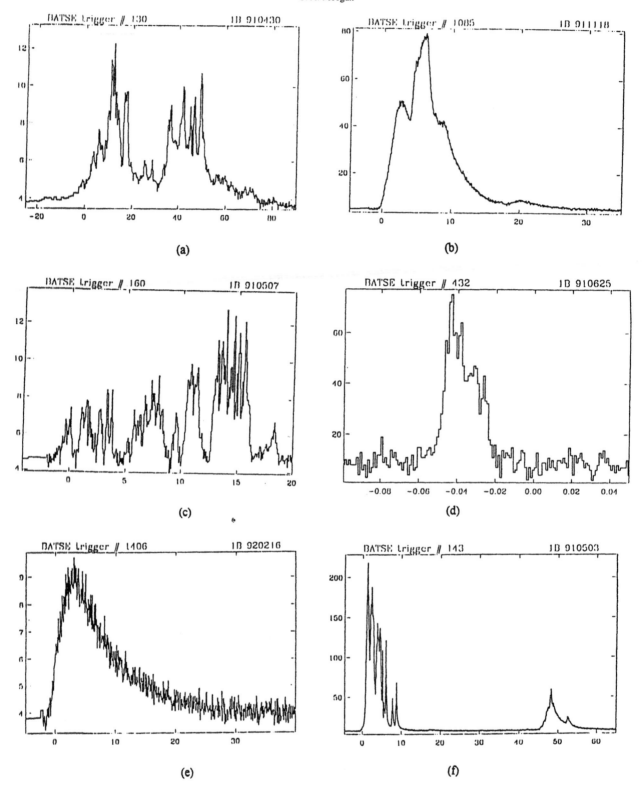

Figure 1. Examples of from BATSE bursts, illustrating the wide range of temporal structure. For each plot, the x-axis is seconds, and the y-axis is 10^3 counts/s.

A number of previous studies have found evidence for a bimodality in the distribution of burst durations, with a division at a few seconds between short and long bursts (see Hurley 1991 and references therein). Figure 2 shows the duration distribution for 427 bursts observed by BATSE that were above threshold on the shortest trigger timescale, 64 ms. The bursts with durations less than 2 seconds have been found to have on average harder spectra (Kouveliotou *et al.* 1993).

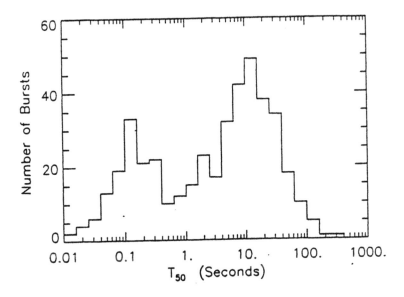

Fig. 2. Burst duration distribution

If bursts are cosmological, their durations would be expected to show the effect of time dilation, with weaker, and presumably more distance, bursts being on average longer. Such a brightness-dependent time stretching, while not requiring a cosmological explanation (Band 1994; Brainerd 1994a), would be supportive of the cosmological hypothesis. Norris *et al.* (1994) have reported that three different techniques show such an effect, with the weakest bursts about a factor of 2 longer than the most intense bursts. On the other hand, Mitrofanov *et al.* (1996), using different selection criteria, find no evidence that the weaker bursts are longer.

SPECTRA

Photon spectra of gamma-ray bursts are well fit by two power laws joined by an exponential, usually referred to as the Band function (Band *et al.* 1993), as follows:

$$N(E) = A E^{\alpha} \exp(-E/E_0), \quad (\alpha-\beta)E_0 > E$$
$$N(E) = B E^{\beta}, \quad\quad\quad\quad (\alpha-\beta)E_0 < E.$$

With proper choice of the constant B, the function and its derivative are continuous. The spectral index α is typically from 0 to -1.5, and β is typically -1.5 to -2.5. The "break" energy E_0 is typically a few hundred keV. Figure 3, adapted from Tavani (1996), shows spectra of several intense bursts observed by multiple instruments on the Compton Gamma Ray Observatory. In this figure, the photon number spectra are multiplied by E^2. The distinguishing feature of such spectra is that most of the energy lies in the low energy gamma-ray band. This raises the question: Why do the spectra of different bursts peak within a relatively narrow range, when there is no reason to expect the Lorentz factors of the emitting regions to be the same for all bursts? To address this difficulty, Brainerd (1994b) proposed that burst spectra are power laws attenuated by Compton scattering in a dense medium at cosmological distances, resulting in observed spectra with characteristic energies of a few hundred keV.

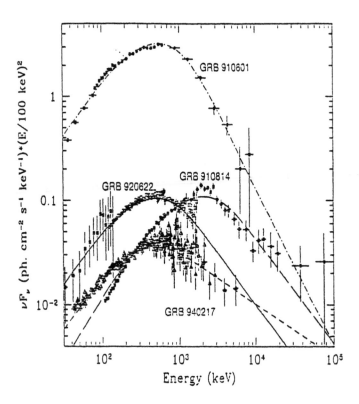

Fig. 3. **Examples of burst spectra (from Tavani 1996). The y-axis (νF_ν) is photon number times E^2.**

Many models of GRBs naturally produce a significant flux of X-rays, for example by reprocessing the gamma rays on the surface of a neutron star. While most GRBs have lower X-ray emission than such models appear to require, some events do show an additional component at low energies. Preece *et al.* (1996) have found that about 15% of the bursts observed by BATSE have emission in the 5 to 10 keV energy range that is significantly above the extrapolation of the Band function to these energies.

Observations at higher γ-ray energies are particularly intriguing. EGRET detected emission from GRB940217 lasting 90 minutes and extending up to 18 GeV (Hurley *et al.* 1994). Several other bursts observed by EGRET (Schneid *et al.* 1995) show no indication of a spectral cutoff up to at least 10 MeV.

We now have a sufficiently large number of intense GRBs to obtain statistical information on the evolution of spectra during a burst and even during individual pulses in a burst. In general, burst spectra evolve from hard to soft (Bhat *et al.* 1994; Ford *et al.*, 1995). A more quantitative study of this effect by Liang and Kargatis (1996) found that for most individual pulses in bursts, the break energy E_0 of the Band function decreases exponentially with photon fluence (the time integrated flux).

A continuing controversy in observations of GRBs is the existence and interpretation of spectral lines. The most compelling case for such features (Murikami *et al.* 1988) were pairs of absorption lines in several bursts observed by *Ginga*. These were interpreted (Fenimore *et al.* 1988) as the fundamental and first harmonic cyclotron absorption lines, indicating magnetic fields of about 10^{12} gauss, typical values for magnetized neutron stars. So far, BATSE has not detected absorption lines (Band *et al.* 1996). A great deal of work has been done in the past few years to compare BATSE and *Ginga* sensitivities to see if there is a significant inconsistency between these instruments (Band *et al.* 1994; 1995). The current status is that the two instruments are consistent at about the 10% probability level, which is uncomfortably low but not low enough to conclude that either BATSE or *Ginga* have instrumental problems.

A necessary consequence of cosmological models is that the spectra of more distant sources should show redshift effects. Mallozzi *et al.* (1996) have fit spectra using the Band function to ~400 BATSE bursts that exhibit curvature in their spectra. They found that there was a correlation between burst peak flux and the characteristic energy of the spectrum, as shown in Figure 4. The correlation can be interpreted as a redshift, with the weakest bursts at z ~1-2.

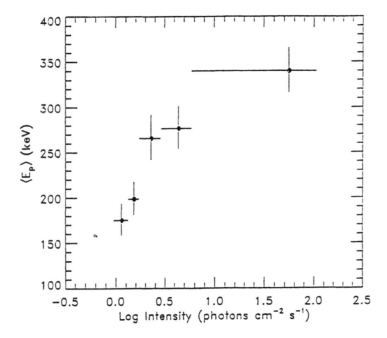

Fig. 4. Correlation between peak energy and intensity (from Mallozzi et al. 1996). If interpreted as a redshift, the weakest bursts are at z ~ 1-2.

COUNTERPART SEARCHES

It is convenient to catagorize the various searches for GRB counterparts as pre-burst, post-burst, and concurrent. In the pre-burst approach, pioneered by Brad Shaefer, archival data are examined for evidence of transient events at the positions of well-located GRBs. Implicit in this approach is the assumption that bursts repeat on time scales of decades or less. Studies of archival plates have found evidence for optical transients (e.g Hudec *et al.* 1994) but the association with the gamma-ray burster is problematic.

In the post-burst approach, deep searches are conducted as soon as possible after a burst is accurately located. Some recent results have been quite encouraging. In X-rays, ASCA observations (Murikami et al. 1996; Hurley et al 1996) have found weak sources at the positions of two well-located bursts, GRB920501 and GRB781119. In both cases, the probability of a chance occurence is < 1%. Vrba, Hartmann, & Jennings (1995) reported an extensive optical search of eight fields, finding no obvious counterparts, but marginal evidence for an excess rate of QSOs. Larsen, McClean, & Becklin (1996) obtained infrared images of 6 IPN error boxes. In each case, they found a bright galaxy, apparently putting to rest the "no-host galaxy" argument against the cosmological scenario. They also found a slightly greater than chance rate of galaxies near GRB error boxes, but the statistical significance is not compelling.

The best evidence for a GRB counterpart would be a simultaneous detection. To this end, Scott Barthelmy has created BACODINE, the BAtse COordinate DIstribution NEtwork, which intercepts the BATSE data stream, computes approximate GRB locations in real time, and transmits them to various observatories worldwide (Barthelmy et al.1995). BACODINE location accuracies range from 6° to 12° (1σ), depending on burst intensity. If a burst happens to be in the COMPTEL field of view and detectable in its energy range, the location uncertainty can be reduced to about a degree, greatly facilitating optical searches (McNamara et al. 1996).

Observations in the TeV energy range provide an important test of the cosmological hypothesis, since these photons would be absorbed by intergalactic radiation fields. Alexandreas et al. (1994), using the CYGNUS extensive air shower array, found no evidence for TeV emission from 58 bursts. For three of the bursts, the upper limit indicates a softening of the spectrum between 2 MeV and 100 TeV.

SPATIAL DISTRIBUTION

The spatial distribution of the sources of GRBs is strongly constrained by the angular and intensity distributions observed by BATSE (Meegan et al. 1992). The angular distribution of 1429 bursts is shown in Figure 5, where the position of each burst is plotted on an Aitoff-Hammer projection using Galactic coordinates. The distribution is isotropic to within the statistical errors. The Galactic dipole moment for

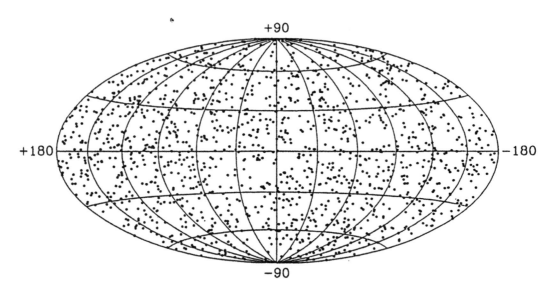

Fig. 5. The angular distribution, in galactic coordinates, of 1429 bursts observed by BATSE.

this distribution, including a correction for anisotropic sky coverage, is <cos θ> = −0.006 ± 0.015. The quadrupole moment is <$\sin^2$ b−1/3> = 0.006 ± 0.008. The intensity distribution is shown in Figure 6, where the differential number of bursts is plotted as a function of peak flux, for 842 BATSE bursts that exceeded the 1024 ms trigger threshold. If the burst sources were distributed homogeneously in Euclidean space, the intensity distribution would follow a −3/2 power law, shown as a dashed line in Figure 6. The deviation of the data from this line requires a deviation from homogeneity. Another test for homogeneity is the V/V_{max} test (Schmidt, Higdon & Heuter 1988). The value of V/V_{max} measured by BATSE is 0.33 ± 0.01, versus 0.5 expected for homogeneity. The intensity distribution of bursts more intense than 20 times the BATSE threshold is consistent with a homogeneous distribution of sources, althought the statistical significance using BATSE data alone is low. Observations by Pioneer Venus Orbiter, however, provide more extensive coverage of the stronger bursts and agree well with BATSE observations in the region of overlap (Fenimore et al. 1993).

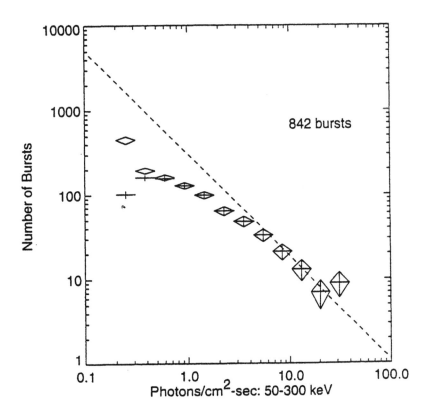

Fig. 6. The intensity distribution for 842 bursts above the BATSE 1024 ms trigger threshold. The dashed line represented the -3/2 power law expected for a homogeneous distribution of sources.

The combination of the isotropy (from the angular distribution) and inhomogeneity (from the intensity distribution) implies a density of sources decreasing with distance, with the Earth close to the center of the distribution. It should be noted that the −3/2 power law intensity distribution of the brighter bursts does not require a constant density of sources nearby, but this is the most straightforward interpretation. The geometrical arguments thus led us to a spatial distribution that is unlike any known Galactic component. In particular, disk geometries are no longer tenable (Briggs *et al.* 1995). If bursts are sampled to distances less than the disk thickness, then the angular distribution will be isotropic and the intensity distribution will exhibit a slope of −3/2. If bursts are sampled to distances larger than the disk thickness, then the angular distribution will anisotropic (having an appreciable quadrupole moment) and the faint end of the intensity distribution will fall below a −3/2 power law extrapolation. The observed combination of isotropy and inhomogeneity does not occur in any disk whose radius is significantly larger than its thickness.

Current thinking has focused primarily on two possibilities for the distance scale: either the burst sources reside in an extended galactic halo, or they lie at cosmological distances. In either case, consideration of time scales and energy requirements lead to very large optical depths in the emitting region, resulting in relativistic thermalized fireballs with high Lorentz factors (Piran & Shemi 1993).

COSMOLOGICAL MODELS

The BATSE observations have created renewed interest in cosmological models of GRBs, first proposed by Prilutski & Usov (1975). The isotropic angular distribution follows naturally, and the apparant inhomogeneity can be ascribed to redshift effects. Several authors have fit cosmological source distributions to the observed intensity distribution (e.g. Mao & Paczynski 1992; Wickramasinghe *et al.* 1993; Fenimore *et al.* 1993; Rutledge, Hui, & Lewin 1995). Good fits are obtained using source luminosities of $\sim 10^{51}$ ergs/s and redshifts of the most distant sources of $\sim$1-2.

Several possibilites for cosmological sources have been proposed, including merging neutron stars (Eichler *et al.* 1989), neutron star-black hole mergers (Mochkovitch *et al.* 1993), an accretion disk around a black hole formed after the collapse of a Wolf-Rayet star (Woosley 1993), and others. A recurring difficulty in most models is how to get a significant fraction of the available energy into gamma-rays with a non-thermal spectrum. Mesząros & Rees (1993) suggested that the kinetic energy of a fireball is converted to gamma radiation via shocks with interstellar matter. Thompson (1994) and Smolski & Usov (1996) employ strong magnetic fields. An interesting recent suggestion by Shaviv & Dar (1996) postulates that the fireball contains heavy ions that absorb and re-emit starlight in dense clusters. Monte-Carlo simulations of this process produce time histories that look remarkably like real GRBs.

EXTENDED GALACTIC HALO MODELS

It is still possible that gamma-ray bursts arise from sources in an extended galactic halo, although the severe constraints on such models are discouraging (e.g. Hakkila *et al.* 1994). The immediate problem for any galactic model is, of course, the isotropic angular distribution. Current upper limits on the dipole moment require that any hypothesized extended galactic halo of GRB sources must extend beyond 100 kpc. At the same time, no excess of bursts has been observed from the giant spiral galaxy M31, 730 kpc distant, effectively limiting the radius of the halo to 350 kpc. Also, Hamilton (1995) reports a null result in a search for GRBs from nearby galaxies using the Imaging Proportional Counter on the *Einstein* Observatory. This instrument is perhaps 10^3 times more sensitive than BATSE for those bursts that are in the IPC field of view and have a soft X-ray component.

The important issue in extended galactic halo models is how to populate the halo with potential sources of GRBs. Eichler and Silk (1992) proposed neutron stars born in the halo. Currently more popular is the idea that high velocity neutron stars (HVNS) might populate an extended halo (Li & Dermer 1992;

Podsiadlowski *et al.* 1995). The HVNS scenario faces several hurdles. First, the bright end of the intensity distribution seems to imply that the closer sources have a homogeneous distribution, but sources emitted from the Galaxy would be expected to show a constantly decreasing density. One clever way around this (Duncan *et al.* 1993) is to have the bursts beamed in a ~20° cone preferentially forwards and backwards along the velocity vector. As sources get farther from the Galaxy, more of them are visible, counteracting the decrease in spatial density. Another possibility is that the bursting phase of the sources is delayed. In either case, the -3/2 slope of the intensity distribution is not a result of homogeneity of the nearby sources, but a coincidental balancing of competing effects. The second hurdle for the HVNS models is that a way must be found to effectively prevent bursting from the neutron stars that remain within the Galactic disk, which would have been easily detected as a second, highly anisotropic component. This issue has yet to be addressed. A third difficulty is that the burst rate is five orders of magnitude larger than the birth rate of HVNSs, requiring many bursts per source, and a total energy reservoir of about 10^{49} ergs, which exceeds the available kinetic energy of rotation (Hartmann and Narayan, 1996).

There has not been a great deal of work yet on the physical mechanisms for generating bursts from high-velocity neutron stars. One possibility is the accretion of planetesimals (Colgate & Leonard 1994), which has the advantage that a delayed turn-on of bursting behavior is a natural consequence of the model.

CONCLUSION

While the overriding issue remains the distance scale to the GRB sources, it is certainly not the case that we will understand bursts once this problem is solved. There are many difficult unanswered questions in both the cosmological and extended galactic halo scenarios. The case for cosmological sources seems overwhelming, but it would be a mistake to abandon alternatives at this stage of our ignorance.

It has become traditional to lament our lack of understanding of gamma-ray bursts. However, the ferment of new ideas and observations gives a glimmer of hope that a solution can be found. On the observational front, the High Energy Transient Explorer (HETE) will soon be launched and Mars 96 will re-establish a complete Interplanetary Network for determining accurate locations. The capabilities of BACODINE sites will continue to improve. BATSE will continue to improve statistics on bursts and observe the rare events whose unusual characteristics may provide important clues.

I believe that further progress will now come primarily from theoretical advances, rather than observational improvements. The role of observations in science has generally been to "constrain" theory; to reduce the number of workable models to one. One could make the case that observations have already reduced the number of workable models to zero. Theoreticians are still wrestling with the gross observational features of GRBs, such as spatial distribution, rates, energetics, and emission mechanisms. A complete theory will have to reproduce the fine points, such as hardness-duration correlations and the complexity of the temporal structure. I suspect that new observations will act not so much to filter existing ideas but to inspire new ones.

REFERENCES

Alexandreas, D. et al., Search for Ultra-High-Energy Radiation from Gamma-Ray Bursts, *ApJ.*, **426**, L1 (1994).

Band, D., Is there Cosmological Time Dilation in Gamma-Ray Bursts?, *ApJ.*, **432**, L23 (1994).

Band, D. *et al.*, BATSE Observations of Gamma-Ray Burst Spectra. I. Spectral Diversity, *ApJ*, **413**, 281 (1993).

Band, D. *et al.*, BATSE Gamma-Ray Burst Line Search II. Bayesian Consistency Methodology, *ApJ.*, **434**, 560 (1994).

Band, D. *et al.*, BATSE Gamma-Ray Burst Line Search III. Line Detectability, *ApJ.*, **447**, 289 (1995).

Barthelmy, S. et al., BACODINE,:The Real-Time BATSE Gamma-Ray Burst Coordinates Distribution Network, *Astrophys. & Space Sci.*, **231**, 235 (1995).

Bhat, P. *et al.*, Spectral Evolution of a Subclass of Gamma-Ray Bursts Observed by BATSE, *ApJ.*, **426**, 604 (1994).

Brainerd, J. J., Mimicking Within Euclidean Space a Cosmological Time Dilation of Gamma-Ray Burst Durations, *ApJ.*, **428**, L1 (1994a).

Brainerd, J. J., Producing the Universal Spectrum of Cosmological Gamma-Ray Bursts with the Klein-Nishina Cross Section, *ApJ.*, **428**, 21 (1994b).

Colgate, S. & Leonard, P., Gamma-Ray Bursts from the Accretion of Solid Bodies onto High-Velocity Neutron Stars, in *AIP Conf. Proc 307, Gamma-Ray Bursts*, ed. G. J. Fishman, J. J. Brainerd, & K. Hurley (New York: AIP), 581 (1994).

Duncan, R., Li, H. & Thompson, C., Halo Beaming Models for Gamma-Ray Bursts, *in AIP Conf. Proc 280, Compton Gamma-Ray Observatory*, ed. M Friedlander, N. Gehrels, D. Macomb (New York: AIP), 1074 (1992).

Eichler, D., Livio, M., Piran, T., & Schramm, D., Nucleosynthesis, neutrino bursts and γ-rays from coalescing neutron stars, *Nature*, **340**,126 (1989).

Eichler, D. & Silk, J. High-Velocity Pulsars in the Galactic Halo, *Science*, **257**, 937 (1992).

Fenimore, E., *et al.*, *ApJ.*, **335**. L71 (1988).

Fenimore, E., *et al.*, The intrinsic luminosity of γ-ray bursts and their host galaxies, *Nature*, **366**, 40 (1993).

Ford, L. *et al.*, BATSE Observations of Gamma-Ray Burst Spectra. II. Peak Energy Evolution in Bright, Long Bursts, *ApJ.*, **439**, 307 (1995).

Hakkila, J. et al., Constraints on Galactic Distributions of Gamma-Ray Burst Sources from BATSE Observations, *ApJ.*, **422**, 695 (1994).

Hartmann, D. & Narayan, R., Are Gamma-Ray Bursts due to Rotation-Powered High-Velocity Pulsars in the Halo?, *ApJ.*, **464**, 226 (1996).

Hudec, R., Dedoch, A., Pravec, P., & Borovicka, J., Optical studies in the fields of gamma ray burst sources II. GRB 910219 and the detection of flashing optical counterpart, *Astron. Astrophys.*, **284**, 839 (1994).

Hurley, K. Gamma-Ray Burst Observations: Past and Future, in *AIP Conf. Proc. 265: Gamma-Ray Bursts*, ed. W. S. Paciesas & G. J. Fishman (NewYork: AIP), 3 (1991).

Hurley, K. *et al.*, Detection of a γ-ray burst of very long duration and very high energy, Nature, 372, 652 (1994).

Hurley, K. *et al.*, Possible Association of a Quiescent X-Ray Source with a Gamma-Ray Burster, *ApJ.*, **464**, 342 (1996).

Klebesadel, R., Strong, I., & Olsen, R., Observations of Gamma-Ray Bursts of Cosmic Origin, *ApJ.*, **182**, 785 (1973).

Kouveliotou, C. *et al.*, Identification of Two Classes of Gamma-Ray Bursts, *ApJ.*, **413**, L101 (1993).

Larsen, S., McLean, I., & Becklin, E., Luminous Galaxies near Gamma-Ray Burst Positions, *ApJ.*, **460**, L95 (1996).

Li, H. & Dermer, C., Gamma-ray bursts from high-velocity neutron stars, *Nature*, **359**, 514 (1992).

Liang, E. & Kargatis, V., Dependence of the spectral evolution of γ-ray bursts on the photon fluence, *Nature*, **381**, 49 (1996).

Link, B., Epstein, R. & Priedhorsky, W., Prevalent Properties of Gamma-Ray Burst Variability, *ApJ.*, **408**, L81 (1993).

Mallozzi, R. *et al.*, The νF_ν Peak Energy Distributions of Gamma-Ray Bursts Observed by BATSE, *ApJ*, **454**, 597 (1996).

Mao, S. & Paczynski, B., On the Cosmological Origin of Gamma-Ray Bursts, *ApJ.*, **388**, L45 (1992).

McNamara, B. *et al.*, Ground-Based Burst Follow-up Efforts: Results of the First Two Years of the BATSE/COMPTEL/NMSU Rapid Response Network, *ApJS.*, **103**, 173 (1996).

Meegan, C. *et al.*, Spatial distribution of γ-ray bursts observed by BATSE, *Nature*, **355**, 143 (1992).

Meszaros, P. & Rees, M., Relativistic Fireballs and Their Impact on External Matter: Models for Cosmological Gamma-Ray Bursts, *ApJ.*, **405**, 278 (1993).

Mochkovitch, R., Hernanz, M., Isern, J., & Martin, X., Gamma-ray bursts as collimated jets from neutron star/black hole mergers, *Nature*, **361**, 236 (1993).

Murikami, T. *et al.*, Evidence for cyclotron absorption spectral features in gamma-ray bursts seen with Ginga, *Nature*, **335**, 234 (1988).

Murikami, T., *et al.*, ASCA Observation of a Possible X-Ray Counterpart of the 1992 May 1 Gamma-Ray Burst , *P.A.S.J.* **48**, L9 (1996).

Narayan, R., Paczynski, B., & Piran, T., Gamma-Ray Bursts as the Death Throes of Massive Binary Stars, *ApJ.*, **395**, L85 (1992).

Nemiroff, R., *et al.*, Gamma-Ray Bursts are TIme-Asymmetric, *ApJ.*, **423**, 432 (1994).

Norris, J. *et al.*, Detection of Signature Consistent with Cosmological Time Dilation in Gamma-Ray Bursts, *ApJ.*, **424**, 540 (1994).

Piran, T. & Shemi, A., Fireballs in the Galactic Halo and Gamma-Ray Bursts, *ApJ.*, **403**, L67 (1993).

Preece, R. *et al.*, BATSE Observations of Gamma-Ray Burst Spectra. III Low-Energy Behavior of Time-Averaged Spectra, *ApJ.*, **473**, in press (1996).

Prilutski, O. & Usov, V., On the Nature of γ-Ray Bursts, *Astrophys. Space Sci.* **34**, 395 (1975).

Rutledge, R., Hui, L., & Lewin, W., Standard cosmology and the BATSE number versus peak flux distribution, *M.N.R.A.S.*, **276**, 753 (1995).

Schmidt, M, Higdon, J. C., & Heuter, G., Application of the V/Vmax Test to Gamma-Ray Bursts, *ApJ.*, **329**, L85 (1988).

Schneid, E., et al., EGRET Measurements of Energetic Gamma Rays from the Gamma-Ray Bursts of 1992 June 22 and 1994 March 1, *ApJ.*, **453**, 95 (1995).

Shaefer, B. *et al.*, BATSE Spectroscopy Catalog of Bright Gamma-Ray Bursts, *ApJ.*, **92**, 285 (1994).

Shaviv, N. & Dar, A., On the Origin of Gamma-Ray Bursts, preprint (1996).

Smolsky, M. & Usov, V., Relativistic Beam-Magnetic Barrier Collision and Nonthermal Radiation of Cosmological γ-Ray Bursters, *ApJ.*, **461**, 858 (1996).

Tavani, M., A Shock Emission Model for Gamma-Ray Bursts, *Phys. Rev. Let.*, **76**, 3478 (1996).

Thompson, C., A Model of Gamma-Ray Bursts, *M.N.R.A.S.*, **270**, 480 (1994).

Usov, V., Millisecond pulsars with extremely strong magnetic fields as a cosmological source of γ-ray bursts, *Nature*, **357**, 472 (1992).

Vrba, F., Hartmann, D., & Jennings, M., Deep Optical Counterpart Searches of Gamma-Ray Burst Localizations, *ApJ.*, **446**, 115 (1995).

Woosley, S., Gamma-Ray Bursts from Stellar Mass Accretion Disks around Black Holes, *ApJ.*, **405**, 273 (1993).

Adv. Space Res. Vol. 22, No. 7, pp. 1077–1082, 1998
© 1998 COSPAR. Published by Elsevier Science Ltd. All rights reserved
Printed in Great Britain
0273-1177/98 $19.00 + 0.00

Pergamon

PII: S0273-1177(98)00198-7

STATISTICS OF TIME HISTORIES AND SPECTRAL VARIABILITY OF GRBS

I.G.Mitrofanov

Institute for Space Research, Moscow 117810, Russia

ABSTRACT

We introduce new parameters for characterizing gamma-ray burst time profiles and energy spectra. Bursts with short and long duration T_{90} are compared, and significant differences are found in the number of pulses and equivalent pulse widths. The trend of the average spectral parameters was resolved at the (α, E_p) plane. The comparison between bright and dim bursts is done, which manifests the difference in the average emissivity curves and in the average spectral hardness.

©1998 COSPAR. Published by Elsevier Science Ltd.

1. INTRODUCTION

The seven years of BATSE measurements have provided an exciting gallery of cosmic gamma-ray bursts (GRBs) (Fishman and Meegan, 1995). So far about 2000 events have been detected with BATSE, measured at similar and rather stable conditions and recorded in the standard data types (Meegan *et al.* 1996). Each GRB is known to be so different from the others that for any two individual events one hardly could conclude whether or not they have the same origin. On the other hand, the large number of BATSE bursts makes it possible to perform statistical studies of their time profiles and energy spectra. One might resolve the most generic signatures of burst flux/spectra evolution in this way and either show the unity of their nature or resolve physically distinguishable groups among them.

Generally, there are three main generic signatures which could be attributed to large samples of cosmic gamma-ray bursts. The first one represents general properties of time histories of bursts in a broad energy range (Section 2), such as the average number of separate pulses, the average pulse duration, the equivalent pulse width and the average interpulse duration. The second signature represents the average spectral properties of bursts and how they change along the burst time history (Section 3). Finally, the third signature corresponds to the most generic differences between intensity groups of bursts, in particular, between bright and dim events. Any difference would be very important for the understanding of the nature of the burst phenomenon. All appropriate signatures have to be used for this comparison (Section 4).

2. GENERAL SIGNATURES OF TIME HISTORIES OF GRBS

The well-known signature of time histories of GRBs is the bimodality of T_{50} and T_{90} distributions (Kouveliotou *et al.* 1993). There are two well resolved peaks in these histograms, separated at ~2 seconds. Generally speaking, the bimodality of the $T_{50/90}$ distributions probably points to two different morphological groups of bursts. On the other hand, it could also result from interference of two different time scales of a single group of bursts with different time profiles. The peak at short T_{90} could comprise of single pulse events, for which $T_{50/90}$ is related with the pulse width. The peak at long T_{90} could be associated with multi-pulse bursts, for which $T_{50/90}$ is determined by interpulse intervals mainly. The choice between these two alternatives is the key question because it deals with the morphological unity of GRBs.

The origin of bimodality may be studied by the statistical method of *Normalized Flux Averaging*, or NFA, when all bursts are normalized to have the same amplitude 1 at the main peaks (Mitrofanov *et al.* 1996a, 1998a). For each burst the NFA provides the number of pulses N_p, pulse duration t_{PD}, equivalent pulse width t_{EPW} and valley duration t_{VD}.

A group of 259 bright BATSE bursts from the 3B Catalog has been analyzed by the NFA method (Mitrofanov *et al.* 1998a) using the data with 64 ms time resolution. The values of t_{EPW} were found for groups short/T_{90} and long/T_{90}. Two well-established peaks were found separated by ~0.2 sec, which correspond to these two groups (Figure 1). It was found that bursts of the long/T_{90} group have rather flat distribution over the number of pulses N_p, while bursts of short/T_{90} group have mainly 1 and much rarely 2 pulses only.

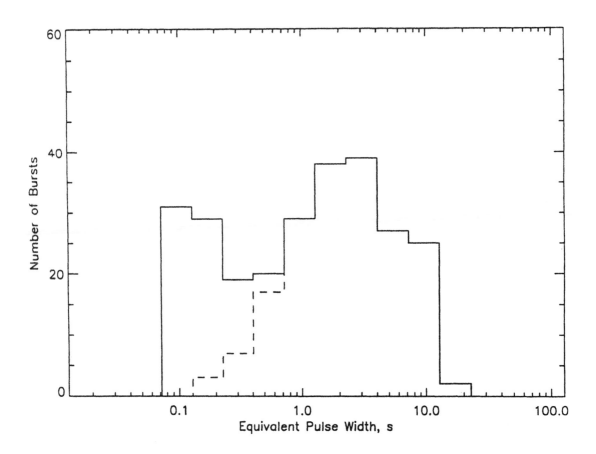

Fig. 1. Bimodal distribution of 259 bright BATSE GRBs over the equivalent pulse width, the contribution of long/T_{90} group is shown by dashed line (Mitrofanov *et al.* 1998a).

Moreover, to exclude the possibility that the difference between t_{EPW} for short/T_{90} and long/T_{90} results from the contribution of secondary pulses for multi-pulse bursts, the comparison has been done between average widths of main pulses for these two groups. The main pulses of short/T_{90} group were shorter than main pulses of long/T_{90} mode; their mean equivalent widths $<t_{EPW}>$ were 0.15 ± 0.01 s and 1.22 ± 0.10 s, respectively. These main pulses have different average hardness ratios also, 1.48 ± 0.07 for short/T_{90} and 1.15 ± 0.03 for long/T_{90}. Therefore, the average parameters of short/T_{90} and long/T_{90} groups of bursts are significantly different, which point out a morphological difference between them.

The NFA analysis might be used for direct verification of theoretical models proposed for GRBs. In particular, recent cosmological models have postulated that GRBs are emitted by highly relativistic flows at cosmological distances. According to this model, one might predict that successive pulses of a burst could have decreasing magnitudes and decreasing spectral hardness. The NFA-based tests of these predictions has been applied to the sub-group of 40 bright long/T_{90} bursts with a large number of separate pulses $N_p>7$. For 7 first successive pulses no significant changes were found for average equivalent width, spectral hardness and dimensionless amplitudes (Table 1).

Table 1. Average parameters of 7 successive pulses of 40 long/T_{90} bright bursts (Mitrofanov *et al.* 1998a).

Successive number of pulses in bursts	#1	#2	#3	#4	#5	#6	#7
Average equivalent pulse width, sec	0.32±0.02	0.31±0.03	0.31±0.05	0.33±0.09	0.21±0.04	0.26±0.01	0.25±0.04
Average amplitude (in fractions of 1)	0.62±0.03	0.64±0.01	0.63±0.02	0.67±0.08	0.56±0.02	0.63±0.04	0.67±0.03
Average spectral hardness ratio	1.47±0.07	1.40±0.06	1.33±0.06	1.31±0.05	1.32±0.06	1.29±0.06	1.33±0.05

3. GENERIC SIGNATURES OF SPECTRAL EVOLUTION OF GRBS

The NFA method permits also to study generic signatures of evolution of GRB energy spectra (Mitrofanov *et al.* 1996b, 1998b). There are 210 sufficiently bright bursts of the 3B Catalog available for spectral studies. The data type CONT with 2,096 sec resolution was used with spectra measured at 16 energy channels from 25 keV up to 2 MeV. Each of these spectra corresponds to ome value of dimensionless flux from 0.0 up to 1.0, and energy spectra for fluxes above the level .1 are not essentially influenced by background fluctuations. Therefore, for any group of bursts, a set of energy spectra could be selected for each particular interval of dimensionless flux.

Using the Band spectral model (Band *et al.* 1993), the photon spectra deconvolution was done from count spectra, and then average spectral parameters were estimated for each interval of dimensionless flux. The average spectral parameters at 2,096 sec interval at the main pulse of 210 bright GRBs are to $<\alpha>$=-0.65 ± 0.04, $<E_p>$=382 ± 28 keV and $<\beta>$=-2.95 ± 0.09 [Mitrofanov *et al.* 1998b]. To study the general trend of spectral evolution along bursts, some selection has to be done between energy spectra measured at different moments at a burst light curve. The simplest approach corresponds to comparison of spectra for intervals with similar dimensionless fluxes, which were accumulated before and after the main pulse of bursts. The significant difference has been resolved between them (Figure 2). When the dimensionless flux is increasing from [0.1-0.4] (point 1) up to [0.4-1.0] (point 2) and up to [1.0] at the main pulse (point 3), and when it decreases from [1.0] down to [0.4-1.0] (point 4) and down to [0.1-0.4] (point 5), the points in the ($<\alpha>$,$<E_p>$) plane move along the "⊃"-like curve.

Therefore, according to some general tendency of spectral evolution, the average value $<\alpha>$ moves down along the total time history of bursts from algebraically larger values to smaller ones. On the other hand, the average peak energy $<E_p>$ shows a quite symmetrical increase before the main pulse and decrease after that, which correlates with changes of dimensionless flux.

This evolution trend of ($<\alpha>$,$<E_p>$) could explain the well-known contradictions concerning the hardness/flux evolution of particular GRBs. Using the simple thermal bremsstrahlung model, Mazets *et al.* (1983) found the positive correlation between exponential cut-off energy kT and photon flux. It might be interpreted now, as resulted from the general correlation between E_p and dimensionless fluxes. On the other hand, the maximal spectral hardness was found at rising fronts of many GRBs (Norris *et al.* 1986), which than went down through the back slops. This peculiarity could be associated with the decrease of $<\alpha>$ along a time history, provided the hardness ratio was mainly influenced by the spectral power index α.

4. MAIN DIFFERENCES BETWEEN GROUPS OF BRIGHT AND DIM GRBS

When a large number of bursts are aligned at moments of main pulses of normalized time histories, the *Average Curve of Emissivity* (ACE) could be built for corresponding time bins before and after main pulses (Mitrofanov *et al.* 1996c). The ACE curve is a generic signature of GRBs which provides the possibility to compare time histories of different groups of GRBs. In particular, the comparison of average time histories for bright and dim bursts could be done, which has the large significance for testing the cosmological models of GRBs.

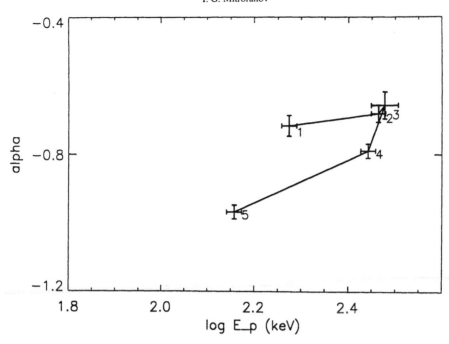

Fig. 2. The spectral evolution trend of bright BATSE GRBs in the ($<\alpha>$,$<E_p>$) plane (Mitrofanov *et al.* 1998b). For each point the values of dimensionless flux is indicated. The rise and decay are shown by arrows.

The most recent comparison between 6 intensity groups with about 150 bursts in each one with $T_{90}>2$ sec was based on the 4B Catalog DISCLA data with 1,024 sec time resolution (Table 2) (Mitrofanov et al 1998c). Two ACE profiles for groups #1 and #5 (Table 2) are presented at Figure 3 for the energy range 50-300 keV.

ACE curves for groups of bright and dim bursts show quite good similarity between them, in particular at the rise parts. Formally speaking, no statistically significant difference is found between rise fronts of ACE for different intensity groups #1-6 at the 1,024 s time scale (Table 2). On the other hand, back slopes of 150 dimmest bursts are stretched in $Y_{BS} \approx 2$ times in respect to 150 the brightest events (Table 2).

The effect of the decrease in spectral hardness with decreasing intensity has been previously seen for different bursts samples (Mitrofanov et al. 1992, Paciesas et al. 1992). It has been presented as the correlation between average peak energy of total spectra and the peak fluxes (Mallozzi et al 1996). It was recently represented as decrease of average peak energy $<E_p>$ at main pulses of dimmer bursts with respect to the similar energy for the brightest bursts (Table 2). The red-shift factor $Y_{RS} \approx$ 3 between 150 dimmest and 150 brightest bursts is larger then the corresponding time-stretching factor Y_{BS} for these groups. However, there are several biases in the comparison of different intensity groups, which have to be studied in more detail and taken into account before final conclusions could be drawn concerning the consistency of time-stretching and energy red-shifting of dimmest bursts.

Table 2. Stretching factors of ACE rise fronts Y_{RF} and back slopes Y_{BS} and red-shift factors Y_{RS} of average peak energy $\langle E_p \rangle$ for groups #2-6 of dimmer bursts in respect with the brightest group #1 (Mitrofanov et al. 1998c).

Intensity groups	Peak flux F_{1024} (ph/cm^{2}s)	Y_{RF}	Y_{BS}	Y_{RS}
#1	>3.8	=1.0	=1.0	=1.0
#2	1.6-3.80	1.12±0.13	1.47±0.16	1.25±0.13
#3	1.0-1.6	1.10±0.12	1.63±0.17	1.83±0.17
#4	0.6-1.0	1.06±0.12	1.41±0.15	2.19±0.20
#5	0.4-0.6	1.21±0.14	1.68±0.18	2.53±0.25
#6	<0.4	0.87±0.11	1.83±0.20	3.02±0.30

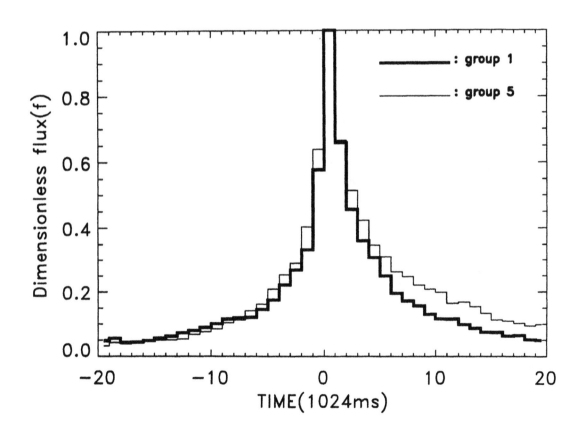

Fig. 3. Comparison of the average curves of emissivity for BATSE bright bursts of group #1 (bold line) and dim bursts of group #5 (thin line) (Mitrofanov 1998c).

5. CONCLUSIONS

1) The physical difference is found between generic signatures of time histories of short/T_{90} and long/T_{90} groups of bursts (Figure 1). Therefore, these groups might be either associated with two different kinds of outbursts, or with two distinct kinds of sources.

2) The general trend of changes of average spectral parameters is found for the group of bright long/T_{90} bursts. It has "⊃"-like shape in the ($<\alpha>$,$<E_p>$) plane (Figure 2): when the average spectral index $<\alpha>$ decreases along the time history of bursts, the average spectral peak energy $<E_p>$ correlates with dimensionless flux variations and has a maximum at the main pulse of bursts.

3) Average curves of emissivity ACE for long bursts with $T_{90}>2.0$ sec are very similar at rise fronts for all intensity groups of bursts (Figure 3). The time-stretching by ~2 times is found at back slope of ACE for 150 dimmest bursts in respect to 150 brightest bursts (Table 3). The energy red-shift by ~3 is found between the average spectral peak energy for these groups. The comprehensive testing of both time-stretching and red-shift effects has to be performed, which will take into account both intrinsic energy-dependent temporal properties of bursts and the noise-produced biases of average parameters. After that, the cosmological models of bursts could be verified using the statistical data.

ACKNOWLEDGMENTS

This paper reviews the joint researches of the BATSE data performed in joint collaboration with Drs. Jerry Fishman and Chip Meegan from NASA/Marshall Space Flight Center, Drs. Michael Briggs, William Paciesas, Geoff Pendelton and Rob Preece from the University at Alabama in Huntsville and with Drs. Anton Chernenko, Alex Pozanenko, Dmitrij Anfimov and Maxim Litvak from the Russian Institute for Space Research. The author thanks all these colleagues for fruitful cooperation.

REFERENCES

Band D. *et al.*, *Astrophys. J.* 413, 281 (1993).

Fishman G. J. and Meegan C.A., *Ann. Rev. Astron. Astrophys.*, 33, 415 (1995).

Kouveliotou, C. *et al.*, *Astrophys. J.*, 413, L101 (1993).

Mallozzi, R.S. *et al.*, *Astrophys. J.*, 454, 597 (1995).

Mazets, E.P. *et al.*, in *Positron-Electron Pairs in Astrophysics*, ed. M.L.Burns, A.K.Harding and R.Ramaty, AIP Conf. #101, pp. 36-53 (1983).

Meegan, C. A. *et al.*, 3rd BATSE Catalog (1996).

Mitrofanov, I.G. *et al.*, in *Gamma-Ray Bursts*, ed. W.S.Paciesas and G.J.Fishman, AIP Conf. #265, pp.195-200 (1992).

Mitrofanov, I.G. *et al.*, in *Gamma-Ray Bursts*, ed. C. Kouveliotou, M.Briggs and G.Fishman, AIP Conf. #384, pp.62-66 (1996a).

Mitrofanov, I.G. *et al.*, in *Gamma-Ray Bursts*, ed. C.Kouveliotou, M.Briggs and G.Fishman, AIP Conf. #384, pp.209-212 (1996b).

Mitrofanov, I.G. *et al.*, *Astrophys. J.*, 459, 570 (1996c).

Mitrofanov, I.G. *et al.*, *Astrophys. J.* in press (1998a).

Mitrofanov, I.G. *et al.*, *Astrophys. J.*, in press (1998b).

Mitrofanov, I.G. *et al.*, *Astrophys. J.*, in press (1998c).

Norris, J.P. *et al.*, *Astrophys. J.* 301, 213 (1986).

Paciesas, W.S. *et al.*, in *Gamma-Ray Bursts*, ed. W.S.Paciesas and G.J.Fishman, AIP Conf. 265, pp.195-200 (1992).

Adv. Space Res. Vol. 22, No. 7, pp. 1083–1092, 1998
© 1998 COSPAR. Published by Elsevier Science Ltd. All rights reserved
Printed in Great Britain
0273-1177/98 $19.00 + 0.00

Pergamon

PII: S0273–1177(98)00199–9

SPECTROSCOPY OF GAMMA-RAY BURSTS: AN OVERVIEW

B. J. Teegarden

NASA/Goddard Space Flight Center, Code 661, Greenbelt, MD 20771, USA

ABSTRACT

More than two decades of accumulated data on the spectra of gamma-ray bursts (GRB's) are now in hand. In general, burst spectra display a wide variety of both spectral and temporal behavior. In essentially all bursts where spectral measurements are possible, the spectrum evolves during the burst. This spectral complexity and variability has presented researchers with a challenging task. Early spectral measurements contained evidence for cyclotron-line-like absorption and emission features. These results were the main anchor for the galactic neutron star paradigm for the origin of gamma-ray bursts. Ginga provided striking confirmation with evidence in three bursts for cyclotron absorption with both first and second harmonics present in two of them. The absorption line energies were consistent with the teragauss magnetic fields expected for neutron stars. However, the BATSE experiment on CGRO has dealt a severe blow to this paradigm. BATSE has found no evidence for the expected galactic anisotropy and also no evidence for cyclotron lines. This paper will: 1) provide an historical overview of GRB spectroscopic measurements 2) a discussion of the BATSE/Ginga controversy and 3) a sampling of the most recent results and analyses of GRB spectral data.   ©1998 COSPAR. Published by Elsevier Science Ltd.

INTRODUCTION

The spectroscopy of gamma-ray bursts began more than 20 years ago with the first spectral observations of Cline *et al.*. (1973) using data from the Imp-6 satellite. These early results showed that most of the energy of the bursts was produced in the few hundred keV range and that the spectra dropped off above a few hundred keV with either exponential or power-law shapes. Since that time a vast number of GRB spectra have been accumulated. The observations have exhibited a very complicated phenomenology, which has made global interpretations very difficult. Understanding the results has been further complicated in several important instances by apparent inconsistencies between the various observations. In this paper I attempt to summarize the current observational situation with regard to GRB spectroscopy with emphasis on the observations as opposed to interpretation. Many interesting models of GRB spectral processes have been proposed, but it remains difficult to recognize which of these are, in fact, truly representative of the real physics of burst emission.

Some general statements can be made with regard to GRB spectra:

1) GRB spectra have been measured over the energy range from a few keV to 20 GeV.

2) GRB spectra generally display featureless smooth continua.

3) The maximum energy output occurs typically between 100 and 500 keV.

4) Most bursts have power law shapes (index ≈ -2) above a few hundred keV.

5) At very low energies (< 10 keV) there is some evidence for departures from a simple spectral form (both deficiencies and excesses).

6) There is a high degree of variability and spectral evolution within bursts and from one to the next.

7) Hard-to-soft spectral evolution is normally present, and there is evidence for hardness-intensity correlations both within bursts and from one to the next.

8) Black-body-like precursors and tails have been reported in the GRB spectra with temperatures of a few keV.

9) Absorption and emission features have been reported in a number of bursts. The former have been interpreted as cyclotron absorption or scattering in teragauss magnetic fields. However, recent experiments (BATSE, TGRS) have found no convincing evidence for such features.

X-RAY OBSERVATIONS

For the purpose of this discussion X-rays will be defined as photons in the range 2-20 keV. This is the most poorly studied of the spectral regions in which observations have been thus far made. It is an observationally difficult region since the signal is weak and the background level from diffuse cosmic emission is very high. The most comprehensive data set is that from the Ginga spacecraft (Strohmayer *et al.*, 1996) which has produced spectral data down to ~ 2 keV for ~ 15 GRB's. Typically, in the X-ray region GRB's display a very hard spectrum (slope = -0.2 to -1.0), however, there is some evidence for turnovers (Strohmayer *et al.*, 1996).

Evidence for soft X-ray precursors and tails has been presented by Murakami *et al.* (1992). In GB 870303 an X-ray tail lasting ~ 50 s with an equivalent temperature of 1.2 keV was observed. Only a few examples of such behavior are now available, due mainly to the lack of appropriate instrumentation. It remains to be determined if this behavior is indeed characteristic of GRB's. If it is, it poses a real challenge for scenarios where the sources are at very large distances.

The possibility of measuring absorption columns in GRB spectra due to intervening material offers an exciting possibility for determining the GRB distance scale. X-ray absorption measurements should be able to fairly easily distinguish between galactic and extra-galactic populations (Owens *et al.*, 1995a). If GRB's are extragalactic the X-ray absorption should be correlated with the known galactic column density. GRB's at low galactic latitude would have easily detectable absorption (column density = $10^{21.7}$ - 10^{23} cm^{-2}). It is less straightforward to distinguish between galactic halo and cosmological populations. The column density through the galactic halo is ~ $10^{20.5}$ cm^{-2}, which would be very difficult, but perhaps not impossible, to measure. If bursts occurred in host galaxies, then there should be measurable absorption depending on the galaxy's angle of inclination. One would expect to observe a range of absorption columns up to ~ 10^{23} cm^{-2} that are uncorrelated with the local galactic column. Absorption within the source, however, could be a complicating factor.

A further exciting possibility in the X-ray region is the measurement of redshifted emission lines or absorption edges. Although no GRB Fe line emission has thus far been observed, the size of the sample is small, and such an observation would surely be a major breakthrough in our understanding of GRB's. Sub-keV absorption edges from abundant elements such as oxygen, silicon, etc. might be produced either by material within the source or within the local galaxy. The determination of a redshift from such an edge would provide a distance determination, but the measurement is certainly very challenging.

20 KEV TO 10 MEV OBSERVATIONS

General Characteristics

The intermediate 20 keV to few MeV energy region is by far the best studied spectroscopically. It is not at all surprising since most of the GRB energy is emitted in this range. Most of the observations have been made with scintillator detectors with typical energy resolutions of 10-100 keV. However, a new instrument, TGRS, using a cooled germanium detector, is now providing burst spectral measurements with a resolution of 2-5 keV. Good spectral measurements exist for many hundreds of GRB's. By far, the most extensive data set is from CGRO/BATSE which has produced high quality spectral data for more than 5 years.

The gross spectral shape can usually be characterized by two different power laws, a relatively flat one at low energies and a steeper one at higher energies with a smooth transition between the two. The following mathematical form from Band et al. (1993) has been quite successful in fitting the spectra most GRB's.

$$N(E) = AE^{\alpha}e^{-E/E_0} \qquad\qquad E \leq (\alpha - \beta)E_0 \qquad\qquad (1a)$$

$$N(E) = A'E^{\beta} \qquad\qquad E > (\alpha - \beta)E_0 \qquad\qquad (1b)$$

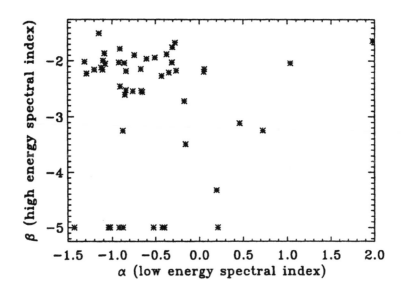

Fig. 1. High energy (β) vs. low energy (α) GRB spectral indices. Cluster of points near β = –5 is not real but is due to inability to perform spectral fit because of lack of counts at higher energies.

Figure 1 (Band et al., 1993) plots the two spectral indices α and β against each other. The data are a sample of 54 GRB's with well-measured spectra from the BATSE Spectroscopy Detectors (SD's). The low energy index α generally falls in the range 0 to -1.5, whereas the values of the high energy index β cluster about a value of -2.0. There is no evidence for a correlation between the two indices. X-ray spectral indices generally fall in the range -0.2 to -1.0 (Strohmayer et al., 1996) which is consistent with the a downward extrapolation of the higher energy data. The group of points with β ≈ -5.0 are from GRB's with insufficient statistics to determine a spectral index. The number of GRB's vs. the transition energy E_0 is plotted in. Figure 2. The most probable value of E_0 lies in the range 100-200 keV but extends up to at least 1 MeV. Some bursts with large values of E_0 have been interpreted as having spectral breaks (Schaefer et al., 1992). Figure 2 shows, however, that these are not anomalous, but fall within the parameter space of the overall burst population.

Spectral Evolution

GRB's display significant spectral evolution within the bursts themselves. One useful way of characterizing the hardness of the burst is the peak energy E_p, defined as the energy at which the power

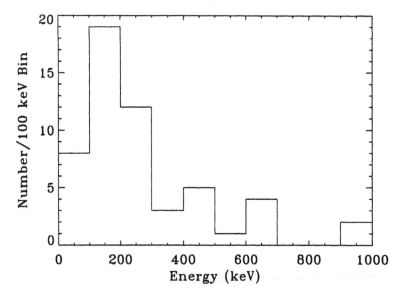

Fig. 2. No. of GRB's vs. E_o.

spectrum νF_ν is a maximum. In the terms of eqn. 1 $E_p = (2+\alpha)E_0$. Ford *et al.* (1995) have used E_p to characterize the spectral evolution within the bursts. Using a sample of 37 bright bursts measured by the BATSE SD's they have tracked the variation of E_p throughout each burst and studied its correlation with burst intensity. They found the following general patterns of behavior (but in all cases there were exceptions):

1) A general correlation of E_p with intensity, especially in the case of major intensity increases. E_p can lead or lag large spikes by a small amount (< 1 s).

2) A general decrease of E_p over the entire burst.

3) A hard-to-soft evolution of E_p within intensity spikes.

4) Later intensity spikes have a smaller value of E_p than earlier ones.

Table 1 (Ford *et al.*, 1995) summarizes the characteristics of this burst sample based on the evolution of E_p.

Table 1. Summary Of Burst Classifications

Number Observed	Number Possible	Property
23	25	E_p - intensity correlation
20	30	E_p softens over whole burst
2	30	E_p hardens over whole burst
18	22	E_p softens within intensity spikes
1	22	E_p hardens within intensity spikes
9	17	Later spikes softer than earlier ones
1	17	Later spikes harder than earlier ones

Peak Energy/Peak Intensity Correlations

In the cosmological scenario for the origin of GRB's it is expected that there will be a correlation between the hardness of a burst and its intensity. Cosmological redshifts will alter the softness/hardness of burst spectra. The weakest bursts in general will be at the greatest distances, have the greatest redshifts, and hence the softest spectra. Using a large sample of 400 bursts from the BATSE Large Area Detectors (LAD's) Mallozzi *et al.* (1995) have explored the correlation of peak intensity with E_p. They have divided the burst sample into five different subsets according to peak intensity. For each sample they have calculated the ensemble average E_p and peak intensity. In Figure 3 the average value of E_p is plotted vs. peak intensity. A rather clearly defined trend is evident with roughly a factor of two increase in peak energy between the dimmest subset and the brightest. It is possible to interpret this result in the context of cosmological models which place GRB's at distances corresponding to Z of order unity. In such scenarios redshifts of order 2 would be expected. This is consistent with similar stretch factors found in the temporal profiles of GRB's (Norris, *et al.*, 1994; Norris, 1995). However, there are several important caveats. The dependence of peak energy on brightness is not a unique indicator of cosmological redshifts. Other scenarios (e. g. those involving energy-dependent beaming) can produce such a correlation. It should also be mentioned that there remains considerable controversy over the interpretations of GRB temporal profiles. For example, using an independent peak-alignment method, Mitrofanov et al. (1995, 1996) find no evidence for time-dilation.

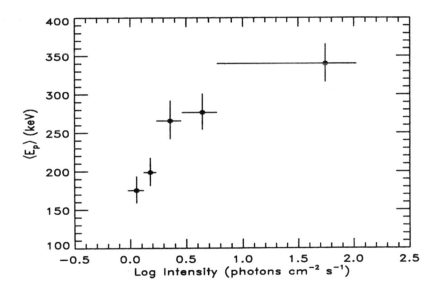

Fig. 3. Peak energy E_p vs. peak flux P. A correlation between the two
    is evident.

Lines in Gamma-Ray Burst Spectra

In the late 1970's the St. Petersburg group under the leadership of E. P. Mazets reported that both emission and absorption lines were common features in GRB spectra (Mazets *et al.*, 1980, 1981). They found, most typically, absorption-like features in the energy range 20-100 keV. They tended to appear early in the burst and were present in ~ 30% of all of those bursts studied. These features were widely believed to be due to cyclotron absorption in strong magnetic fields. The line energy is given by $E(keV) = 11.6\ B_{12}$ (where $B_{12} = $ (mag. field)$/10^{12}$ gauss). This data provided the strongest evidence for a galactic neutron star origin for GRB's. The St. Petersburg group also reported observations of emission lines at higher energies (300-500 keV) which were associated with red-shifted annihilation radiation.

Subsequently, two other groups reported similar features in the spectra of GRB's. The UCSD experiment on HEAO-I (Hueter *et al.,* 1982) found evidence for an absorption feature at 55 keV in a single event GB780325. Perhaps the most striking results came from Ginga (Fenimore *et al.,* 1988; Murakami *et al.,* 1988) . Absorption features at ~ 20 and ~ 40 keV were found in three different events. In two of these the two lines were simultaneously present. These were naturally interpreted as 1st and 2nd harmonics of cyclotron absorption.

BATSE, launched on CGRO in April 1990, has essentially ruled out the nearby neutron star scenario for the origin of GRB's. The BATSE GRB sky map is isotropic and inconsistent with a galactic disk population. Interest in the question of the reality of cyclotron lines, however, remains quite high. This is because in certain galactic halo scenarios, which are consistent with the BATSE GRB sky map, it is possible to produce cyclotron lines. Thus far BATSE, has found no concrete evidence for any line-like features in GRB spectra. With more than five years of accumulated data, the BATSE SD's have produced the most extensive currently available data set of GRB spectra. More than 250 bursts have been visually examined (Palmer, 1996), without finding any concrete examples of line-like features. There have been some tantalizing candidates, where an apparently significant feature was found in one detector, but none in a second detector which, in principle, should have seen it. It is possible that these candidates are simply statistical fluctuations. The number of independent spectra examined is large, and one must exercise extreme caution in deciding whether or not a candidate is real. Even with the large BATSE sample the formal disagreement between BATSE and Ginga is not yet so large as to rule out the possibility of a statistical explanation (Palmer *et al.,* 1994; Band *et al.,* 1996). The probability that the BATSE and Ginga observations are consistent lies in the few to 10% range depending the particular manner in which the question is posed (Band *et al.,* 1996).

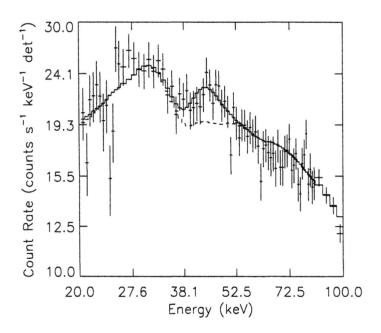

Fig. 4. BATSE SD spectrum of GB940703. There is a possible emission
line at ~ 44 keV.

In addition to the visual search, the BATSE team has begun an automated search for spectral features (Briggs *et al.,* 1996). Spectra are generated for essentially all possible combinations of reasonable time intervals during each burst. Using the Band function to describe the continuum, the algorithm tests for the presence of a line at each independent energy over the entire recorded spectrum. A set of 42 bright bursts has now been analyzed. The number of trials is quite large ~ 8 x 10^5, but these are clearly not all independent. Figure 4 shows the most interesting candidate thus far found. Although, at first glance, it appears to be an absorption feature, it is, in fact, more consistent with an emission feature at 44 keV. The enhancement at ~ 30 keV is due to the particular properties of the detector (absorbing material, NaI light output). A formal significance of 10^{-6} is given, but the actual probability of chance occurrence taking into account the number of trials is much more difficult to calculate and is not given by the authors. Such automated searches represent a much more powerful method for finding spectral features and eventually they may turn up some interesting results. However, it will always be necessary to take extreme care in evaluating these results and to take into account properly the effective number of independent trials.

High Resolution GRB Spectroscopy

Until recently, nearly all GRB spectral measurements were made using scintillator detectors having typical energy resolutions of 10 - 60 keV (FWHM) in the energy range of interest. There is now in operation a new experiment TGRS (Transient Gamma-Ray Spectrometer) that uses cooled germanium technology to achieve significantly superior spectral resolution (2 - 5 keV FWHM) for gamma-ray bursts (Owens et al.., 1995b). It is the first experiment to produce a sizable data set of GRB spectra with an energy resolution significantly superior to that of previous detectors. TGRS was launched on the WIND satellite on 1 Nov. 1994 and has been in continuous operation since 1 Jan. 1995. Its energy resolution at launch varied between 2 and 4 keV over the range 20 - 1500 keV, which is typically a factor of 5 - 20 better than the BATSE SD's. The TGRS detector is smaller than BATSE (~ 1/4 x BATSE area), so that only bright bursts can be studied in great detail. However, if spectral features exist that are narrower than the BATSE resolution it is possible that TGRS could see them and that BATSE could not.

Since 1 Jan 1996 TGRS has detected 59 GRB's, ~ 15 of which were bright enough to carry out sensible spectroscopy . The spectra of these events have been visually examined and no evidence for any spectral features has been found. The energy range covered varies depending on burst brightness. For the brightest bursts the spectral coverage extends down to 20 keV. More typically the lower threshold is 30 - 35 keV, which means that most of the time TGRS would be unable to detect the lower harmonic of the Ginga line pairs (E ≈ 20 keV). However, TGRS should have been able to detect the higher energy features such as reported by the St. Petersburg group.

There are now a significant number of GRB's for which simultaneous TGRS and BATSE measurements exist. One of these is shown in Figure 5. A best-fit model to the BATSE data using the Band function was determined. This model was then folded through the TGRS response matrix. The amplitude was allowed to vary since the time intervals over which the BATSE and TGRS spectra were accumulated did not exactly coincide. The BATSE model is in excellent agreement with the TGRS data to within a single

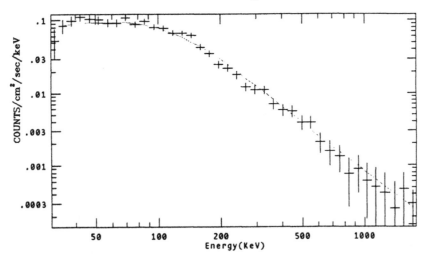

Fig. 5. TGRS data from GB 950425 (crosses). The best fit Band model to the BATSE data for the same burst folded through the TGRS response matrix is also shown (dashed line).

constant factor. The maximum adjustment of the amplitude was only ~ 15%. This provides a valuable confirmation of the BATSE performance using an entirely different detector technology.

TGRS has the ability to detect spectral fine structure in GRB's down to ΔE = few keV. Most bursts are not bright enough to produce enough counts in such small energy bins to allow a search for such fine structure. It is, however, possible to do so for the few brightest bursts each year. The burst GB 950822 is the brightest thus far detected by TGRS. In an initial search a spectrum was created for the main pulse of the burst and fit with the Band function. The residuals were calculated and found to be small, but non-negligible. This is entirely expected since the Band function is not a perfect representation of GRB spectra. Since the interest is in fine structure in the few to ~ 50 keV range, a filter was applied to remove variations on scales larger than 50 keV. The results of this procedure are shown in Figure 6 Two different binnings of the data are displayed (ΔE = 2 keV and 10 keV). No evidence is present for any spectral fine structure, or, in other words, to the best of our ability to measure, GRB spectra are smooth.

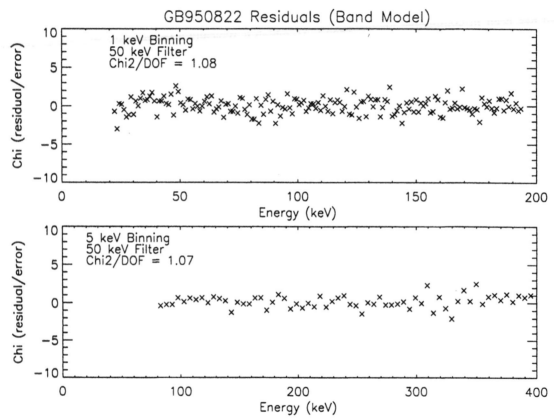

Fig. 6. Residuals of fit of Band model to spectrum of GB960822. A filter has been applied to remove variations on scales larger than 50 keV. Top panel: 2 keV binning. Bottom panel: 10 keV binning.

HIGH ENERGY OBSERVATIONS (E > 10 MeV)

With the launch of CGRO it has become possible to measure GRB spectra to much higher energies. The EGRET experiment has now detected > 10 MeV emission in at least 5 bursts (Hurley *et al.*, 1994; Sommer *et al.*, 1994) . Table 2 (Hurley *et al.*, 1994) summarizes the results. The most dramatic of these was GB 940217 (see Figure 7). The main part of the burst lasted for ~ 3 min., during which several GeV photons were detected. Shortly after the end of the main portion of the burst the source was eclipsed by the earth. When the satellite came back into operation 90 min. later, more high energy photons from the direction of the burst were observed. One of these had the highest energy (18 GeV) of any detected photon from any

Table 2. EGRET High Energy Gamma-Ray Burst Observations

Date	Max. Energy (GeV)	Duration of High Energy Emission
1991 May 3	10	84 s
1991 June 1	0.314	200 s
1993 Jan 31	1.2	100 s
1994 Feb. 17	18	1.5 h
1994 Mar 1	0.16	30 s

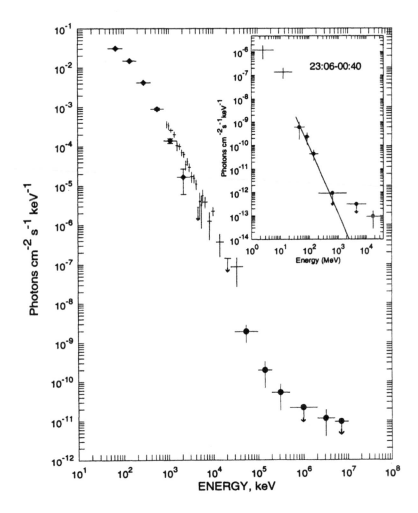

Fig. 7. Spectrum of GB940217. Main panel is accumulated for the 3 min. period of the main burst emission. Crosses with diamonds, BATSE SD's; plain crosses, EGRET TASC; crosses with circles, EGRET spark chamber. Inset shows spectrum of full 90 min. period.

burst. The probability that this is a chance occurrence of a background event is very low. The main panel of Figure 7 shows the spectrum of the burst during the 3 min. main emission episode. A clear excess above the power law tail is seen above 100 MeV. The inset shows the spectrum of the full 90 min. period (including the 18 GeV photon). It also shows a clear excess providing evidence for a new high-energy component that lasts much longer than the main phase of the event.

CONCLUSIONS

After more than two decades of investigation the spectral behavior of gamma-ray bursts remains poorly understand. Further observations with more sensitive instruments may shed further light and lead to greater understanding, but, there is no guarantee. Probably the most fruitful areas for further investigation are at the extremes of the current range, i.e. the X-ray and high-energy gamma-ray regions. The former is challenging, since wide fields-of-view are needed and the diffuse cosmic background is bright in the X-ray range. It may, however, be highly rewarding since there is the possibility of establishing a distance scale through the detection of galactic absorption. A further, but more remote, possibility is detection of X-ray absorption edges, which offers the exciting prospect of an unambiguous redshift measurement.

Only a handful of high-energy (> 1 GeV) GRB observations are now available. The fact that GRB's can produce photons at such extreme energies, and that the emission can be delayed by more and an hour is indeed fascinating. With the limited lifetime of EGRET, the prospects for further observations in the near future are not good, . A future mission such as GLAST is probably the best hope for significant progress in this area.

REFERENCES

Band, D. L. *et al.*, *Ap. J.*, 413, 281 (1993).
Band, D. L. *et al.*, *Ap. J.*, 458, 746 (1996).
Briggs, M. S. *et al.*, *these proceedings* (1996).
Cline, T. L. *et al.*, *Ap. J.*, 185, L1, (1973).
Fenimore, E. E. *et al.*, *Ap. J.*, 335, 71L (1988).
Ford, L. A., *et al.*, *Ap. J.*, 439, 307 (1995).
Hueter, G. J. *et al.*, in *Accreting Neutron Stars*, ed. W. Brinkmann and J. Trumper (MPE Report 177: Garching), 213 (1982).
Hurley, K. H. *et al.*, *Nature*, 372, 652 (1994).
Mallozzi, R. *et al.*, *Ap. J.*, 454, 597 (1995).
Mazets, E. P. *et al.*, *Astrophys. Space Sci.*, 80, 3 (1980).
Mazets, E. P. *et al.*, *Nature*, 290, 378 (1981).
Mitrofanov, I. G. *et al.*, *Adv. in Space Res.*, 5, 131 (1995).
Mitrofanov, I. G. *et al.*, *Ap. J.*, 459, 570 (1996).
Murakami, T. *et al.*, *Nature*, 335, 234 (1988).
Murakami, T. *et al.*, , in *Gamma-Ray Bursts*, ed. C. Ho et al. (Cambridge Univ. Press: Cambridge), 239 (1992).
Norris, J. P., *et al.*, *Ap. J.*, 423, 432 (1994).
Norris, J. P., *Astrophys. Space Sci.*, 231, 95 (1995).
Owens, A., Schaeffer, B. E. and Sembay, S., *Ap. J.*, 447, 279 (1995a).
Owens, A. *et al.*, *Spa. Sci. Rev.*, 71, 273 (1995b).
Palmer, D. M. *et al.*, *Ap. J.*, 433, L77 (1994).
Palmer, D. M., private communication (1996).
Sommer, M. *et al.*, *Ap. J.*, 422, L63 (1994).
Schaefer, B. E., *Ap. J.*, 393, L51 (1992).
Strohmayer, T. E. *et al.*, submitted to *Ap.J.* (1996).

Pergamon

Adv. Space Res. Vol. 22, No. 7, pp. 1093–1096, 1998
© 1998 COSPAR. Published by Elsevier Science Ltd. All rights reserved
Printed in Great Britain
0273-1177/98 $19.00 + 0.00

PII: S0273-1177(98)00200-2

THE SPECTRAL EVOLUTION OF CLASSICAL GAMMA-RAY BURSTS

D. Band and L. Ford

CASS 0424, UC San Diego, 9500 Gilman Dr., La Jolla, CA 92093-0424, USA

ABSTRACT

Gamma ray burst spectral evolution is characterized by two trends: on time scales of ~ 1 s the hardness and intensity rise and fall together, but there is hard-to-soft evolution within and between intensity spikes. The major issues in the study of spectral evolution are the universality of this characterization, and whether the observed broad-band spectra are truly broad-band or whether they consist of many short-lived narrow-band (e.g., black body) flares. We present results from techniques which sacrifice spectral resolution for temporal resolution.

©1998 COSPAR. Published by Elsevier Science Ltd.

INTRODUCTION

Even a cursory inspection of gamma-ray burst time profiles at different energies reveals that these profiles are not identical, and therefore the spectrum must change over the course of the burst. Prominent low energy spikes late in the burst are often absent at high energies, and temporal structure appears broader at low energy. Variations in the spectrum as a burst progresses reflect the evolution of the emitting region; although there is currently no consensus on burst origin, the observed phenomena provide a powerful constraint on future physical models. Two major trends were identified by early missions. By comparing the rates in the *Veneras 13* and *14* KONUS detectors with different gain settings (and therefore energy ranges), Golenetskii *et al.* (1983) found that the spectrum hardens and softens as the intensity increases and decreases. On the other hand, using *SMM* spectra Norris *et al.* (1986) found hard-to-soft spectral evolution within and between intensity spikes. Subsequent work using SIGNE (Kargatis *et al.* 1994) and BATSE data (Bhat *et al.* 1994; Ford *et al.* 1995) demonstrated that both trends are often present: the hardness is correlated with the count rate on time scales of a second, but the hardness peaks earlier than the intensity, and the maximum hardness decreases with successive intensity spikes. These studies usually compare the intensity and hardness time profiles, where the hardness is a measure derived from fitting series of spectra accumulated during the burst. Thus this method of studying spectral evolution requires spectra with sufficient statistics to be fit properly. A common hardness measure is E_p, the energy of the peak of $E^2 N_E \propto \nu F_\nu$; the $E^2 N_E(E)$ curve indicates the energy band in which the source radiates. Note that we use "hard" and "soft" to refer to the average photon energy, not the spectral slope.

This spectral behavior is not surprising: although we do not know the precise physical processes, we expect that the average radiating particle, and thus the emitted photons, is more energetic when energy is first injected into the emitting region. However, that the hardness usually decreases from spike to spike indicates that the emitting region retains a memory of previous energy injection events, if there is only one region, or that the emitting regions communicate, if there are more than one such region. By and large this characterization of spectral evolution has been qualitative,

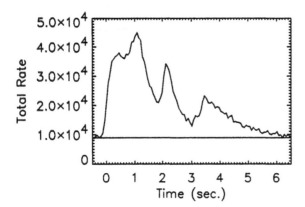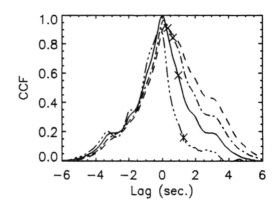

Fig. 1. Time history (left) and CCFs (right) for GRB 910717. The CCFs compare BATSE LAD discriminators: 1 with 3 (dashes), 2 with 3 (dot-dashes), 3 with 3 (ACF—solid curve) and 4 with 3 (3 dots-dash). The shifts between CCFs at small values of the lag indicate hard-to-soft evolution within the intensity spikes, while the relative values of the CCFs at a lag of 3 s reveals hard-to-soft evolution between spikes. Note the difference between positive and negative lags.

and not very constraining. Recently Liang and Kargatis (1996) found that the logarithm of the hardness fell on a straight line for each major intensity spike when plotted against the photon fluence since the beginning of the burst; the slope is the same for each spike. This suggests a model where a fixed number of radiating particles are energized at the beginning of each major intensity spike and then lose their energy through the observed radiation.

UNIVERSALITY

Spectral evolution is usually studied by fitting series of spectra. However, because of telemetry limits and the limited number of counts from which spectra can be formed, the number of fittable spectra across a burst is usually small, and typically only a few intense bursts can be analyzed. Other methods are therefore necessary to determine the spectral evolution in a large number of bursts. In general these methods sacrifice spectral resolution for temporal resolution.

We have been comparing the autocorrelations of, and crosscorrelations between, time histories in different energy bands. These are the Fourier transforms of the cross-spectra studied by Kouveliotou *et al.* (1992). During a burst the BATSE Large Area Detectors (LADs) provide the count rates every 64 ms in 4 channels: 30–50 keV, 50–100 keV, 100–300 keV, and 300–2000 keV. The crosscorrelation of time histories v and u is $CCF(\tau; v, u) = \langle v(t)u(t + \tau)\rangle / \sigma_v \sigma_u$ where $\sigma_v^2 = \langle v^2 \rangle$; the autocorrelation is $ACF(\tau; v) = CCF(\tau; v, v)$, and thus $ACF(\tau = 0) = 1$. Because bursts are transient events, we use background-subtracted time histories instead of zero-mean time histories. In addition we correct $\sigma_{v,u}$ for Poisson noise which would otherwise reduce the ACFs and CCFs (Link *et al.* 1993; Fenimore *et al.* 1995).

If u lags v, then $CCF(\tau; v, u)$ will be shifted to positive lags τ relative to $ACF(\tau; v)$. If v and u represent the time histories of high and low energy channels, respectively, from a burst characterized by hard-to-soft spectral evolution, then the bulk of counts in u will be shifted to later times relative to v; this will be evident from a comparison of $ACF(\tau; v)$ and $CCF(\tau; v, u)$. Specifically, the shift of the peak of the CCF relative to the ACF at τ near 0 reveals the spectral evolution within a spike. The relative heights of secondary peaks in the ACF and CCF (when one intensity spike has been shifted to coincide with a second spike) indicate the relative intensities between spikes in the different energy bands, and therefore is a measure of the spectral evolution between spikes.

Thus far we have analyzed 118 of the more intense bursts observed by BATSE, all but one with durations greater than 1 s. We compared the 4 LAD discriminator rates to the rate in discriminator 3. Of these bursts 101 (86%) showed hard-to-soft evolution within a spike, and only 2 (less than 2%) showed the opposite trend (the rest could not be classified). The shift of the low energy discriminators relative to the fiducial discriminator 3 was typically 1/3 s. The secondary peaks in the ACFs and CCFs showed hard-to-soft evolution between intensity spikes in 82 bursts (69%) and the opposite trend in 7 (6%); in the rest the trend was ambiguous or the structure was too simple. In most bursts the low energy structure is clearly broader. These results are preliminary, but the qualitative conclusion is secure: hard-to-soft spectral evolution within and between intensity spikes is the norm, although counterexamples exist.

LIMITS ON NARROW-BAND EMISSION

The time-integrated burst spectrum has a broad-band shape which can be described from ~ 20 keV to 100s of MeV by the simple 4 parameter "GRB" form: a low energy power law with an exponential cutoff which merges with a high energy power law (e.g., Band et $al.$ 1993 for BATSE spectra, Greiner et $al.$ 1995 for the BATSE-Ulysses-COMPTEL spectrum of GRB 920622). It is remarkable that the simple concave-down "GRB" function also describes burst spectra on shorter time scales. Spectral evolution indicates that the parameters of the "GRB" form must change during a burst, yet when all the spectra are summed, the result can be described by the same functional form. This can be attributed to the broad-band spectral shape, the continuous evolution of the spectral parameters and NaI's moderate spectral resolution.

But is the instantaneous burst spectrum truly broad-band, or does it consist of short-lived narrow-band events which sum rapidly to a broad-band spectrum? For example, is the observed spectrum the result of a large number of short black body flares? The early relativistic fireball models (Paczyński 1986; Goodman 1986) predicted quasi-black body spectra when the expanding fireball became optically thin. A few hundred counts are required to form a spectrum, and thus spectra must be accumulated over a finite time. The shortest accumulation time for the BATSE Spectroscopy Detectors (SDs) is 0.128 s, but usually of order a second is required. The BATSE SDs also provide a list of counts with their energy channel and arrival time (to 128 μs), but the count rate is insufficient to form usable spectra on millisecond time scales. However, if the spectrum averaged over time scales longer than the difference Δt between the arrival times of two photons is narrow-band, then the difference in energy between these two photons will be smaller than if the spectrum is broad-band. Thus a relevant diagnostic is the average energy separation between photons as a function of their temporal separation Δt.

However, the SDs report counts, not photons. Not all photons incident on the detector will be detected, and more significantly, because of the detector's moderate energy resolution, the true photon energy is uncertain. The detector response can be inverted to produce a probability distribution for the true photon energy given the recorded energy channel; the comparison between photon energies becomes the correlation between these probabilities for pairs of counts. We use as the relevant energy separation statistic $A(\Delta t)$ the fraction of pairs whose fractional energy separation (i.e., $2\Delta E/(E_1 + E_2)$) is less than a constant chosen so that the fraction for a narrow-band spectrum will be significantly larger than for a broad-band spectrum.

The result is the energy separation statistic $A(\Delta t)$ as a function of the temporal separation Δt between counts. If the spectrum is truly broad-band down to the shortest time separations, then this statistic will be constant (to within the expected sampling fluctuations). However, if the spectrum is composed of narrow-band events with a duration t_d, such as black body flares, then the statistic $A(\Delta t)$ will be constant for $\Delta t \geq t_d$, and increase with decreasing Δt for $\Delta t \leq t_d$. We use a Bayesian formalism to quantify the significance of changes in $A(\Delta t)$ for small Δt. In the absence of a significant detection of short-lived narrow-band emission we can set limits on the fraction of narrow-band emission as a function of the duration of this emission.

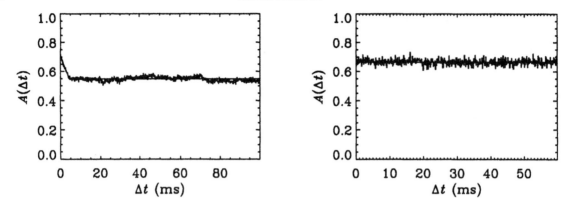

Fig. 2. The energy separation statistic $A(\Delta t)$ as a function of the time separation. On the left is a simulation where the spectrum is formed from black bodies with a distribution of temperatures and durations, while on the right is data from GRB 911210. Note the upturn in $A(\Delta t)$ for the simulation with narrow-band emission, and the absence of an upturn in actual data.

We have applied this analysis to 20 short (duration less than 2 s), intense bursts (Ford 1996). We do not find any indication of narrow-band emission at the few percent level down to time scales of 1 ms. In a few cases $A(\Delta t)$ is not constant, but has a small slope over all Δt which appears to result from spectral evolution of the broad-band spectrum. We therefore conclude that there is no evidence that the observed broad-band spectra consist of narrow-band events with durations greater than 1 ms.

ACKNOWLEDGEMENTS

We thank the members of the BATSE spectroscopy team for their assistance in studying burst spectra. BATSE-related research at UCSD is supported by NASA contract NAS8-36081.

REFERENCES

Band, D., J. Matteson, L. Ford, B. Schaefer, D. Palmer, *et al.*, *Ap. J.*, **413**, 281 (1993).
Bhat, N., G. Fishman, C. Meegan, R. Wilson, C. Kouveliotou, *et al.*, *Ap. J.*, **426**, 604 (1994).
Fenimore, E., J. In't Zand, J. Norris, J. Bonnell and R. Nemiroff, *Ap. J. Lett.*, **448**, L101 (1995).
Ford, L., D. Band, J. Matteson, M. Briggs, G. Pendleton, *et al.*, *Ap. J.*, **439**, 307 (1995).
Ford, L., Ph.D. Thesis, UCSD (1996).
Golenetskii, S., E. Mazets, R. Aptekar, and V. Ilyinskii, *Nature*, **306**, 451 (1983).
Goodman, J., *Ap. J. Lett.*, **308**, L47 (1986).
Greiner, J., M. Sommer, N. Bade, G. Fishman, L. Hanlon, *et al.*, *Astron. Astrophys.*, **302**, 121 (1995).
Kargatis, V., E. Liang, K. Hurley, C. Barat, E. Eveno, and M. Niel, *Ap. J.*, **422**, 260 (1994).
Kouveliotou, C., J. Norris, G. Fishman, C. Meegan, R. Wilson, and W. Paciesas, *Gamma-Ray Bursts, AIP Conference Proceedings 265*, eds. W. Paciesas and G. Fishman, pp. 299-303, AIP, New York, NY (1992).
Liang, E., and V. Kargatis, *Nature*, **381**, 49 (1996).
Link, B., R. Epstein, and W. Priedhorsky, *Ap. J. Lett.*, **408**, L81 (1993).
Norris, J., G. Share, D. Messina, B. Dennis, U. Desai, *et al.*, *Ap. J.*, **301**, 213 (1986).
Paczyński, B., *Ap. J. Lett.*, **308**, L43 (1986).

Pergamon

Adv. Space Res. Vol. 22, No. 7, pp. 1097–1100, 1998
© 1998 COSPAR. Published by Elsevier Science Ltd. All rights reserved
Printed in Great Britain
0273-1177/98 $19.00 + 0.00

PII: S0273-1177(98)00201-4

CHARACTERISTICS OF GAMMA-RAY BURSTS AT MeV ENERGIES MEASURED BY COMPTEL

R. M. Kippen[1,5], J. M. Ryan[1], A. Connors[1], C. Winkler[2], V. Schönfelder[3], L. Kuiper[4], M. McConnell[1], M. Varendorff[3], W. Hermsen[4] and W. Collmar[3]

[1] *Space Science Center, University of New Hampshire, Durham, NH 03824, USA*
[2] *Astrophysics Division, ESA/ESTEC, NL-2200 AG Noordwijk, NL*
[3] *Max-Planck-Institut für Extraterrestrische Physik, D-85740 Garching, Germany*
[4] *SRON-Utrecht, Sorbonnelaan 2, NL-3584 Utrecht, NL*
[5] *Presently at: CSPAR, University of Alabama in Huntsville, Huntsville, AL 35899, USA*

ABSTRACT

We present results from the spectral analysis of 29 gamma-ray bursts measured by the COMPTEL *Telescope* instrument (0.75–30 MeV) during the period from April 1991 through May 1995. The time-averaged spectra of these events are consistent with a simple power law model with spectral index in the range 1.5–3.5, whereas simple thermal models are statistically inconsistent with the burst sample. We find good agreement between burst spectra measured simultaneously by BATSE, COMPTEL and EGRET, which typically show a spectral transition or "break" in the BATSE energy range around a few hundred keV followed by simple power law emission extending to hundreds of MeV. Observations of rapid intensity variations up to a few MeV in the stronger bursts are used to investigate the importance of photon-photon opacity and to place limits on the source distance and bulk Lorentz factor.

©1998 COSPAR. Published by Elsevier Science Ltd.

INTRODUCTION

Most of our knowledge of gamma-ray bursts (GRBs) comes from measurements made in the energy range around a hundred keV. Data from several instruments (most notably BATSE) have shown that burst emission in this energy regime is well described by models which predict a break or transition in the photon flux spectrum connecting two power laws (e.g., Band *et al.*, 1993). The distribution of transition energies for large ensembles of BATSE bursts has provided some insight into the GRB phenomenon (e.g., Mallozzi *et al.*, 1995), but other potentially important spectral characteristics have yet to be investigated in detail. Unfortunately, observational data at lower ($\lesssim 10$ keV) and higher ($\gtrsim 1$ MeV) energies are limited. In this study, a sample of 29 bursts observed by the COMPTEL *Telescope* aboard the *Compton* Gamma Ray Observatory is used to investigate GRB spectral properties in the 0.75–30 MeV energy band. It was discovered by *SMM* instruments that MeV emission is a common and energetically important feature of burst spectra (Matz *et al.*, 1985). The COMPTEL data presented here supplement the earlier *SMM* results and complement contemporaneous *Compton–* BATSE and EGRET observations.

COMPTEL TELESCOPE BURST SPECTRA

In the main *Telescope* instrument mode, COMPTEL measures the direction, energy (FWHM ~10% at 1 MeV) and arrival-time ($125\mu s$ tagging accuracy) of photons originating within a ~1 sr field-of-view. Gamma-ray bursts are identified by searching for excess numbers of measured photons at the times of known BATSE GRBs using a maximum likelihood imaging technique (Kippen *et al.*, 1995 and references therein). During the period from April 1991 through May 1995, 29 significant detections have been identified. Several of these events have been described earlier (see Winkler *et al.*, 1995 and references therein). Here, we use a larger sample of bursts to investigate the global characteristics of MeV emission with enhanced sensitivity (cf. Kippen *et al.*, 1996).

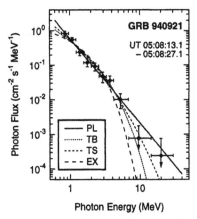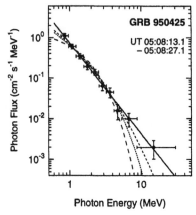

Fig. 1. *Time-averaged, deconvolved photon flux–energy spectra of GRB 940921 and GRB 950525 with best-fit model functions (cf. Table 1).*

A detailed description of the spectral analysis will be given elsewhere. Briefly, individual time-averaged count spectra are obtained by combining photons from the approximate full duration of burst emission. The count spectra, combined with livetime, instrument response and background information, are "deconvolved" into photon spectra using a direct inversion technique. General characteristics of the individual COMPTEL spectra are examined by modeling the observed data. A "forward folding" maximum likelihood technique is used since small numbers of photons make the standard χ^2 fitting method inapplicable. The likelihood statistic is evaluated between an observed count spectrum and the counts predicted by a model (convolved with the instrument response function and corrected for background and livetime). Best-fit model parameters and confidence limits are determined by a search through likelihood values, analogous to a search of χ^2. However, the acceptability of models must be evaluated through Monte Carlo simulation (see Winkler *et al.*, 1995).

Table 1. Spectral Models and Global Fit Results.

Model	Differential Spectrum	Average Spectral Shape[a]	Q[b]
PL (Power Law)	$N_E(E) \propto E^{-\alpha}$	$\langle \alpha \rangle = 2.53$	9.9×10^{-1}
TS (Thermal Synchrotron)	$N_E(E) \propto e^{-(4.5E/E_c)^{1/3}}$	$\langle E_c \rangle = 30.29$ keV	1.3×10^{-2}
TB (Thermal Bremsstrahlung)	$N_E(E) \propto E^{-1} e^{-E/kT}$	$\langle kT \rangle = 2.12$ MeV	$< 10^{-6}$
EX (Photon Exponential)	$N_E(E) \propto e^{-E/E_o}$	$\langle E_o \rangle = 0.93$ MeV	$< 10^{-6}$

[a] Average of all 29 best-fit spectral shape parameters.

[b] Probability that all 29 burst spectra can be explained by random deviations of the model.

TIME-AVERAGED SPECTRAL MODELING

Each of the 29 time-averaged burst spectra were fit with four simple model functions, chosen for their mathematical behavior rather than for their physical meaning (Table 1). As an example, Figure 1 shows the best-fit models and directly inverted photon spectra of two relatively strong bursts. For these particular events, the PL and TS models provide adequate fits to the data, while the TB and EX models can be rejected with >99% confidence. This is typical of large bursts, where the number of high-energy counts (above a few MeV) is sufficient to rule out soft emission models. For the majority of bursts, however, the number of detected photons is smaller—making it difficult to distinguish between different models.

The limitations of studying individual bursts are overcome by examining the *distribution* of goodness-of-fit estimates (or "acceptability") from the full sample of 29 bursts. This technique allows us to investigate the underlying properties of all bursts with greater sensitivity—assuming that all bursts have similar MeV spectral shape characteristics. The acceptability of a single model/burst combination is evaluated by simulating random Poisson deviations of the model and determining how often the best-fit model is within the range predicted by those deviations. The distributions of model acceptability for all 29 bursts are shown in the histograms of Figure 2, where, for convenience, acceptability is

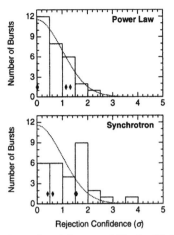
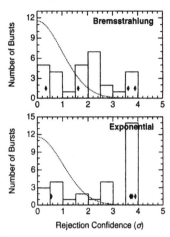
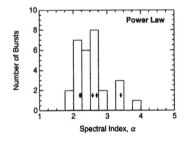
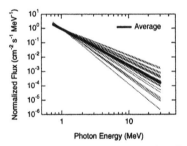

Fig. 2. *Distributions of model acceptability for 29 COMPTEL bursts (solid histograms) and the distributions expected from random statistical deviations (dotted curves). Diamonds indicate values from the five largest bursts.*

expressed as the number of Gaussian standard deviations σ corresponding to the goodness-of-fit probability (i.e., large σ indicates a bad fit). If all 29 burst spectra were drawn from the same parent model, the distribution of acceptability in this representation should follow a Gaussian with zero mean and unit variance (dotted curves in Figure 2).

The histograms of Figure 2 indicate that the power law is the only model tested which yields a distribution that is consistent with random fluctuations only. This indication is quantified in Table 1 by the probability Q (determined by Monte Carlo simulation) that the observed distributions could arise from the expected Gaussians. With this test, it is clear that all models except the power law can be rejected with a high level of confidence. Further evidence of this result is seen by examining the largest bursts (smallest statistical uncertainties), which are well fit by a power law and poorly fit by the TB and EX models (see Figure 2).

Fig. 3. *(Left) Distribution of best-fit power law indices for 29 COMPTEL bursts (diamonds indicate values from the five largest bursts). (Right) Best-fit power law spectra normalized to the average observed intensity.*

Since the observed MeV burst emission is globally well described by the power law model, we can investigate the entire burst population via power law fit parameters. The *average* burst spectrum measured by COMPTEL has a highly constrained intensity $I = 0.91 \pm 0.04$ (photons cm^{-2} s^{-1} MeV^{-1} at 1 MeV) and spectral index $\alpha = 2.53 \pm 0.05$. However, there is a clear distribution about this average. The distribution of spectral hardness (Figure 3), indicates a preference for $2.0 \lesssim \alpha \lesssim 2.6$, with fewer events on either side. This is generally consistent with *SMM* results and with the high-energy end of spectra measured by BATSE (e.g., Band *et al.*, 1993).

RAPID MeV INTENSITY VARIATIONS

The intensity of MeV burst emission measured by COMPTEL is often observed to vary rapidly on short time-scales. While this is not always the case, several of the bursts with sufficient counts exhibit short, high intensity pulses of emission that are observed up to several MeV. The most striking examples are shown in Figure 4. Few counts, electronics deadtime and telemetry gaps severely limit COMPTEL's ability to measure such pulses. However, using the available measurements we are able to put conservative limits on the intensity variation time-scale ($\delta t \lesssim 50$–100 ms),

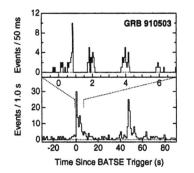 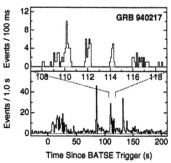

Fig. 4. COMPTEL intensity–time profiles (>1 MeV) of GRB 910503 and GRB 940217 showing rapid intensity variations. Gaps in the profiles are due to the limited capacity of on-board telemetry storage.

the instantaneous peak flux ($F_p \gtrsim$ 10–20 photons cm^{-2} s^{-1} MeV^{-1}) and the maximum energy at which short pulses are observed ($E_{\max} \gtrsim$ 2–5 MeV).

DISCUSSION

The COMPTEL measurements confirm earlier *SMM* findings that MeV-range emission is a common and energetically important feature of GRB spectra. The time-averaged MeV emission is well described by a simple power law model and inconsistent with models predicting substantial curvature or cutoff above a few MeV. When these results are combined with contemporaneous BATSE and EGRET data, a compelling wide-band view of burst energetics emerges: most burst spectra exhibit a turnover in the BATSE energy range around a few hundred keV, but then follow a simple power law out to at least 100 MeV (see e.g., Winkler *et al.*, 1995). The measured distribution of spectral hardness presented earlier provides a characteristic signature of the underlying non-thermal burst emission process.

A power law spectrum extending to high energies, *combined* with observations of rapid, high-intensity flux variations at high energies implies that γ–γ pair production is an inefficient attenuation mechanism (e.g., Baring 1993). If MeV burst emission is isotropic, the COMPTEL measurements alone require that the sources lie well within a distance of 1 kpc in order to avoid γ–γ attenuation. For sources at greater distances, significant anisotropic beaming of the photons is required. Beaming due to bulk relativistic motion of the emitting plasma requires bulk Lorentz factors $\Gamma_{\min} \gtrsim 10^2$ for cosmological distances ($\sim$1 Gpc) and $\Gamma_{\min} \gtrsim 3$ for Galactic Halo distances ($\sim$100 kpc).

ACKNOWLEDGMENTS

COMPTEL is supported through NASA grant NAS5-26645, DARA grant 50 QV 90968 and by the Netherlands Organization for Scientific Research (NWO).

REFERENCES

Band, D., J. Matteson, L. Ford, B. Schaefer, D. Palmer, *et al.*, BATSE Observations of Gamma-Ray Burst Spectra. I. Spectral Diversity, *Astrophys. J.*, **413**, 281 (1993).

Baring, M.G., γ–γ Attenuation and Relativistic Beaming in Gamma-Ray Bursts, *Astrophys. J.*, **418**, 391 (1993).

Kippen, R.M., J. Ryan, A. Connors, M. McConnell, C. Winkler, *et al.*, Analysis of COMPTEL Gamma-Ray Burst Locations and Spectra, *Astrophys. Space Sci.*, **231**, 231 (1995).

Kippen, R.M., J. Ryan, A. Connors, M. McConnell, C. Winkler, *et al.*, COMPTEL Measurements of MeV Gamma-Ray Burst Spectra, in Proc. of the Third Huntsville Gamma-Ray Burst Symposium, in press (1996).

Mallozzi, R.S., W.S. Paciesas, G.N. Pendleton, M.S. Briggs, R.D. Preece, *et al.*, The $\nu \mathcal{F}_\nu$ Peak Energy Distributions of Gamma-Ray Bursts Observed by BATSE, *Astrophys. J.*, **454**, 597 (1995).

Matz, S.M., D.J. Forrest, W.T. Vestrand and E.L. Chupp, High-Energy Emission in Gamma-Ray Bursts, *Astrophys. J. Lett.*, **288**, L37 (1985).

Winkler, C., R.M. Kippen, K. Bennett, W. Collmar, L.O. Hanlon, *et al.*, COMPTEL Observations of the Strong Gamma-Ray Burst GRB 940217, *Astron. Astrophys.*, **302**, 765 (1995).

Pergamon

Adv. Space Res. Vol. 22, No. 7, pp. 1101–1104, 1998
© 1998 COSPAR. Published by Elsevier Science Ltd. All rights reserved
Printed in Great Britain
0273-1177/98 $19.00 + 0.00

PII: S0273-1177(98)00202-6

HIGH ENERGY PHENOMENA ASSOCIATED WITH GAMMA RAY BURST SOURCES

A.R. Rao and M.N. Vahia

Space Physics Group, Tata Institute of Fundamental Research, Homi Bhabha Road, Colaba, Mumbai 400 005, INDIA

ABSTRACT

Gamma-ray burst (GRB) spectrum is known to extend upto GeV energies, indicating the occurrence of transient high energy phenomena near GRB sources. Here we present evidence of statistical association between the locations of ultra high energy cosmic rays and GRBs. We examine our earlier finding that bright GRBs are associated with high energy cosmic rays (Energy $> 3 \times 10^{19}$ eV). Additionally we find a significant positional association between faint GRBs and cosmic rays. If confirmed, these results have very far-reaching implications for GRB source models: GRB sources have to be nearby; GRBs should repeat with a time scale of about 10 years; GRBs are heterogeneous and not mono-luminous. We discuss these results in the light of our earlier suggestion that GRBs and cosmic rays originate from Magnetically Active Stellar Systems.

©1998 COSPAR. Published by Elsevier Science Ltd.

INTRODUCTION

The general similarities of the properties of gamma ray bursts (GRBs) and ultra high energy cosmic rays (UHECR) like the typical occurrence rate, homogeneity etc. have prompted several workers (Waxman, 1995; Vietri, 1995) to suggest that these two enigmatic astrophysical events may have the same origin. Though some evidence was presented for a direct one to one association between individual GRB and UHECR sources (Milgrom and Usov, 1995), an attempt to associate a larger population of sources did not yield any positive results (Stanev *et al.*, 1996). For any reasonable source distance the Galactic magnetic field will deflect the UHECRs and smear out any positional or temporal association with GRBs.

In Vahia and Rao (1991) a statistical approach was taken and GRBs and UHECRs were found to be associated at a confidence level of 99.5%. Cosmic rays of energy $> 3 \times 10^{19}$ eV were found to be associated with highly localized GRBs. This result was explained using the hypothesis that GRBs and UHECRs originate from the inter-binary regions of Magnetically Active Stellar Systems (MASS). In this work we extend this correlation using the *CGRO/BATSE* 3B catalog of GRBs.

MAGNETICALLY ACTIVE STELLAR SYSTEMS

Solar-like flaring activity due to magnetic interaction occurs in most of the late type stars and these activities are in the most intense form in the Magnetically Active Stellar Systems (MASS). The stars classified in this category are flare stars, RS CVn binaries and Cataclysmic Variables (Rao and Vahia, 1987, Vahia and Rao, 1988; 1990). Most of the Fast Transient X-ray (FTX) sources detected by the Ariel V satellite were identified as the most intense form of stellar X-ray flares with energies in the range of $10^{36} - 10^{37}$ ergs (Pye and McHardy, 1975; Rao and Vahia, 1987). Though these identifications are of statistical nature, the conclusions are vindicated by the recent direct detection of soft X-ray and UV flares of similar energies in such stars (Ottmann and Schmitt, 1996; Pagano et al., 1996).

Some of the FTX sources are identified with GRBs and there appears to be a causal relationship between MASS and GRBs. The energetics of such systems are discussed in Vahia and Rao (1988) and some GRBs are associated with MASS. In a later work (Vahia and Rao, 1990) the inter-binary regions of MASS were shown to be an attractive site for particle acceleration and a direct association between UHECRs and GRBs was obtained at a confidence level of 99.5%. For this study 76 GRBs with positional error less than 1 square degrees were selected from the pre-GRO observations (Atteia et al., 1987; Golenetskii et al., 1987) and UHECR data was taken from the observations of SUGAR array (Winn et al., 1986), Volcano Ranch (Linsley 1980) and Haverah Park (Reid and Watson 1980). The sky around each GRB was divided into annular bins of equal angle and the number of UHECRs in each bin was counted. Data for all GRBs are co-added and a significant excess in the first bin was detected.

ASSOCIATION OF UHECRS WITH GRBS

Since positional association between UHECRs and GRBs are possible for only nearby sources (due to the deflection caused by the Galactic magnetic field) we have selected the brightest GRBs (fluence $> 5 \times 10^{-5}$ ergs s^{-1}) from the CGRO/BATSE data base which, according to our hypothesis, originate from the nearby Cataclysmic Variables (CVs). We adopt the same procedure for comparing the GRB and UHECR catalogues that is discussed in detail in Vahia and Rao (1990). For 76 UHECRs and 43 GRBs we find an association at a significance level of about 80%.

This significance is much lower compared to the earlier work of pre-GRO data. This can partly be explained by the lower number of GRBs in the data used for the association. Further, the UHECR data and the pre-GRO GRB data is almost contemporaneous and if GRBs are indicative of enhanced magnetic activity, one can expect cosmic ray acceleration during that time. This is not the case for the CGRO/BATSE observations which are about a decade later than the UHECR observations.

Detailed analysis of the comparison of GRBs with MASS showed that while the most energetic of the GRBs would occur on CVs, the flare stars would contributed significantly to the lower fluency end of GRBs (Rao and Vahia, 1994). On the other hand, the GRBs on flare stars would be of much more local origin (distance of the order of a few parsecs). Hence while looking for GRBs we specifically look for weak GRBs in the catalogues for comparison with the cosmic ray catalogue. The results of the comparison are shown in Figure 1. The observed integrated number of UHECRs of energy greater than 10^{18} eV (a total of 943) in the vicinity of the 353 weak GRBs (fluence $< 2 \times 10^{-7}$ erg) are expressed per unit solid angle and plotted against the angle. A significant peak is seen in the first bin (corresponding to an angular deviation less than 5°). We estimate the significant level to be 99.8%. For the intermediate fluence GRBs (between 10^{-6} ergs/s and 5×10^{-5} ergs/s) we do not find any significant association with UHECRs. According to our hypothesis these intermediate energy GRBs originate from very distant CVs and RS CVn binaries and hence significant association with UHECRs is not expected.

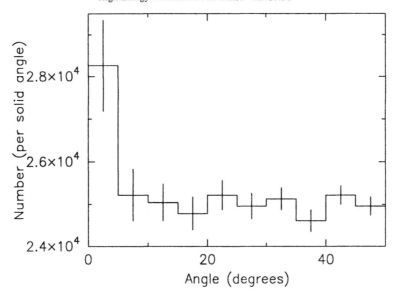

Fig. 1. The number of UHECRs per unit solid angle around
faint GRBs plotted as a function of angle.

DISCUSSION

Both the cosmic ray and GRB positions have a few degrees uncertainty. More than 65% of the sources in both the catalogues have $< 5°$ uncertainty, justifying the choice of $5°$ as the bin size. For the weak GRBs, however, the positional uncertainties are larger (65% of the sources have $< 8°$ uncertainty). Since the association presented here are of statistical nature, this larger uncertainty in weak GRBs does not significantly affect the results. The very fact that there is a significant positional association between UHECRs and GRBs with angular separation $< 5°$ puts a limit on the distance to these sources as < 1.5 kpc for an interstellar magnetic field of 2 μG. Further, the UHECR and GRB observations refer to two different times of observations. In particular, while the GRB observations are for the last five years or so, the UHECR catalogues refer to observations from 1960 onwards. Though the sample size of GRBs for the association is larger than the pre-GRO study (Vahia and Rao 1991), the significance level of association has not increased very much. Nevertheless, the observed association implies that both the GRB and the UHECR sources must be repeating to obtain the observed association. We estimate a repetition rate of about once in 10 years. A similar value for the repetition rate was derived for the GRBs using the GRB and FTX association (Vahia and Rao, 1990). The independent significant association between the faint and bright GRBs shows that GRBs as a class are heterogeneous and consists of several classes with widely separated intrinsic luminosities.

According to our hypothesis, GRBs originate from widely different classes of objects with the same inherent production mechanism: magnetic reconnection on inter binary scale. Using diverse range in the intrinsic luminosity it is possible to explain the observed isotropy (Rao and Vahia, 1994a). There are, however, direct evidences to support this hypothesis. One of the highly localized GRB error box included an active star (Rao and Vahia 1994b). Another highly localized GRB was identified with a Galactic stellar object (Hurley et al. 1996) Further confirmation of our hypothesis would come from 1) improved statistics of brighter GRBs with *CGRO/BATSE* which should improve the significance of the association between GRBs and UHECRs and 3) the large duty cycle *RXTE/ASM* should reveal many more FTX like events coincident with GRBs.

REFERENCES

Atteia J.L. *et al.*, A Second Catalogue of Gamma-Ray Bursts: 1978 − 1980 Localization from the Interplanetary Network, *ApJS*, 64, 305 (1987).

Golenetskii, S.V. *et al.*, A.F. Ioffe Physical-Technical Institute Preprint Number 1026, (1987).

Hurley, K. *et al.*, Possible Association of a Quiescent X-ray Source with a Gamma-ray Burster, *Ap.J.*, 464, 342i (1996).

Linsley, J, Catalogue of Highest Energy Cosmic Rays, ed. WDC-C2 for cosmic rays, p5 (1980).

Milgrom, M. and V. Usov, Possible Association of Ultra-High-Energy Cosmic Ray Events with Strong Gamma-Ray Bursts, *Ap.J.*, 449, L37 (1995).

Ottmann, R. and J.H.M.M. Schmitt, Rosat Observation of a Giant X-ray Flare on Algol: Evidence for Abundance Variations ?, *Astron. Astrophys.*, 307, 813 (1996).

Pagano, I., R. Ventura, M. Rodono, G. Peres, and G. Micela, A Major Optical Flare on the Recently Discovered X-ray Active dMe Star G102-21, *Astron. Astrophys.*, submitted (1996).

Pye, J.P. and I.M. McHardy, Fast Transient X-ray Sources, *MNRAS* 205, 875 (1983).

Rao, A.R. and M.N. Vahia, Fast Transient X-rays From Flare Stars and RS CVn Binaries *Astron. Astrophy.*, 188,109 (1987).

Rao, A.R. and M.N. Vahia, Observational Constraints on the Inter-Binary Stellar Flare Hypothesis for the Gamma-Ray Bursts *Astron. Astrophy.*, 281,L21 (1994a).

Rao, A.R. and M.N. Vahia, On the Possible Source of GRB 930131, *Astron. Astrophy.*, 287,L34 (1994b).

Reid, R.J.O., A.A. Watson, Catalogue of Highest Energy Cosmic Rays, ed. WDC-C2 for cosmic rays, p71 (1980).

Stanev, T., R. Schaefer, and A. Watson, Search for Correlation Between the Highest Energy Cosmic Ray Showers and Gamma Ray Bursts, *Astroparticle Physics*, 5,75 (1996).

Vahia, M.N. and A.R. Rao, Origin of the Gamma Ray Bursts, *Astron. Astrophy.*, 207,55 (1988).

Vahia, M.N. and A.R. Rao, Cosmic Rays from Magnetically Active Stellar Systems *Astron. Astrophy.*, 250,424 (1991).

Vietri, M., *Ap.J.*, 453, 883 (1995).

Waxman, E., *Phys. Rev. Lett.*, 75, 386 (1995).

Winn, M.M. J. Ulrichs, L.S. Peak, C.B. McCusker, L. Horton, Catalogue of Highest Energy Cosmic Rays, ed. WDC-C2 for cosmic rays (1986).

Pergamon

Adv. Space Res. Vol. 22, No. 7, pp. 1105–1110, 1998
© 1998 COSPAR. Published by Elsevier Science Ltd. All rights reserved
Printed in Great Britain
0273-1177/98 $19.00 + 0.00

PII: S0273-1177(98)00203-8

ON DETECTION OF GAMMA-RAY BURSTS FROM ANDROMEDA

T. Bulik[1,2], and D. Q. Lamb[1]

[1]*Department of Astronomy and Astrophysics, University of Chicago, 5640 South Ellis Avenue, Chicago, IL 60637, USA*
[2]*Nicolaus Copernicus Astronomical Center, Bartycka 18, 00-716 Warsaw, Poland*

ABSTRACT

If gamma-ray bursts originate in a corona of high velocity neutron star around the Milky Way, it should also be possible to detect them from a similar corona around Andromeda. Adopting a simple model of such a corona, we evaluate the ability of instruments on existing and on some proposed missions to detect an excess of bursts toward Andromeda. We investigate the optimal properties of an instrument designed to detect such an excess. We find that if the bursts radiate isotropically, an experiment with a sampling distance $d_{max} > 600$ kpc is required in order to detect the expected excess of bursts in the direction of Andromeda in ≈ 1 year of observation. If the bursts are beamed forward and backward along the direction of the neutron star's kick velocity, an experiment with $d_{max} > 800$ kpc is required in order to detect the expected excess in a similar amount of time. Detection of an excess of bursts toward Andromeda would constitute definitive evidence that the bursts are Galactic in origin. Conversely, the observation of no excess toward Andromeda would strongly suggest that the bursts are cosmological in origin, provided that the observation is made by an instrument at least orders of magnitude more sensitive than BATSE, given current constraints on the BATSE sampling distance.   ©1998 COSPAR. Published by Elsevier Science Ltd.

INTRODUCTION

Gamma-ray bursts continue to puzzle astrophysicists nearly a quarter century after their discovery (Klebesadel *et al.*, 1973). Before the launch of the *Compton Observatory*, most scientists thought that gamma-ray bursts came from magnetic neutron stars residing in a thick disk having a scale height of up to ~ 2 kpc in the Milky Way (see, e.g., Higdon & Lingenfelter, 1990; Harding, 1991). It was expected that the Burst and Transient Source Experiment (BATSE) on the *Compton Observatory* would find that the sky distribution of faint bursts is concentrated in the Galactic plane, and would thus confirm that the burst sources lie in a thick Galactic disk.

Instead, the data gathered by BATSE confirmed the existence of a rollover in the cumulative brightness distribution of gamma-ray bursts and showed that the sky distribution of even faint bursts is consistent with isotropy (Meegan *et al.*, 1992; Briggs *et al.*, 1996). The rollover in the cumulative brightness distribution and the isotropic sky distribution imply that we are at, or very near, the center of the spatial distribution of burst sources and that the intrinsic brightness and/or spatial density of the sources decreases with increasing distance from us. Therefore the bursts cannot come from the Galactic disk (Mao & Paczyński, 1992; Hakkila *et al.*, 1994; Smith, 1994).

Recent studies (Lyne & Lorimer, 1994; Frail *et al.*, 1994) have revolutionized our understanding of the birth velocities of radio pulsars. They show that a substantial fraction of neutron stars have velocities that are high enough to escape from the Galaxy. Consequently, Galactic models attribute the bursts primarily to high-velocity neutron stars (HVNSs) in an extended Galactic corona (Li & Dermer, 1992; Duncan *et al.*, 1993; Li *et al.*, 1994; Podsiadlowski *et al.*, 1995; Lamb *et al.*, 1996; Bulik & Lamb, 1995; Bulik & Lamb, 1996; Bulik & Lamb, 1998; Coppi *et al.*, 1996).

CORONA SOURCE SPATIAL DISTRIBUTION AND BURSTING RATES

We assume that the gamma-ray bursts seen by BATSE come from HVNSs in a distant Galactic corona, and that they have velocities high enough that their motions are unaffected by the gravitational potential due to the Milky Way and Andromeda (M31). Then the HVNSs move along straight lines at constant velocities. This is true to a good approximation, since HVNS corona models of the bursts require that the velocities of the neutron stars are large in comparison with the escape velocity from the Milky Way, i.e. $v > 800$ km s^{-1} $> v_{esc} \approx 550$ kms $^{-1}$ (Li & Dermer, 1992; Duncan et al., 1993; Li et al., 1994; Podsiadlowski et al., 1995; Lamb et al., 1996; Bulik & Lamb, 1995; Bulik & Lamb, 1996). Further, we neglect the circular velocity around the center of the Milky Way or Andromeda which the neutron stars have at birth. This is also a good approximation, since the circular velocity $v_{\mathrm{circ}} \approx 220$ km s$^{-1} \ll 800$ km s^{-1}. We take the center of the Andromeda galaxy to lay 700 kpc from the center of the Milky Way , and neglect the relative motion of the two galaxies. The latter is a good approximation, since the relative velocity of the two galaxies $v_{\mathrm{M31-MW}} \approx -123$ km s^{-1} (Peebles, 1993).

We assume that over the past few billion years the birth rate of HVNSs in Andromeda is the same as in the Milky Way. The mass of the Andromeda galaxy is not accurately known, but is estimated to be $1 - 2$ times that of the Milky Way. However, the current neutron star birth rate in Andromeda (as measured by the number of young supernova remnants observed) is less than that of the Milky Way (B. Burke, private communication). Thus, while our assumption that over the past few billion years the birth rate of HVNSs in Andromeda and in the Milky Way are the same is conservative, it may be approximately correct.

We neglect the gradual or delayed turn-on required in isotropic emission models (see, e.g., Lamb et al., 1996). This is a good approximation because the number of bursts from sources within several tens of kpc of the center of the Milky Way and of the center of M31 (i.e., during the turn-on phase) is small compared to the total number of bursts from the HVNS coronae around the Milky Way and M31 seen by any instrument designed to detect an excess of bursts toward M31. We assume that the rate of bursting is constant and that bursting continues indefinitely (i.e., for times long compared to the time for a HVNS to travel from the center of the Milky Way to M31, or vice versa).

The number of bursts per year seen by BATSE is $N_{\mathrm{BATSE}} \approx 800$ yr^{-1} in the 1B catalogue and comparison of Galactic corona models with the BATSE 3B data constrain the current BATSE sampling distance to lie between 130 and 350 kpc (see, e.g., Bulik & Lamb, 1996).

If bursts radiate isotropically, then the rate-density at a point $\vec{r}$ of gamma-ray bursts is

$$n(r,\mu) = \mathcal{N} \left(\frac{1}{r^2} + \frac{1}{r^2 - 2rd_{M31}\mu + d_{M31}^2} \right) \text{ bursts yr}^{-1}\text{kpc}^{-3}, \tag{1}$$

where $\mathcal{N}$ [bursts yr^{-1} sr^{-1} kpc^{-1}] is the burst rate per steradian per unit distance, $r = |\vec{r}|$ is the distance from the Galactic center, $\mu = \cos(\vec{r}, n\vec{M31})$ is the cosine of the angle between the direction to Andromeda and $\vec{r}$, and $d_{M31} \approx 700$ kpc is the distance to M31.

The beaming model assumes that there is a correlation between the direction of a neutron star's kick velocity at birth, and the direction in which the bursts are emitted (Li et al., 1994). Since we neglect the circular velocity of the neutron star around the center of either Andromeda or the Milky Way when it is born, the effects of the gravitational potentials of the two galaxies, and their relative motion, the kick velocity and the velocity of motion of the neutron star coincide. Consequently, bursts can be observed only in a cone of angular radius θ_b along the direction of motion of a neutron star and opposite to the direction of motion of a neutron star. This reduces substantially the number of bursts from M31 that can be seen, since there exist large regions of space from which we are unable to observe bursts.

In the beaming model, we only see bursts from stars born in M31 for which the angle between the direction of motion and direction from us is smaller than the beaming angle θ_b. Let us denote $\mu_b = \cos\theta_b$. Thus the rate density of gamma-ray bursts now becomes

$$n_b(r,\mu) = \mathcal{N}_b \left(\frac{1}{r^2} + \frac{\Theta(|\vec{n}(\vec{n} - \vec{n}_{M31})| - \mu_b)}{r^2 - 2rd_{M31}\mu + d_{M31}^2} \right) \text{ bursts yr}^{-1}\text{kpc}^{-3}, \tag{2}$$

where $\Theta(x)$ is the Heaviside function, $\vec{n}$ is a unit vector, and $\vec{n}_{M31}$ is the unit vector pointing in the direction of M31. Note that the normalization $\mathcal{N}_b$ is different than $\mathcal{N}$.

The bursting rates seen by a detector are obtained by integrating the bursting rates $n(r, \mu)$, or $n_b(r, \mu)$ over a volume V seen by a detector. Discussion of the integration regions and the calculation of the rates is presented by Bulik et al. (1998).

SAMPLING DISTANCE OF A MODEL INSTRUMENT

The sampling distance of an instrument is a key factor in its ability to detect (or rule out) an excess of gamma-ray bursts toward Andromeda. In this section, we determine the sampling distance of a model instrument, taking into account the diffuse X-ray and particle backgrounds and using the spectral and temporal properties typical of gamma-ray bursts.

For a burst of a given luminosity L, the limiting flux, and therefore the sampling distance, of an instrument can be increased by: a. increasing the area A; b. collimating the detector, and reducing the particle background; c. lowering the detection criterion; d. optimizing the triggering timescale $t_{trigger}$; and e. optimizing the energy range of the detector. Increasing the detector area A is a straightforward but not particularly efficient way of increasing the sampling distance of an instrument because d_{max} increases only as $A^{1/4}$. Collimating the detector (i.e., decreasing θ_d) greatly increases the sampling distance of the instrument, but at the cost of reducing the rate at which bursts are detected. Lowering the detection threshold κ is also an effective way of increasing the sampling distance of an instrument, provided that this does not lead to a large increase in the rate of false triggers. Ways of avoiding this include placing the instrument in far Earth orbit or in interplanetary space, environments in which the diffuse X-ray and particle backgrounds are stable; or incorporating a coded aperture mask in the instrument, which allows continuous, simultaneous measurements of the background and any signal (Harrison et al., 1995; Harrison & Thorsett, 1996).

RESULTS

We present our results in Figures 1 and 2 for the case of the isotropic emission and the beaming model with the beaming angle of 20° respectively. Each figure consists of three panels corresponding to the BATSE sampling distances: 100, 200, and 300 kpc. The horizontal axis is the area of a detector, while the vertical axis is the collimation half-angle. For simplicity we assume that all other properties of the detectors in question are similar to these of BATSE LADs. The thin lines are the curves of constant sensitivity, and are labeled by the sampling distance at the center of the detector. The sampling distance increases with increasing area. Collimating the detector decreases the diffuse X-ray background and the sampling distance increases with decreasing collimation angle. However, this relation becomes weaker at small collimation angles because particle background starts to dominate and further collimating the instrument does not decrease the total background significantly.

Given a detectors sampling distance and collimation angle one can integrate the bursting rate-densities of equations (1), and (2) to obtain predicted rate when looking at M31 and in the opposite direction. In the calculation presented here we have assumed that the sampling distance is the same for each direction, that is there are no effects of shadowing of the detecting area by the collimator. A full discussion of these effects is given by Bulik et al., 1998. Given these rates we calculate the time required to detect a 5σ excess in number of bursts towards M31. The thick lines correspond to different timescales: short-dashed line: 0.1 year, long-dashed line: 1 year, and solid line: 10 years. Most of the curves indicate that there exists a clear optimal design of a detector. Realistically, one looks at the curve corresponding to the timescale of one year. It is best to consider a detector with a smallest possible area, however one would like to have the largest possible field of view to detect a relatively large number of bursts. The most difficult case is when the current BATSE sampling distance is smallest, thus we look at the top panels, corresponding to $d_{BATSE} = 100$ kpc. For the isotropic emission model one requires $A\tau \sim 40 \times (A\tau)_{BATSE}$ and a collimation half angle of 8°, while for the beaming model the requirement is stronger $A\tau \sim 140 \times (A\tau)_{BATSE}$ and the collimation half angle of 5°. The main reason for the difference is that the beaming model makes

No beaming

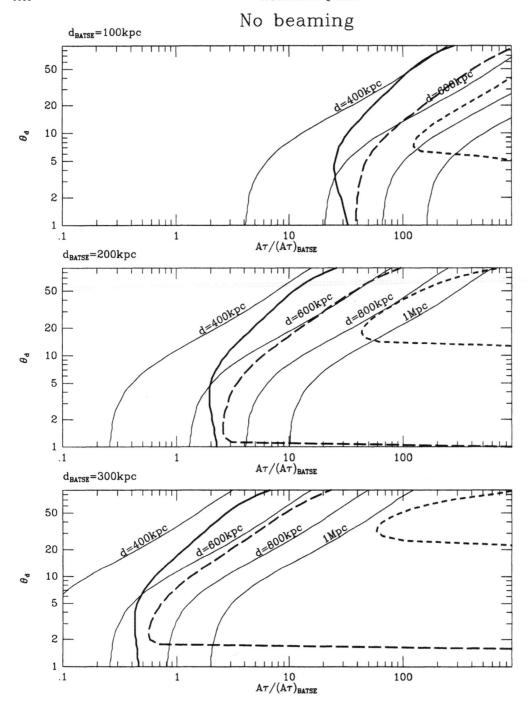

Fig. 1: Contours of constant sampling distance d (or sensitivity) for a focusing instrument, and the constant time T to detect or rule out an excess of bursts toward M31 for the isotropic emission model ($T = 0.1$yr short dashed line, $T = 1$ yr long dashed line, $T = 10$ yr solid line). The three panels correspond to current BATSE sampling distances of 100, 200, and 300 kpc.

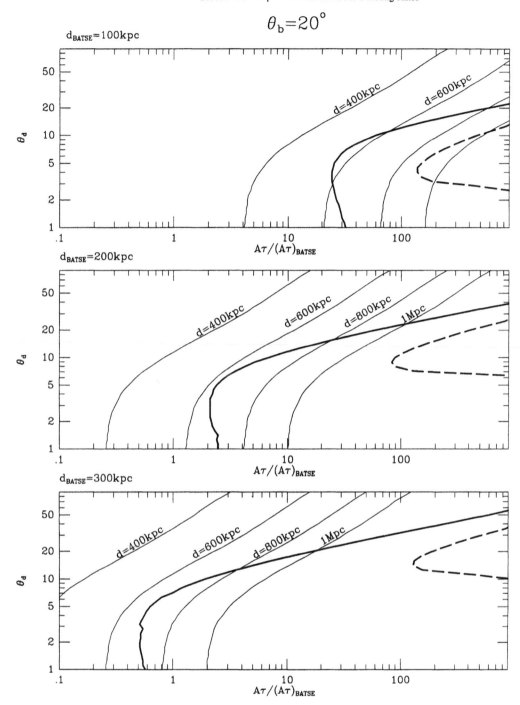

Fig. 2: Contours of constant sampling distance d (or sensitivity) for a focusing instrument, and the constant time T to detect or rule out an excess of bursts toward M31 for the beamed emission model ($T = 1$ yr long dashed line, $T = 10$ yr solid line). The three panels correspond to current BATSE sampling distances of 100, 200, and 300 kpc

detection of gamma-ray bursts from M31 more difficult because of geometrical effects, (Li *et al.*, 1996). In the isotropic emission model we need to reach the distance of ≈ 600 kpc, while in the beaming model we need to see beyond the center of M31, out to the distance of ≈ 900 kpc Thus in the framework of the isotropic emission model one requires an increase on sensitivity by a factor of ≈ 36 over BATSE, while in the framework of the beaming model the increase in sensitivity required is a factor of ≈ 90.

Acknowledgements. This research was supported by NASA grant NAG 5-2868 and KBN grant 2P03D00911

.

REFERENCES

Briggs, M. S., *et. al.* ApJ 459, 40 (1996)

Bulik, T. and Lamb, D. Q., Astrophys. Space. Sci. 231, 373 (1995)

Bulik, T. and Lamb, D. Q., in Proceedings of the 3rd Gamma-Ray Burst Workshop, Univ. of Alabama, Huntsville, M. Briggs, C. Kouveliotou, and G. Fishman (eds.), (New York: AIP) (1996)

Bulik, T. and Lamb, D. Q., in preparation (1998)

Bulik, T., Coppi, P .S., Lamb, D. Q., ApJ in press (1998)

Coppi, P. S., Bulik, T., and Lamb, D. Q., in Proceedings of the 3rd Gamma-Ray Burst Workshop, Univ. of Alabama, Huntsville, M. Briggs, C. Kouveliotou, and G. Fishman (eds.), (New York: AIP) (1996)

Duncan, R., Li, H., and Thompson, C., in Compton Gamma Ray Observatory, M. Friedlander, N. Gehrels, and D. J. Macomb (eds.), p. 1074, (New York: AIP) (1993)

Frail, D. A., Goss, W. M., and Whiteoak, J. B. Z., ApJ 437, 781 (1994)

Hakkila, J., *et al.* ApJ 422, 659 (1994)

Harding, A. K., Science 251, 1033 (1991)

Harrison, F. A., *et al.* Proc. SPIE 2518, 223 (1995)

Harrison, F. A. and Thorsett, S. E., ApJ 460, L99 (1996)

Higdon, J. C. and Lingenfelter, R. E., ARA&A p. 401 (1990)

Klebesadel, R. W., Strong, I. B., and Olson, R. A., Rev. Astron. 182, L85 (1973)

Lamb, D. Q., Bulik, T., and Coppi, P. S., in Proceedings of the Conference on High Velocity neutron Stars and Gamma-Ray Bursts, R. E. Rotschild (ed.), (New York: AIP) (1996)

Li, H. and Dermer, C. D., Nature 359, 514 (1992)

Li, H., Duncan, R., and Thompson, C., in Gamma-ray bursts; Proceedings of the 2nd Workshop, Univ. of Alabama, Huntsville, G. J. Fishman, J. J. Brainerd, and K. Hurley (eds.), p. 600, (New York: AIP) (1994)

Li, H., Fenimore, E. E., and Liang, E. P., ApJ 461, L73 (1996)

Lyne, A. G. and Lorimer, D. R., Nature 369, 127 (1994)

Mao, S. and Paczyński, B., ApJ 389, L13 (1992)

Meegan, C. A., *et al.*, Nature 355, 14 (1992)

Peebles, P., Principles of Modern Cosmology, Princeton: Princeton University Press, p. 18 (1993)

Podsiadlowski, P., Rees, M. J., and Ruderman, M., MNRAS 273, 755 (1995)

Smith, I. A., ApJ 429, L65 (1994)

Adv. Space Res. Vol. 22, No. 7, pp. 1111–1114, 1998
© 1998 COSPAR. Published by Elsevier Science Ltd. All rights reserved
Printed in Great Britain
0273-1177/98 $19.00 + 0.00

PII: S0273-1177(98)00204-X

Pergamon

GAMMA-RAY BURSTS FROM THE FINAL STAGE OF PRIMORDIAL BLACK HOLE EVAPORATIONS

A. A. Belyanin, V. V. Kocharovsky, and Vl. V. Kocharovsky

Institute of Applied Physics, 46 Ulyanov st., 603600 Nizhny Novgorod, Russia

ABSTRACT

It is now accepted that within the Standard Model of particles the evaporating primordial black holes cannot produce the detectable gamma-ray bursts because the expected photon flux from black-hole explosions is too weak, and consists mainly of GeV photons. Contrary to this verdict, we put forward a scenario, in which a large fraction of black-hole power is converted into the photon luminosity in the MeV spectral range, producing a burst of duration $10^{-1} - 10^3$ s. We show that some of the gamma-ray bursts detected by BATSE can be associated with evaporating black holes. ©1998 COSPAR. Published by Elsevier Science Ltd.

INTRODUCTION

The idea of a possible connection between γ-ray bursts (GRBs) and the final stage of evaporating primordial black holes (PBHs) was considered in a number of papers (Page & Hawking 1976, MacGibbon & Webber 1990; Halzen et al. 1991; Cline & Hong 1992; Semikoz 1994). The result of these works can be summarized as follows: (i) within the standard quantum chromodynamics (QCD) the individual PBH explosions *cannot* be detected by modern γ-ray telescopes; (ii) the evaporating PBHs *cannot* be associated with the observed GRBs because the typical photon energies at the final stage of PBH evaporation lay in the range $\gtrsim 100$ MeV (MacGibbon & Webber 1990), in contrast to the observed GRBs in which the main energy release occurs in the $0.1 - 1$ MeV range.

Contrary to this verdict, the main statement of this paper can be put in the following way:
1) There exists at least one mechanism which allows to convert a large fraction of power of PBH explosion into the photon luminosity in the "proper" GRB spectral range $0.1 - 1$ MeV during the time interval $10^{-1} - 10^3$ s. Namely, we show below that when the PBH temperature exceeds ~ 10 GeV, the charged particle outflow from a black hole forms a well-defined plasma. In this case the magnetohydrodynamical (MHD) regime may be realized in the expanding particle wind, and the kinetic energy of particles is converted into the soft γ-rays due to the synchrotron radiation and electromagnetic cascade in the turbulent magnetic field.
2) *If* the significant fraction of the PBH energy at the final stage of its evaporation (when the PBH temperature exceeds ~ 1 TeV) is deposited in the "proper" GRB range, then individual PBH explosions should already have been detected by BATSE instrument. Therefore, at least, some of the GRBs observed by BATSE could originate from evaporating black holes.

MHD WIND SCENARIO OF PBH EVAPORATION

Outline of PBH Evaporation

Black holes of small masses may have been formed in the early Universe by several mechanisms including initial density and gravitational-wave perturbations, and more exotic possibilities, see MacGibbon & Carr (1991), Halzen et al. (1991) for the review. According to the Hawking's results (Hawking 1974), an isolated black hole radiates particles approximately like a black body of temperature $\tilde{T} \simeq 10^{10}/M(\text{g})$, where M is the PBH mass, $\tilde{T}$ the temperature in TeV. The rate of mass loss is (Halzen et al. 1991) $dM/dt \simeq -8 \times 10^6 \tilde{T}^2$ g/s. The lifetime of a hole with temperature T before evaporation is given by

$$\Delta \tau \simeq 5 \times 10^2 / \tilde{T}^3 \text{ s.} \tag{1}$$

When the PBH temperature exceeds the QCD scale, ~ 100 MeV, a PBH begins emitting quarks and gluons that subsequently develop into hadron jets (MacGibbon & Webber 1990). Starting from GeV temperatures, the total particle flux from the evaporating black hole is dominated by products of the jet fragmentation, mainly pions. Finally, pions and other unstable jet products decay into the stable particles.

Validity of Hydrodynamical Approximation

Using, e. g., the results of MacGibbon & Webber (1990), one can easily find that at all stages of jet fragmentation particle-particle collisions cannot provide the hydrodynamical approximation. Nevertheless, the hydrodynamical regime *is* possible, if the particle wind forms a well-defined plasma in the usual sense that the local value of the Debye radius r_D at a given distance r from the black hole is less than r. In this case plasma-wave turbulence may have time to develop during the expansion, and the hydrodynamical regime is supported in the collisionless plasma by the collective wave-particle interactions and/or due to the turbulent magnetic field.

One can show that an outflow emanating from a black hole becomes a plasma at the final stage of the evaporation, both for $\pi^-\pi^+$ wind and farther from the black hole, when pions and muons decay finally into e^-e^+ pairs (Belyanin et al. 1996). Consider, e. g., the $\pi^-\pi^+$ wind. Convolving the Hawking emission spectrum with the jet fragmentation function, we find that most of the number flux of pions is contained in particles with energies $E_\pi/m_\pi c^2 - 1 \sim 2$, where $m_\pi c^2 \sim 140$ MeV is the pion rest energy. At the same time, for sufficiently hot black holes the main *power* is carried by ultrarelativistic pions, and the mean energy of jet-produced pions is therefore $\bar{E}_\pi \sim 20\sqrt{\tilde{T}}$ GeV (MacGibbon & Webber 1990). Calculating the Debye radius $r_{D\pi}$ in a $\pi^-\pi^+$ wind at a given distance r from the black hole (Belyanin et al. 1996), we obtain that the inequality $r_{D\pi} < r$ is satisfied when the PBH temperature exceeds ~ 5 GeV. This estimation is only weakly sensitive to the value of E_π. Similar estimation can be obtained for the e^-e^+ stage of the wind.

The resulting plasma outflow should be strongly non-equilibrium and anisotropic. This may naturally give rise to the development of various plasma instabilities and a high level of wave turbulence, thus establishing the fluid-like expansion and supporting the turbulent magnetic field. The timescale for the magnetic field generation is probably determined by a turbulent diffusion. Estimations of the diffusion coefficient for $\pi^-\pi^+$ and e^-e^+ stages of the wind (Belyanin et al. 1996) show that the diffusion time is much greater than the expansion time r/c, but much smaller than the lifetime of a black hole with temperature $T \sim 10$ GeV corresponding to the beginning of MHD expansion. Therefore, there should be time delay between the beginning of the hydrodynamical expansion regime (when, say, $T \sim 10$ GeV), and the generation of turbulent magnetic field of enough strength to trigger the pair cascade considered below.

Soft Gamma-Ray Luminosity of the Evaporation Wind

An efficiency of the conversion of particle kinetic energy into the radiation in the $0.1 - 1$ MeV range depends crucially on whether or not the energy density of the magnetic field can be close to the equipartition with the pressure of charged pions that constitute roughly 2/3 of the total PBH power. Suppose that the energy density of a magnetic field is a fraction λ of the equipartition value

$$B_{\text{eq}} \simeq \left(8\pi (2/3) |dM/dt| c^2 / 4\pi r^2 c \right)^{1/2} \simeq 7 \times 10^8 \tilde{T}/r \quad \text{G.} \tag{2}$$

For $\lambda = \text{const}$, or more generally, slowly varying function of r, the synchrotron losses, $d\gamma_\pi/dr = -\gamma_\pi/l_\pi(r)$, by a charged pion propagating in a radial direction in a chaotic magnetic field of strength $B = \lambda^{1/2}B_{\text{eq}} \ll m_\pi^2 c^3/e\hbar = 3 \times 10^{18}$ G are characterized by the pion free path $l_\pi \simeq 5 \times 10^{26}/\gamma_\pi B^2$ cm $\propto r^2/\lambda$. This gives the following radial dependence of a pion Lorentz factor:

$$\gamma_\pi(r) \simeq \frac{\gamma_i}{1 + (r_{\rm syn}/r_i)(1 - r_i/r)} \; ; \quad r_{\rm syn} \simeq 10^{-9}\lambda\tilde{T}^2\gamma_i \quad {\rm cm}. \tag{3}$$

Here $\gamma_i \equiv \gamma_\pi(r_i)$ is a pion Lorentz factor at some initial distance r_i from the black hole, $r_0 < r_i < r_d$; r_0 and r_d are the distances where the pions are created and decay into muons: $r_0 \sim 10^{-10}\tilde{T}$ cm, and $r_d \simeq 10^3\gamma_\pi$ cm. It is clear from (3) that when $r_i \ll r_{\rm syn}$, the charged pions should radiate all their energy during much shorter time than the expansion time r_i/c. For pions with the mean energy, $\bar{\gamma}_\pi = \bar{E}_\pi/m_\pi c^2 \sim 10^2\tilde{T}^{1/2}$ we have $\bar{r}_{\rm syn} \simeq 10^{-7}\lambda\tilde{T}^{5/2}$ cm.

The synchrotron photons have energies well beyond the one-photon e^-e^+ pair-production threshold, $E_{\rm syn} \simeq 0.4\hbar\omega_{B\pi}\bar{\gamma}_\pi^2 \simeq 1 \, (\lambda(r_i))^{-1/2}\tilde{T}^{-1/2}$ GeV, and initiate the electromagnetic cascade: photons create e^-e^+ pairs, and electrons (positrons) radiate photons in a strong magnetic field. The cascade develops if the energy E_0 of an initial particle (photon or electron) satisfies the condition $\xi = (B/B_{\rm cr})(E_0/m_e c^2) \gg 1$ (Akhiezer et al. 1995). This requirement together with inequality $r_0 \ll r_i$ gives the following interval of parameters:

$$10^{-3}\tilde{T}^{1/2} \ll \lambda(r_i)\tilde{T}^2 \ll 3 \times 10^5. \tag{4}$$

The mean free path l of a photon of energy $\xi \gg 1$ with respect to one-photon pair production in a chaotically oriented magnetic field is given by (Akhiezer et al. 1995) $l \simeq 7 \times 10^{-8}(B_{\rm cr}/B)\xi^{1/3}$ cm. The cascade develops until the photon energies degrade to the threshold value $E_c = \max\{m_e c^2(B_{\rm cr}/B), 2m_e c^2\}$. The integration of the inverse free path over the radius, from r_i where $B(r_i) \gg B_{\rm cr}$ to $r_c \simeq 4 \times 10^{-5}\,(\lambda(r_c))^{1/2}\,\tilde{T}$ where $B(r_c) \simeq B_{\rm cr}/2$, gives the necessary condition for the *complete* cascade developing to $E_c \sim 1$ MeV: $10^{-3}q^{3/2}\tilde{T}^{1/4} \lesssim \lambda(r_c)\tilde{T}^2$, which should be fulfilled together with (4). For $\lambda \sim 1/30$ the cascade starts at $r_i \sim 3 \times 10^{-9}$ cm, and terminates at $r_c \sim 6 \times 10^{-6}$ cm. Note that due to the growth of the PBH temperature with time, the final energies of the cascade photons decrease with time from GeV values when $\tilde{T} \ll 1$ to ~ 1 MeV in the complete cascade. Therefore, the spectrum should exhibit a hard-to-soft evolution with a timescale defined by (1).

The second possibility to convert $\sim 30\%$ of the PBH energy into soft γ-rays is related to the electromagnetic cascade initiated by π^0 decays. At the radius $r_{\rm ph} = c\tau_0\gamma_\pi \simeq 3 \times 10^{-6}\gamma_\pi$ cm the neutral pions with Lorentz factor γ_π decay into photons with energies of order $E_{\rm ph} \sim E_\pi/2$. The cascade parameter $\xi = (B/B_{\rm cr})(E_{\rm ph}/m_e c^2)$ for pion-produced photons is $\xi(r_{\rm ph}) \simeq 7 \times 10^2\lambda^{1/2}\tilde{T}$, and does not depend on the π^0 Lorentz-factor. Therefore, in the case $\lambda^{1/2}\tilde{T} \gg 2 \times 10^{-3}$, when $\xi(r_{\rm ph}) \gg 1$, the situation is similar to that described above for the $\pi^-\pi^+$ wind: the photons produced by π_0 decays initiate a cascade. The final energies of particles, produced in the cascade, are given by $E_c({\rm MeV}) \sim B_{\rm cr}/2B(r_{\rm ph}) \simeq 15\lambda^{-1/2}\tilde{T}^{-1/2}$. The final e^-e^+ pairs of energies $\gamma_c \sim E_c/m_e c^2 \gg 1$ emit all their kinetic energy to the synchrotron radiation, and the corresponding break in the synchrotron spectrum is at the energy $E_b \simeq 0.4\hbar\omega_{Be}\gamma_c^2 \sim 5\lambda^{-1/2}\tilde{T}^{-1/2}$ MeV. Note again a hard-to-soft spectral evolution with time.

DISCUSSION

Suppose that the detector is triggered at the moment of time $t = 0$ when the PBH temperature $T = T_0$. The total duration of the burst is determined by the lifetime of a hole with temperature T_0, $\Delta\tau(T_0)$, see Eq. (1). The mean luminosity of the burst cannot be higher than $\bar{L} \sim \beta M(T_0)c^2/\Delta\tau(T_0) \sim \bar{\beta}10^{28}\tilde{T}_0^2$ erg/s, where β is a fraction of the PBH rest mass converted into the radiation in the $0.1 - 1$ MeV interval, expected to be of order $1/3$. The source of luminosity $\bar{L}$ will be detected by BATSE from the maximum distance $d_{\rm max} \sim \bar{L}/(4\pi F_{\rm th})^{1/2} \simeq 10^{17}\beta^{1/2}\tilde{T}_0$ cm. Here $F_{\rm th} \simeq 10^{-7}$ erg/cm^{2}s is the threshold peak flux which triggers BATSE instrument (Fishman et al. 1994). The value of $\bar{L}$ rapidly grows with the PBH temperature, while the burst duration decreases. The optimal conditions for triggering BATSE by the faintest (most distant) bursts are realized when $\Delta\tau(T_0)$ is of the order of the smallest accumulation time of the detector, 64 ms. This gives $T_{\rm max} \simeq 20$ TeV and $d_{\rm max} \sim 0.7\beta^{1/2}$ pc. The rate of the PBH explosions in the effective volume scanned by BATSE is $\dot{N} \simeq (dn/dt)(4/3)\pi d_{\rm max}^3$. Here dn/dt is the volume density of individual PBH explosion rate, restricted by observed γ-ray background (MacGibbon & Carr 1991; Halzen et al. 1991): $dn/dt \lesssim 10$ pc^{-3} yr^{-1}, provided PBHs are clustered to the same degree as the other matter in the Galactic halo. Accepting this estimation, we arrive at ~ 10 detectable bursts per year.

To increase the predicted number of bursts detectable by BATSE, we should either suppose that the peak intensity during the final stage of the evaporation can be much higher than the mean value, or assume that a local enchancement of the PBH density is greater than that for other matter in the Galactic halo. The latter would be possible, for example, if PBHs had a heliocentric distribution. This would simultaneously solve the $\log N - \log S$ problem, inevitable in local models (Horack et al. 1994). Note that the total energetics of PBH explosions is consistent with current BATSE statistics of all GRBs. Indeed, the evaporation of a black hole of temperature $\tilde{T}$ TeV at the distance d from the Earth will produce a burst with a MeV fluence $f = \beta M c^2/(4\pi d^2) \simeq 8 \times 10^{-8} \beta \tilde{T}^{-1} (d/1\mathrm{pc})^{-2}$ erg/cm^2. For the sensitivity threshold $f_{\min} \sim 10^{-8}$ erg/cm^2, the maximum distance scanned by BATSE is $d_{\max} = (\beta M c^2/4\pi f_{\min})^{1/2} \sim 3\beta^{1/2}\tilde{T}^{-1/2}$ pc, leading to $\sim 10^3$ detected bursts per year. This is in agreement with the all-sky GRB rate.

In conclusion, note some spectral peculiarities of PBH-originated bursts:

1) Since evaporating black holes are more or less standard candles (at least, with respect to their fluences and mean luminosities $\bar{L}$), the brighter bursts are simply located closer to the Earth than the fainter bursts. The closely located evaporating black holes (bright bursts) should trigger BATSE detector at lower initial PBH temperatures than more distant bursts. According to the expressions for break energies of photons in the cascade, this means that bright bursts should be harder than faint ones. Also, the hard-to-soft evolution of the spectrum follows directly from the above analysis. Both facts are in a qualitative agreement with observations; see Kouveliotou (1994) for review.

2) Pair cascades in a strong magnetic field may produce spectral breaks around 1 MeV (Baring 1990), unless two-photon processes smear out this feature. Curiously, several GRBs exhibiting such spectral breaks have been identified (Kouveliotou 1994).

3) The PBH evaporation must be accompanied by flux of primary particles of energy $\sim 5T$, and neutrinos originated from pion decays. However, the probability of observing these events with modern generation of detectors is very low (Halzen et al. 1991; LoSecco 1994).

ACKNOWLEDGEMENTS

This work has been supported by the Commission of the European Communities – DG III/ESPRIT Projects CTIAC and ACTCS 9282, by Russian Foundation for Fundamental Research, and by International Science Foundation through grant R 93300.

REFERENCES

Akhiezer, A. I., N. P. Merenkov, and A. P. Rekalo, *Nucl. Phys.*, **58**, 491 (1995).
Baring, M. G., *MN RAS*, **244**, 49 (1990).
Belyanin, A.A., V. V. Kocharovsky, and Vl. V. Kocharovsky, *MN RAS*, in press (1996).
Cline, D. V., and W. Hong, ApJ, **401**, L57 (1992).
Fishman, G. J., C. A. Meegan, R. B. Wilson, M. N. Brock, J. M. Horack et. al., ApJS, **92**, 229 (1994).
Halzen, F., E. Zas, J. H. MacGibbon, and T. C. Weekes, Nat, **353**, 807 (1991).
Hawking, S. W., Nat, **248**, 30 (1974).
Horack, J. M., T. M. Koshut, R. S. Mallozzi et al., ApJ, **429**, 319 (1994).
Kouveliotou, C., ApJS, **92**, 637 (1994).
LoSecco, J. M., ApJ, **425**, 217 (1994).
MacGibbon, J. H., and B. J. Carr, ApJ, **371**, 447 (1991).
MacGibbon, J. H., and B. R. Webber, Phys. Rev. D, **44**, 3052 (1990).
Page, D. N., Hawking S. W., ApJ, **206**, 1 (1976).
Semikoz, D. V., ApJ, **436**, 254 (1994).

Pergamon

Adv. Space Res. Vol. 22, No. 7, pp. 1115–1119, 1998
© 1998 COSPAR. Published by Elsevier Science Ltd. All rights reserved
Printed in Great Britain
0273-1177/98 $19.00 + 0.00

PII: S0273–1177(98)00205–1

PAIR PRODUCTION OPACITIES FOR GAMMA-RAY BURST SPECTRA IN THE 10 GEV–1 TEV RANGE

Matthew G. Baring* and Alice K. Harding

NASA Goddard Space Flight Center, Code 661, Greenbelt, MD 20771, U.S.A.

ABSTRACT

Recent calculations of pair production opacities for gamma-ray bursts detected by EGRET have suggested that relativistic bulk motion with large Lorentz factors must exist in these sources, whether of galactic or cosmological origin, to permit the emergence of a high energy continuum. The extant calculations usually assume an infinite power-law source spectrum for simplicity, an approximation that is quite adequate for most bursts detected by EGRET. However, for a given bulk Lorentz factor, photons above the EGRET range can potentially interact with sub-MeV photons in pair production opacity calculations, so it is important to accurately include the spectral curvature in bursts seen by BATSE. In this paper we modify our previous work to include such spectral forms, well-represented by broken power-laws, determining opacity as a function of energy for different source Lorentz factors Γ. The comparative depletion of low energy photons below 1 MeV turns out to be immaterial to considerations of bulk motion in cosmological bursts, but is crucial to estimates of source transparency in the 1 GeV – 1 TeV range for sources located in the galactic halo. We find that broad absorption dips can arise at these energies for suitable expansion parameters. Whipple, MILAGRO and GLAST will play an important role in the observation of the presence or absence of such distinctive structure in GRB spectra.   ©1998 COSPAR. Published by Elsevier Science Ltd.

INTRODUCTION

The relationship between gamma-ray burst observability at energies around or above 1 MeV and pair production ($\gamma\gamma \to e^+e^-$) attenuation was first explored by Schmidt (1978), who concluded at the time that the detection of MeV photons limited bursts to being closer than a few kpc. The basic premise of this proposition was that the source produced a quasi-isotropic radiation field that was optically thin at the maximum observed energy, observing that the optical depth scales linearly with the photon density and that source flux is proportional to this density divided by the square of the distance to the burst. As time passed and instrumental capabilities improved, bursts began to be seen at higher energies, thereby providing more severe constraints to the source distance. With the launch of the Compton Gamma-Ray Observatory (CGRO), the discovery by BATSE that the spatial distribution of gamma-ray bursts (GRBs) is isotropic and non-homogeneous (Meegan et al. 1991) suggested that these sources are either in an extended halo or at cosmological distances. Consequently their intrinsic luminosities (and therefore their optical depths to pair production) would be much higher than previously believed, and completely incompatible with the assumption of isotropic radiation at the source. Hence it was natural to suggest (e.g. Fenimore et al. 1992) that GRB photon angular distributions were highly beamed and produced by a relativistically moving (or expanding) plasma.

Determinations of the bulk Lorentz factor Γ of the medium radiating the GRB emission soon followed, first focusing (e.g. Krolik and Pier 1991, Baring 1993, Baring and Harding 1993) on the simplest case where the angular extent of the source was of the order of $1/\Gamma$. Assuming an infinite power-law GRB burst spectrum,

*Compton Fellow, USRA

they deduced that gamma-ray transparency up to the maximum energy detected by the EGRET instrument aboard CGRO required $\Gamma \gtrsim 5$ for galactic halo sources and $\Gamma \gtrsim 250$ for cosmological bursts. These limits were reinforced by a much more detailed consideration of a wide range of source geometries (Baring and Harding 1995). While the expedient assumption of an infinite power-law source spectrum is adequate for most EGRET-detected bursts (six in all), the spectral curvature seen in most GRBs by BATSE (e.g. Band et al. 1993) is expected to play an important role in reducing estimates for Γ for potential TeV emission from these sources. This paper investigates the effects of introducing spectral curvature into pair production opacity considerations, and we find that these effects are minimal for bursts of cosmological origin. However, for galactic halo sources, we observe that for realistic bulk motion parameters, source opacity may arise only in a portion of the 1 GeV – 1 TeV range, resulting in the appearance of distinctive broad absorption dips. Such features may provide a unique identifier for bursts in halo locales, with current and future ground-based initiatives such as Whipple, MILAGRO and space missions such as GLAST perhaps playing a key role in determining the distance scale for gamma-ray bursts.

γ-γ ATTENUATION AND RELATIVISTIC BEAMING

For photon isotropy, the optical depth to pair production has been calculated by Gould and Schreder (1967). However, $\gamma\gamma \to e^-e^+$ has a threshold energy ε_1 that is strongly dependent on the angle Θ between the photon directions: $\varepsilon_1 > 2m_e^2c^4/[1 - \cos\Theta]\varepsilon_2$ for target photons of energy ε_2. Hence radiation beaming can dramatically reduce $\tau_{\gamma\gamma}$ and suppress γ-ray spectral attenuation turnovers, blueshifting them out of the observed spectral range; GRBs can then be more distant from earth. The simplest picture of beaming has $\Theta \sim 1/\Gamma$ (Krolik and Pier 1991, Baring 1993, Baring and Harding 1993) in "blobs" with line-of sight and transverse dimensions differing by a factor of Γ. The source size can be inferred from time variability arguments, and can be estimated to be $R \sim R_v\,\Gamma$ (this is model-dependent), where $R_v \sim 10^7$ cm is the variability size of the emission region. Furthermore, a source of size R that is a distance d from earth and has a density n_γ of photons, yields a flux f observed at earth of $f = 4\pi n_\gamma c\,R^2/d^2$. Typically $f \gtrsim 1$ cm^{-2} sec^{-1} at 511 keV for BATSE bursts. Then assuming an infinite power-law spectrum $n(\varepsilon) = n_\gamma\,\varepsilon^{-\alpha}$, the optical depth to pair creation assumes the form $\tau_{\gamma\gamma}(\varepsilon) \propto n_\gamma\sigma_{\rm T}R\varepsilon^{\alpha-1}\Gamma^{-2\alpha}$ for $\Gamma \gg 1$. Setting this equal to unity then gives an estimate for the minimum bulk Lorentz factor $\Gamma_{\rm MIN}$ compatible with source transparency out to a maximum observed energy $\varepsilon_{\rm MAX}$ (Baring and Harding 1993, Harding 1994):

$$\Gamma_{\rm MIN}^{1+2\alpha} \gtrsim \frac{(3.83)^\alpha}{\alpha^{5/3}(4/3+\alpha)^{27/11}}\,d_{\rm kpc}^2\left[\frac{\varepsilon_{\max}}{1\text{ MeV}}\right]^{\alpha-1}f_{\rm MeV}\left(\frac{R_v}{10^7\text{cm}}\right)^{-1}. \qquad (1)$$

Here $d_{\rm kpc}$ represents the distance d to the source in kpc, and $f_{\rm MeV}$ is the flux f at 1 MeV. Evaluations of this criterion for all the EGRET bursts and using only EGRET data are presented in Harding (1994) and Baring and Harding (1997a, in preparation), which can be summarized as $\Gamma_{\rm MIN} \gtrsim 5$ for galactic halo sources and $\Gamma_{\rm MIN} \gtrsim 100$ for cosmological bursts. Note that such estimates can be more or less accommodated by a wide range of source geometries (Baring and Harding 1995), from narrow beams to thin spherical shells.

The estimates for $\Gamma_{\rm MIN}$ in Eq. (1) are strongly dependent on $\varepsilon_{\rm MAX}$. With the Whipple instrument producing exciting TeV detections in other astrophysical contexts, and the new water Cerenkov detector MILAGRO coming on line soon (see, for example, Yodh 1996, for a description) it is natural to extend these estimates up to the TeV energy range. Fig. 1a shows $\Gamma_{\rm MIN}$, computed from Eq. (1), as a function of high energy (i.e. EGRET-Whipple range) power-law index α_h for maximum energies $\varepsilon_{\rm MAX}$ up to the TeV range where Whipple and MILAGRO are sensitive. Results are depicted for burst distances that are typical of galactic halo and cosmological scenarios. Due to the larger $\varepsilon_{\rm MAX}$ considered here, the $\Gamma_{\rm MIN}$ are generally larger than those in Harding (1994). The flux f is scaled to typical BATSE sensitivity, so that only sources with spectra flatter than around 2.6 will be simultaneously detectable by MILAGRO (and perhaps slightly steeper for Whipple). Conversely, sources with given Γ are less likely to be observable by Whipple or MILAGRO since attenuation is more probable.

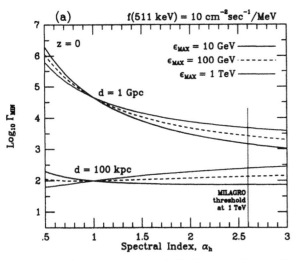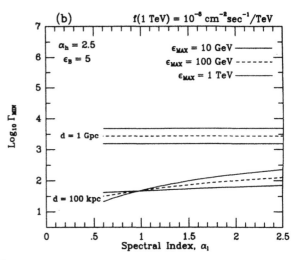

Fig. 1. (a) The minimum bulk Lorentz factor Γ_{MIN} that guarantees source transparency up to energy ε_{MAX} for two different source distances d as labelled. Infinite power-law source spectra are assumed. The source flux f at 511 keV is typical of BATSE burst detections; for this flux, Whipple and MILA-GRO will be sensitive to bursts with $\alpha_h \lesssim 2.6$. The cosmological redshift was taken to be $z = 0$ for simplicity. (b) The effect of breaking the power-law at energy $\varepsilon_B m_e c^2$, with a low-energy index α_l flatter than $\alpha_h = 2.5$. Here $\varepsilon_B = 5$ is at the upper end of the range of break energies for EGRET-detected bursts, and significant reductions in Γ_{MIN} below the values in Fig. 1(a) occur only for galactic halo ($d = 100$ kpc) sources.

Despite the usefulness of approximating the source photon spectrum by an infinite power-law, most bursts detected by BATSE show significant spectral curvature in the 30 keV–500 keV range (e.g. Band et al. 1993). Furthermore, BATSE sees MeV spectral curvature in 4–5 bright bursts (Schaefer et al. 1992) from 18 selected for brightness at 1 MeV, and EGRET observes three of these sources with such "high energy" breaks (Schneid et al. 1992; Kwok et al. 1992) — for parameters, see Table 1. Hence, EGRET detections seem correlated with spectral breaks at the upper end of the BATSE energy range, which is very probably a selection effect for the observability of bursts by EGRET: GRBs with breaks at higher energies tend to be more luminous in the super-MeV range. These breaks and those generally seen at lower energies in the BATSE data for many GRB spectra could, in principle, reduce the opacity of potential TeV emission from these sources, so it is important to generalize the pair production opacity/relativistic beaming analysis to include the effects of spectral curvature. The principal effect is expected to be a possible reduction of Γ_{MIN} from the above power-law estimates at energies above 1 GeV, due to the relative deficiency of interacting photons at low energies.

Table 1: GRB Breaks

		BATSE		EGRET
	ε_B	α_l	α_h	α_h
GRB 910503	1.0 MeV	1.0	1.6 ± 0.2	1.2 ± 0.2
GRB 910601	0.7 MeV	0.4	2.4 ± 0.3	2.7 ± 0.2
GRB 910814	1.1 MeV	0.0	1.6 ± 0.3	1.8 ± 0.2

The values of the break energy ε_B, and spectral index below (α_l) and above (α_h) the break, as determined by BATSE (see Schaefer et al. 1992), for the first three bursts detected by EGRET, together with their high-energy spectral indices as determined by EGRET.

The effects of a depletion of low energy photons in the observed spectra can quickly be determined by taking the simplest approximation to spectral curvature, namely a power-law broken at an energy ε_B. The optical depth determination for such a distribution follows the above description for pure power-laws, and utilises results obtained in Gould and Schreder (1967) and Baring (1994) for truncated power-laws. Deferring the details of our calculations to Baring and Harding (1997b, in preparation), the optical depth $\tau_{\gamma\gamma}(\varepsilon)$ for pair production attenuation of a broken power-law photon distribution with $n(\varepsilon) \propto \varepsilon^{-\alpha_l}$ for $\varepsilon < \varepsilon_B$ and $n(\varepsilon) \propto \varepsilon^{-\alpha_h}$ for $\varepsilon > \varepsilon_B$ is found to be

$$\frac{\tau_{\gamma\gamma}(\varepsilon)}{n_\gamma \sigma_T R} \approx \frac{\mathcal{A}(\alpha_l)}{1+\alpha_l}\left\{\mathcal{H}(\alpha_l, 1) - \mathcal{H}(\alpha_l, \eta)\right\}\frac{\varepsilon^{\alpha_l-1}}{\Gamma^{2\alpha_l}} + \frac{\mathcal{A}(\alpha_h)}{1+\alpha_h}\mathcal{H}(\alpha_h, \eta)\frac{\varepsilon^{\alpha_h-1}}{\Gamma^{2\alpha_h}} \quad , \tag{2}$$

where $\eta = \max\{1, \sqrt{\varepsilon_B\varepsilon}/\Gamma\}$ and

$$\mathcal{H}(\alpha, \eta) = \frac{4}{\sigma_T}\int_1^\infty d\chi\, q^{2(1+\alpha)}\chi^{1-2\alpha}\sigma_{\gamma\gamma}(\chi) \quad , \quad q = \min\left(1, \frac{\chi}{\eta}\right) \quad , \tag{3}$$

and $\sigma_{\gamma\gamma}$ is the pair production cross-section. The parameter η governs the importance, or otherwise, of spectral curvature effects. Clearly, if either ε_B or ε is low enough that $\eta = 1$, then the pure power-law result emerges from Eq. (2). Setting $\alpha_l = \alpha_h$ also yields the previous infinite power-law dependence: $\tau_{\gamma\gamma}(\varepsilon) \propto \varepsilon^{\alpha_h-1}\Gamma^{-1-2\alpha_h}$ (remembering that $n_\gamma \propto f/R^2$ with $R \propto \Gamma$), which arises because of pair production kinematics. Note that $\mathcal{A}(\alpha)$ is a function that arises from the integrations over photon angles, and is approximately given by (Baring 1994) $2/(4/3+\alpha)^{27/11}$, a coefficient that appears in Eq. (1).

Fig. 1b illustrates the behaviour of Γ_{MIN} as determined by Eq. (2), depicting it as a function of α_l for α_h fixed at 2.5. Results are displayed for source distances typical of galactic halo and cosmological populations. The break energy $\varepsilon_B \approx 2.5\,\mathrm{MeV}$ is high enough (at the upper end of the range appropriate for EGRET bursts) that deviations from the infinite power-law case are very significant for the galactic halo distance scale. In contrast, spectral curvature obviously can be neglected for cosmological sources. The key parameter in determining spectral curvature effects is $\sqrt{\varepsilon_B\varepsilon_{MAX}}/\Gamma_{MIN}$, which is clearly less than unity for $d = 1\,\mathrm{Gpc}$, but can be significantly greater than unity for the chosen halo distance, thereby reducing $\tau_{\gamma\gamma}$. Note that general (i.e. more gradual) spectral curvature can be treated by fitting the continuum with piecewise continuous power-laws, generalizing the structure inherent in Eq. (2); our technique is well-suited to this adaptation. However the qualitative nature of the reduction of Γ_{MIN} for general spectral curvature can be deduced from these broken power-law results. Note that reducing ε_B diminishes the dependence of Γ_{MIN} on α_l so that sensitivity to α_l is lost when $\varepsilon_B \lesssim 250\,\mathrm{keV}$.

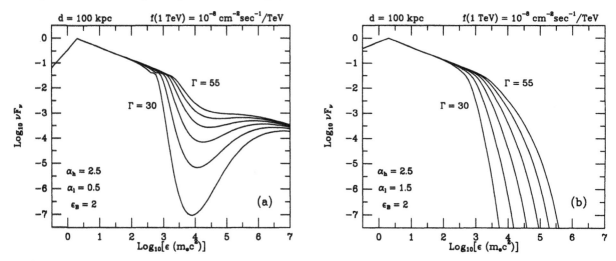

Fig. 2. Sample broken power-law source spectra for galactic halo bursts emanating from regions of different bulk Lorentz factors Γ. The spectra are plotted in νF_ν format, are of unspecified normalization, and are attenuated by pair creation according to the factor $\exp\{-\tau_{\gamma\gamma}\}$ (i.e. no pair cascading is included). The spectral break is more severe in (a) where $\alpha_l = 0.5$, resulting in absorption dips, while the more moderate break in (b), for which $\alpha_l = 1.5$, yields the more familiar exponential turnovers. Cosmological source spectra would appear much like Fig. 2b, with d and Γ scaled up accordingly.

A remarkable feature can be deduced from the cross-over of the Γ_{MIN} curves for the $d = 100\,\text{kpc}$ case in Fig. 1b. For a source with $\alpha_l \lesssim 0.8$ with a bulk Lorentz factor lying between the values of Γ_{MIN} for the $\varepsilon_{\text{MAX}} = 10\,\text{GeV}$ and $\varepsilon_{\text{MAX}} = 1\,\text{TeV}$ curves, opacity to pair creation is clearly achieved at the lower energy but not the higher. The natural consequence of this inversion, which is contrary to normal situations where optical depths are increasing functions of ε_{MAX}, is that a broad absorption feature appears in the sub-TeV range. This is evident in Fig. 2a, where attenuated broken power-law spectra are exhibited for a range of galactic halo source bulk Lorentz factors. These distinctive features arise only for large spectral breaks (i.e. $|\alpha_h - \alpha_l| \gtrsim 1.5$), and disappear in Fig.2b, where $\alpha_l = 1.5$, and also MeV-type break energies, in favour of the more familiar exponential turnover. Note that details of pair cascading have been neglected; their inclusion would generate a reprocessing of attenuated photons to energies just below the dip, rendering the absorption feature slightly more pronounced.

These absorption dips result directly from a significant depletion of sub-MeV photons, and can only occur for galactic halo burst scenarios since their photon fields are less dense than for cosmological bursts, and hence the photons at specified energy ε_{MAX} must interact with photons at lower (i.e. sub-MeV) energies where depletion relative to the EGRET power-laws occurs. For cosmological bursts, the interacting photons are generally above the break and the old unbroken power-law opacity calculations suffice (see Fig. 1b), yielding exponential turnovers similar to those depicted in Fig. 2b. Therefore the absorption dips in Fig. 2a are a clear marker of a galactic halo burst population, and consequently a potentially powerful observational diagnostic. By inspection of Table 1, the bursts GRB910601 and GRB910814 could possibly exhibit this behaviour, contrasting GRB910503, whose break is too moderate to produce anything but an exponential cutoff. Note also that the structure in the EGRET spectrum of GRB940217 (Hurley et al. 1995) is somewhat suggestive of a GeV dip. Observations by Whipple and MILAGRO at the TeV range, and perhaps even HEGRA at higher energies, combined with a probing of the 100 MeV – 100 GeV range by future instrumentation such as GLAST or Celeste could confirm or deny the existence of such absorption features. While the observation of exponential cutoffs would not distinguish between cosmological or galactic burst hypotheses, the detection of broad absorption dips would be a very strong indicator that gamma-ray bursts are of galactic origin.

REFERENCES

Band, D., et al.: 1993 *Ap.J.* **413**, 281.

Baring, M. G.: 1993 *Ap.J.* **418**, 391.

Baring, M. G.: 1994 *Ap.J. Suppl.* **90**, 899.

Baring, M. G. and Harding, A. K.: 1993 in *Proc. 23rd ICRC* **1**, 53.

Baring, M. G. and Harding, A. K.: 1995 *Adv. Space Res.* **15(5)**, 153.

Fenimore, E. E., Epstein, R. I. and Ho, C.: 1992 in *Gamma-Ray Bursts*, AIP Conference Proceedings 265, eds. Paciesas, W. S. and Fishman, G. J., (AIP, New York) p. 158

Gould, R. J. and Schreder, G. P.: 1967 *Phys. Rev.* **155**, 1404.

Harding, A. K.: 1994 in *The Second Compton Symposium*, eds. Fichtel, C., Gehrels, N. & Norris, J., (AIP Conf. Proc. 304, New York) p. 30.

Hurley, K., et al.: 1994 *Nature* **372**, 652.

Krolik, J. H. and Pier, E. A.: 1991 *Ap.J.* **373**, 277.

Kwok, P. W. et al.: 1993 in *Compton Gamma-Ray Observatory*, eds. Friedlander, M., Gehrels, N., and Macomb, D. (AIP Conf. Proc. 280, New York) p. 855

Meegan, C. A., et al.: 1991 *Nature* **335**, 143.

Schaefer, B. E., et al.: 1992 *Ap.J. (Lett.)* **393**, L51.

Schmidt, W. K. H.: 1978 *Nature* **271**, 525.

Schneid, E. J., et al.: 1992 *A&A Lett.* **255**, L13.

Yodh, G. B.: 1996 *Space Science Reviews* **75**, 199.

Pergamon

Adv. Space Res. Vol. 22, No. 7, pp. 1121–1124, 1998
© 1998 COSPAR. Published by Elsevier Science Ltd. All rights reserved
Printed in Great Britain
0273–1177/98 $19.00 + 0.00

PII: S0273–1177(98)00206–3

NUMERICAL EVALUATION OF MAGNETIC PHOTON SPLITTING RATES VIA A PROPER-TIME TECHNIQUE

Matthew G. Baring* and Alice K. Harding

NASA Goddard Space Flight Center, Code 661, Greenbelt, MD 20771, U.S.A.

ABSTRACT

The results of our numerical computation of magnetic photon splitting rates according to the proper-time formulation developed by Stoneham are presented in this paper. Calculations were performed for both the total rate of splitting and the differential distributions for the polarization mode $\perp \to \parallel \parallel$ of splitting. Analytic asymptotic limits obtainable from the triple integral expressions of Stoneham's formalism are also presented, and these are reproduced by the full computation of the integrals. The rates produced here agree with numerous previous analyses of photon splitting but fall somewhat below the S-matrix calculation of Wilke and Wunner. The numerical determination of these rates is of great use to current topical models of gamma-ray pulsars and magnetar scenarios for soft gamma repeaters.

©1998 COSPAR. Published by Elsevier Science Ltd.

INTRODUCTION

The exotic QED process of the splitting of a photon into two ($\gamma \to \gamma\gamma$) in the presence of a strong magnetic field has recently become of interest in the study of neutron star sources such as soft gamma repeaters (SGRs) and gamma-ray pulsars. The astrophysical consequences of splitting in these objects have been explored in considerable detail by the authors, and other groups. In particular, splitting can act (Baring 1995, Baring and Harding 1995) to soften SGR spectra to the observed range if they originate from regions of $B \gtrsim 10^{14}$ Gauss, and can inhibit emission above 10 MeV in high field pulsars such as PSR1509-58 (Harding, Baring and Gonthier 1997). This research used relatively simple approximations to the splitting rates that were derived by Adler (1971, see also Bialynicka-Birula & Bialynicki-Birula 1970), using effective Lagrangian techniques, which are valid for low photon energies $\varepsilon m_e c^2$ satisfying $\varepsilon B \ll B_{cr}$, where $B_{cr} = 4.413 \times 10^{13}$ Gauss is the quantum critical field strength. Restriction to this asymptotic regime has been necessitated by the total lack of more general evaluations of splitting rates that are sufficiently amenable for use in astrophysical models. This is a real deficiency, since the $\varepsilon B \gtrsim B_{cr}$ regime can easily be realized in both SGRs and pulsars, and is precipitated by the inherent difficulty in generating numerical rates for this third-order QED process.

The most general calculation of photon splitting rates was performed by Stoneham (1979), who used Schwinger's proper-time method to determine formal expressions for the splitting rate for ε below pair creation threshold and arbitrary B. Although Stoneham's analysis included dispersion in the birefringent, magnetized vacuum, they reduced in the zero-dispersion limit to triple integrals that involve only moderately complicated combinations of elementary functions (see Adler 1971, Papanyan & Ritus 1972, and Baïer,

*Compton Fellow, USRA

Mil'shteĭn, & Shaĭsultanov 1986 for alternative but less convenient presentations). We have developed Stone-ham's triple integral expressions considerably (Baring, Harding and Weise, in preparation), specifically to facilitate numerical evaluations, and in this paper we present the results of our numerical computation of splitting rates, at a wide range of energies, for the polarization mode $\perp\to\|\|$ permitted in the limit of weak dispersion by energy-momentum kinematic selection rules (see Adler 1971; $\perp\to\perp\perp$ and $\|\to\perp\|$ are the other two modes permitted by QED in the non-dispersive limit). Here $\perp$ and $\|$ denote the orientation of the photon electric vector relative to the plane containing its momentum and the field line. Our computed rates vary continuously from the low energy limit, and reproduce analytic asymptotic forms at both low and highly supercritical field strengths; they differ dramatically from the recent numerical evaluation of the S-matrix determination of the splitting rate by Mentzel, Berg and Wunner (1994) that is orders of magnitude greater than the earlier proper-time determinations.

PHOTON SPLITTING RATES

While Adler's (1971) use of the effective Lagrangian technique was tailored to low energies or low field strengths, using these quantities as QED expansion parameters, the more general "Schwinger" proper-time analysis by Stoneham (1979) is historically the most palatable presentation of the splitting reaction rates for the purposes of numerical evaluation that encompasses a broad range of parameter space, with Baĭer, Mil'shteĭn, & Shaĭsultanov (1996) recently producing another alternative facile presentation. The differential attenuation coefficient of the photon splitting mode $\perp\to\|\|$ is

$$T_{\perp\to\|\|} \;=\; \frac{\alpha_f^3}{2\pi^2}\,\frac{1}{\lambda}\,B^6\,\sin^6\theta\,\varepsilon_1^2\,(\varepsilon-\varepsilon_1)^2\,\mathcal{M}_{\perp\to\|\|}^2 \;\;, \tag{1}$$

where $\mathcal{M}$ denotes the (scaled) scattering amplitude; other polarization modes are similar. Here θ is the angle of photon propagation relative to the field lines. Throughout this paper, B is expressed in units of B_{cr} and photon energies ε (initial) and ε_1 (final) are scaled by $m_e c^2$. Note that α_f is the fine structure constant and λ is the electron Compton wavelength over 2π. The triple integral expressions obtained by these authors for the scattering amplitudes $\mathcal{M}$ are too lengthy for presentation here. When the incident photon energy is low, namely $\varepsilon \ll 1$ (i.e. 511 keV), analytic simplification is possible:

$$\mathcal{M}_{\perp\to\|\|} \approx \mathcal{M}_1(B) = \frac{1}{B^4}\int_0^\infty \frac{ds}{s}\,e^{-s/B}\left\{\left(-\frac{3}{4s}+\frac{s}{6}\right)\frac{\cosh s}{\sinh s}+\frac{3+2s^2}{12\sinh^2 s}+\frac{s\cosh s}{2\sinh^3 s}\right\} \tag{2}$$

which specializes to $\mathcal{M}_1(B) \approx 26/315$ when $B \ll 1$, and $\mathcal{M}_1(B) \approx 1/6B^3$ when $B \gg 1$. These results were first obtained by Adler (1971), and are those used in the recent astrophysical models of SGRs (Baring and Harding 1995) and gamma-ray pulsars (e.g. Harding, Baring and Gonthier 1997). Recently, Baĭer, Mil'shteĭn, & Shaĭsultanov (1996) have derived the high field limit (i.e. $B \gg 1$) for splitting below pair creation threshold:

$$\mathcal{M}_{\perp\to\|\|} \approx \frac{1}{B^3\varepsilon\varepsilon_1\varepsilon_2}\left\{\frac{4\varepsilon_1}{\varepsilon_2\sqrt{4-\varepsilon_2^2}}\arcsin\left(\frac{\varepsilon_2}{2}\right)+\frac{4\varepsilon_2}{\varepsilon_1\sqrt{4-\varepsilon_1^2}}\arcsin\left(\frac{\varepsilon_1}{2}\right)-\varepsilon\right\}, \tag{3}$$

for $\varepsilon_2 = \varepsilon - \varepsilon_1$. This form has also been reproduced from Stoneham's (1979) formulae (Baring, Harding and Weise, in preparation).

Motivation for numerical computation of proper-time results has been amplified by a new result on the rates of photon splitting: Mentzel, Berg & Wunner (1994) presented an S-matrix calculation of the polarized split-ting rates. While their formal development is comparable (Weise, Baring and Melrose 1996, in preparation) to an earlier S-matrix formulation of splitting in Melrose & Parle (1983), their presentation of numerical results (see also Wunner, Sang & Berg 1995) appeared to be in violent disagreement with the splitting results obtained by Adler (1971) and Stoneham (1979), which are embodied in Eqs. (1) and (2). The numerical computation of the S-matrix formalism is a formidable task. The proper-time analysis, though difficult, is more amenable, and has been reproduced in the limit of $B \ll B_{cr}$ by numerous authors. No analytic reduction of the S-matrix formalism to directly compare with extant proper-time results was presented in Mentzel, Berg & Wunner (1994). As the S-matrix and proper-time techniques should produce equivalent results, and indeed have done so demonstrably in the case of pair production (see Daugherty & Harding 1983; Tsai & Erber 1974), the numerical work of Mentzel, Berg and Wunner seemed highly questionable; they have since retracted these results in Wilke and Wunner (1996).

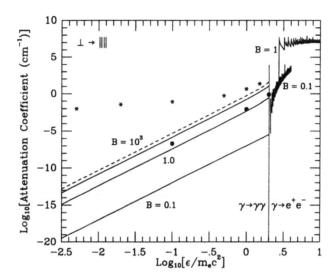

Fig. 1. The computed photon splitting attenuation coefficients for polarization mode $\perp \rightarrow \| \|$ for field strengths $B = 0.1, 1, 10$ (solid lines below $2m_e c^2$, in ascending order), and the asymptotic rate in Eq. (3) labelled as $B = 10^3$ (dashed curve). For comparison, the pair creation rates for $\perp$ photons at two different field strengths are depicted above $2m_e c^2$. The stars denote the corresponding numerical evaluation that Mentzel, Berg and Wunner (1994) performed of their S-matrix formulation for $B = 0.1$; their $B = 1$ data points are more or less coincident with those shown. The filled circles denote the numerical revision in Wilke and Wunner (1996) for $B = 1$.

Our numerical computations of Stoneham's (1979) general formulae are presented in Fig. 1 (total attenuation coefficient obtained by integrating Eq. (1) over ε_1) and Fig. 2 (normalized differential spectra). In Fig. 1, the numerical results for the polarization mode $\perp \rightarrow \| \|$ indicate that Adler's (1971) simple analytic forms [see Eqs. (1) and (2)], which have an ε^5 energy dependence (i.e. when integrating Eq. (1) over produced energies ε_1), are quite accurate for all energies below pair creation threshold when $B \lesssim 0.4$, and in the limit of $B \gg 1$ they merge smoothly into the asymptotic limit in Eq. (3). Furthermore, the rates are strongly dependent on B, and are substantially smaller than the coefficients for the first order QED process of pair creation $\gamma \rightarrow e^+ e^-$. Our numerical splitting rates are clearly less than those of Mentzel, Berg and Wunner (1994) by orders of magnitude (see also the recent work of Baïer, Mil'shteĭn, & Shaĭsultanov 1996) for their chosen range of energies and field strengths, namely $\varepsilon B / B_{cr} \lesssim 1.5$, and far more sensitive to the value of B. The revised computations of Wilke and Wunner (1996) still exceed our proper-time rates by factors of 2–3. In Fig. 2, the differential spectra produced by Stoneham's expressions are considerably narrower than those obtained by Mentzel Berg and Wunner, and reproduce the dependence implied by Eq. (1) when $\varepsilon \ll 1$ or $B \ll 1$; the revisions of Wilke and Wunner (1996) approximate this differential shape quite well.

In summary, we find our numerical results quite consistent with several analytic limits obtained by various groups, comparable to the numerical computations of Baïer, Mil'shteĭn, & Shaĭsultanov (1996), and well-behaved in the light of physical intuition related to strong field QED processes. We believe that such computations can be reliably used in astrophysical models, clearly stand in strong contradiction to the numerical aspects of the work of Mentzel, Berg and Wunner, and still represent a significantly more accurate evaluation than the revised work of Wilke and Wunner.

We thank Ramin Sina for providing us with his pair creation code.

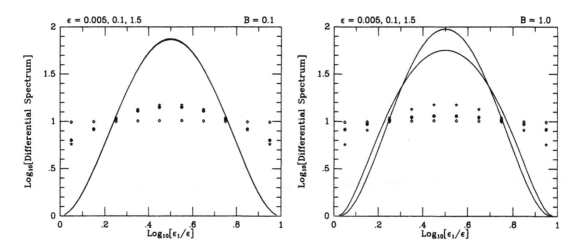

Fig. 2. The computed photon splitting differential rates, normalized to unity, for polarization mode $\perp \rightarrow \|\|$ for field strengths (a) $B = 0.1$ and (b) $B = 1$. Spectra are depicted for three different initial photon energies ε (in units of $m_e c^2$) as functions of the energy ε_1 of one of the produced photons, and become narrower as ε is increased. The $B = 0.1$ curves closely resemble the asymptotic low energy form of $\varepsilon_1^2 (\varepsilon - \varepsilon_1)^2 / 30$, which can be deduced from Eq. (1). For comparison, the corresponding numerical evaluation that Mentzel, Berg and Wunner (1994) performed of their S-matrix formulation is shown as open squares ($\varepsilon = 0.005$), filled circles ($\varepsilon = 0.1$) and stars ($\varepsilon = 1.5$).

REFERENCES

Adler, S. L. 1971: *Ann. Phys.* **67**, 599.

Baĭer, V. N., Mil'shteĭn, A. I., & Shaĭsultanov, R. Zh.: 1986 *Sov. Phys. JETP* **63**, 665.

Baĭer, V. N., Mil'shteĭn, A. I., & Shaĭsultanov, R. Zh.: 1996 *Phys. Rev. Lett.* **77**, 1691.

Baring, M. G.: 1995 *Ap.J. (Lett.)* **440**, L69.

Baring, M. G. & Harding, A. K.: 1995 *Astr. Sp. Sci.* **231**, 77.

Bialynicka-Birula, Z., & Bialynicki-Birula, I.: 1970 *Phys. Rev. D* **2**, 2341.

Daugherty, J. K. & Harding, A. K.: 1983 *Ap.J.* **273**, 761.

Harding, A. K., Baring, M. G. & Gonthier, P. L.: 1997 *Ap.J.* in press.

Melrose, D. B. & Parle, A. J.: 1983 *Aust. J. Phys.* **36**, 775, 799.

Mentzel, M., Berg, D., & Wunner, G.: 1994 *Phys. Rev. D* **50**, 1125.

Papanyan, V. O., & Ritus, V. I.: 1972 *Sov. Phys. JETP* **34**, 1195.

Stoneham, R. J.: 1979 *J. Phys. A* **12**, 2187.

Tsai, W.-Y. & Erber, T.: 1974 *Phys. Rev. D.* **10**, 492.

Wilke, C. & Wunner, G. 1996, *Phys. Rev. D* submitted.

Wunner, G., Sang, R., & Berg, D.: 1995 *Ap.J. (Lett.)* **455**, L51.

Pergamon

Adv. Space Res. Vol. 22, No. 7, pp. 1125–1128, 1998
© 1998 COSPAR. Published by Elsevier Science Ltd. All rights reserved
Printed in Great Britain
0273-1177/98 $19.00 + 0.00

PII: S0273–1177(98)00207–5

VERIFYING THE ACCURACY OF THE THIRD INTERPLANETARY NETWORK: LOCALIZATION OF THE BURSTING PULSAR GRO J1744-28 BY TRIANGULATION

K. Hurley[1], C. Kouveliotou[2], A. Harmon[2], G. Fishman[2], M. Briggs[2], J. van Paradijs[3], J. Kommers[4], W. Lewin[4], T. Cline[5], M. Boer[6], M. Niel[6]

1. UC Berkeley Space Sciences Laboratory, Berkeley CA 94720-7450, USA
2. NASA Marshall Space Flight Center, Huntsville, AL 35812, USA
3. University of Alabama, Huntsville, AL 35889, USA
4. Massachusetts Institute of Technology, Center for Space Research, Cambridge, MA 02139, USA
5. NASA Goddard Space Flight Center, Greenbelt, MD 20771, USA
6. Centre d'Etude Spatiale des Rayonnements, 31029 Toulouse Cedex, France

ABSTRACT

The bursting pulsar GRO J1744-28 is the first repeating source of intense bursts whose position is accurately known, which has become active since the establishment of the third interplanetary network. We show how observations of this object with Ulysses and BATSE can be used to verify many aspects of the triangulation method. Analysis of the preliminary data indicates that there are no unknown sources of error in this method. ©1998 COSPAR. Published by Elsevier Science Ltd.

INTRODUCTION

It is generally agreed that deep counterpart searches are a promising avenue to explore in the solution of the gamma-ray burst mystery. Such searches require small ($\lesssim 1'$) error boxes, and there is presently only one way of obtaining them, namely triangulation using a network of widely spaced detectors. The accuracy of this method depends on the accuracy to which the time and coordinates are known for the spacecraft observing the burst. Here we review the methods which can be used to verify the timing and ephemeris of the Ulysses spacecraft. The presence of a repeating source of intense bursts whose position is well known (GRO J1744-28: Kouveliotou et al. 1996a) is an essential element in this verification.

We note that one verification method which involves no cosmic sources is an "end-to-end timing test": commands are sent to the Ulysses GRB experiment at known times and their arrival times are measured. Using this method, it has been shown that the timing accuracy of the Ulysses spacecraft may be in the millisecond range (Hurley & Sommer 1994). For technical reasons, however, assumptions must be made about possible sources of systematic errors; if they are incorrect, the timing accuracy is only verifiable to about 100 ms. Thus it is interesting to search for other methods of verifying the timing accuracy.

THE METHOD

Repeating burst sources can be localized by triangulation using Ulysses and BATSE, by several methods. For very weak sources which are not *directly* detectable by the small Ulysses instrument, the "network synthesis" method may be used. It consists of defining a grid of α, δ values, and for each BATSE detection of a burst from a soft gamma repeater, predicting the arrival time of the burst at Ulysses, and co-adding the Ulysses data rephased so that the burst signals are aligned and produce a detectable pulse. This method has been successfully applied to the soft repeaters SGR1806-20, SGR1900+14 (Hurley et al. 1994) and to GRO J1744-28 (Hurley et al. 1996). Because many bursts, occurring over a period of weeks - months, must be co-added to improve statistics and produce a detectable signal, this method does not allow the timing or ephemeris for any one burst to be verified individually, although it does verify the average timing to ~100 ms, and the average ephemeris to several arcminutes.

For slightly stronger sources, such as GRO J1744-28, bursts *are* detected individually by Ulysses. They may either be detected in the real-time data (time resolution 0.25 - 2 s) or, if they are intense enough, in the triggered data (8 - 32 ms time resolution). A burst is found in the real-time data by searching a time window which is defined by the burst arrival direction, the arrival time at BATSE, and the burst/BATSE/Ulysses angle. Since there are very few triggers/week on Ulysses, and almost all are for cosmic bursts, there is no corresponding search procedure required for these events. Both the real-time and triggered events are triangulated by cross-correlating the Ulysses and BATSE time histories. A knowledge of the Ulysses time and ephemeris is needed for both. In both cases, the burst time is ultimately derived from the spacecraft clock. However, as the triggered bursts are timed to an accuracy of several ms on board, fine timing circuits are involved for them, but not for the real-time events. Table 1 summarizes what each method confirms, and the approximate accuracy to which it confirms it using GRO J1744-28. Here R refers to the spacecraft distance, and α, δ are the right ascension and declination. 'Ave' means that the average value is verified.

Table 1. What is verified with various tests

Method	Spacecraft timing, ms	Fine timing, ms	R	α,δ
End-to-end test	100 or 3	100 or 3	100 or 3 light-ms	NO
Network synthesis	100 (ave.)	NO	100 light-ms (ave.)	3' (ave.)
Real-time bursts	<300	NO	<300 light-ms	<1'
Triggered bursts	<300	<300	<300 light-ms	<1'

By cross-correlating the Ulysses and BATSE time histories, we can obtain an annulus of possible arrival directions. This is shown in figure 1. We have assumed that the location of GRO J1744-28 is the one found by the ROSAT HRI observation (Kouveliotou et al. 1996b). An 8" radius error circle is shown for this source. The annulus center (dashed line) represents the most probable location derived from triangulation. Taking an estimated 3σ uncertainty of ±1 s in the cross-correlation gives a ±3' annulus width. Because the bursts from GRO J1744-28 are quite weak for Ulysses, statistical errors dominate; they are of order 300 ms (1σ). Thus as Table 1 indicates the systematic errors in the timing can only be verified to about this accuracy.

RESULTS

We have triangulated 58 real-time bursts and 7 triggered bursts observed by Ulysses and BATSE between January 1 and 29, 1996. In each case, we calculate the error in the triangulated result, i.e. the distance between the center of the triangulation annulus and the center of the ROSAT position. In most cases, the data are preliminary, and this work will be redone not only with the final data for these events, but also, for more bursts.

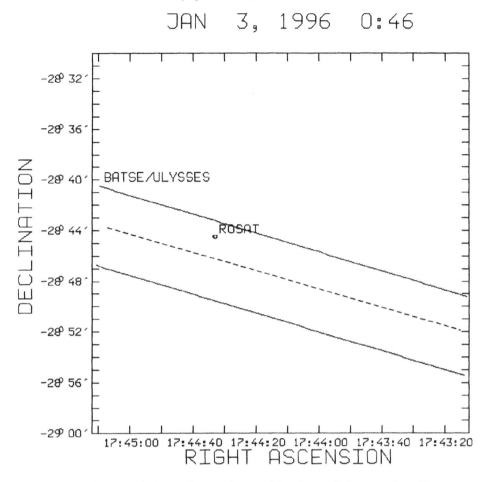

Fig. 1. The result of triangulating a burst observed in the real-time mode. The center of the BATSE/Ulysses annulus is shown as a dashed line, and the solid lines give the estimated 3σ confidence annulus width. The angular distance between the dashed line and the ROSAT source is the actual error for this event, in this case about 1.6'.

We have calculated the mean error and the sample standard deviation for the triggered and real-time events separately. The results are 0.16'±1.03' and 0.22'±1.24' respectively. (The corresponding timing uncertainties are 53±343 and 73±413 ms, respectively). In figure 2, we have plotted the distribution of the errors for the 58 real-time events and compared it with a Gaussian distribution having mean zero, but the same standard deviation. Although the number of events is still too small to allow a detailed comparison, the distributions appear to be consistent with one another. Moreover, they appear to be consistent with the estimated annulus width (±3', 3σ), and they are fully consistent with the earlier, but less accurate, network synthesis results.

CONCLUSION

Some of the Ulysses data used in these comparisons were preliminary (quick-look), and some timing uncertainty can be expected in them. Also, there are numerous events yet to be analyzed. Study of them should reduce the uncertainties to a still lower value. However, the results at this point indicate that both the procedures used in triangulation, such as searching for weak events in the real-time data and cross-correlating time histories, as well as the spacecraft timing and ephemeris information, are accurate. The errors found are dominated by statistical errors in the cross-correlation of the time histories, and no unknown sources of sytematic errors appear to be present. Because classical gamma-ray bursts are generally much more intense, the uncertainties in their localization will be much less than the 3' uncertainty found here.

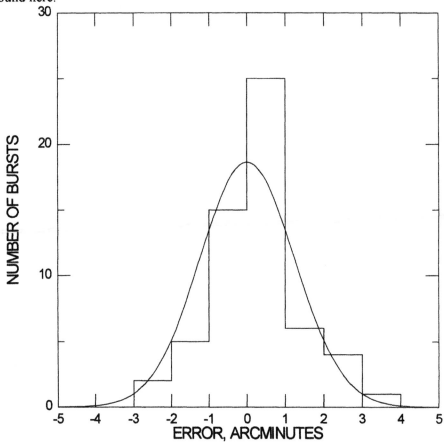

Fig. 2. The observed distribution of errors compared with a Gaussian.

ACKNOWLEDGMENTS

The Ulysses GRB experiment is supported under JPL Contract 958056, and collaborative Ulysses/BATSE data analysis is supported under NASA Grant NAG5-1560.

REFERENCES

Hurley, K., and Sommer, M., Timing Accuracy of the Ulysses GRB Experiment, in *Gamma-Ray Bursts, 2nd Workshop*, edited by. G. Fishman, J. Brainerd, and K. Hurley, AIP 307, AIP Press, New York, NY, 682 (1994)

Hurley, K., Sommer, M., Kouveliotou, C., Fishman, G., Meegan, et al., Network Synthesis Localization of Two Soft Gamma Repeaters, *Ap. J.* 431, L31 (1994)

Hurley, K. on behalf of the Ulysses Gamma-Ray Burst Team, Kouveliotou, C., Harmon, A., Fishman, G., et al., Transient X-Ray Burster GRO J1744-28, *IAUC* 6286, January 12 (1996)

Kouveliotou, C., van Paradijs, J., Fishman, G., Briggs, M., Kommers, J., et al., A New Type of Transient High-Energy Source in the Direction of the Galactic Center, *Nature* 379, 799 (1996a)

Kouveliotou, C., et al., GRO J1744-28, *IAUC* 6369, April 1 (1996b)

Pergamon

Adv. Space Res. Vol. 22, No. 7, pp. 1129–1132, 1998
© 1998 COSPAR. Published by Elsevier Science Ltd. All rights reserved
Printed in Great Britain
0273-1177/98 $19.00 + 0.00

PII: S0273-1177(98)00208-7

THE GAMMA-RAY BURSTS MONITOR ONBOARD SAX

E. Costa[1], F. Frontera[2,3], D. Dal Fiume[3], L. Amati[1], M.N. Cinti[1], P. Collina[2], M. Feroci[1], L. Nicastro[3], M. Orlandini[3], E. Palazzi[3], M. Rapisarda[4], and G. Zavattini[2].

[1] *Istituto di Astrofisica Spaziale, CNR, Via E.Fermi 21, I-00044 Frascati, Italy*
[2] *Dipartimento di Fisica, Universita' di Ferrara, Via Paradiso 12, I-44100 Ferrara, Italy*
[3] *Istituto Tecnologie e Studio Radiazioni Extraterrestri, CNR, Via P.Gobetti 101, I-40129 Bologna, Italy*
[4] *Divisione di Neutronica, C.R.E., ENEA, Via E.Fermi 27, I-00044 Frascati, Italy*

ABSTRACT

The four Lateral Shields of the PDS experiment aboard the Beppo-SAX satellite are equipped with dedicated electronics and used as a Gamma Ray Burst Monitor (GRBM). On a suitable trigger time histories are recorded and transmitted. The large geometric area and the low background of these detectors provide a remarkable capability which is limited, on the other hand, by the no uniform exposure to different direction, due to the shielding of the other experiments of SAX. Here we discuss the on-ground calibrations and the performance simulations in progress, aimed to reconstruct the cosmic X-ray fluxes impinging on the satellite from the recorded counting rates.

©1998 COSPAR. Published by Elsevier Science Ltd.

GAMMA RAY BURST MONITOR OF SAX

SAX (Satellite di Astronomia X) is a joint venture of Italian and Dutch Space Agencies (ASI and NIVR) for a broad band (0.1-300 keV) study of the X-ray sky (Boella et al. 1996). The high energy part of this band is covered by the Phoswich Detection System (PDS) designed by ITESRE-CNR, IAS-CNR and University of Ferrara. It consists (Frontera et al. 1996) of four Phoswich detectors [NaI(Tl)/CsI(Na)], with a pair of passive collimators that can be rocked in order to monitor the background while observing a source. The whole assembly is surrounded by four active shields, disposed as a square well with the goal of reducing the background and its modulation along the orbit. Each shield is made of a large area CsI(Na) detector (about 275 x 413 x 10 mm^3 size) seen by two Hamamatsu R2238 photomultipliers (PMT). For each lateral shield a light pulser, made with an Am^{241} α source on a NaI(Tl) crystal, is mounted amid the two PMT and used as a gain monitor during the mission. The signals from the two PMT of each shield are amplified and analogically summed before being fed into a single-channel discriminator that generates the veto signal for the main detectors and, in parallel, to a window discriminator that generates the enable signal for the ratemeter counting and the AD conversion, and possibly the selection logic of the GRBM trigger. The anticoincidence threshold (ACT) and the lower (LLT) and upper (ULT) thresholds of the GRBM function are independently programmable on each shield (ACT: 7 steps from nominal 100 to 300 keV; LLT: 15 steps from nominal 20 to 90 keV; ULT: 7 steps from nominal 200 to 600 keV) .

THE TRIGGER AND DATA ACCUMULATION

The GRBM Dedicated Channel

The counts from signals within the LLT and ULT of each shield are used to detect the presence of a gamma-ray burst (GRB) candidate event and trigger the production of the time history records. All the relevant parameters to determine the trigger conditions can be set from the ground station by telecommand. The count rate of each detector is accumulated on a short integration time, programmable from 4.8 to 4000 ms. The background level is continuously computed by the moving average of the counting rate on a fixed time interval, programmable from 8 to 200 s. If the counting rate exceeds, simultaneously on two shields, the level of $n \cdot \sigma$ (n programmable to 4, 8 or 16 by command), the GRBM logic is triggered. The temporal profile of the GRB candidate is then stored in the onboard memory with the following structure of the time bins: 7.8 ms for the last 10 s before the trigger, 0.48 ms for the first 10 s after the trigger and 7.8 ms for the following 110 s.

The Relevant Housekeeping Data

Useful additional information is included in some housekeeping data. Ratemeters of GRB counts and anticoincidence counts for each shield are transmitted with a 1 s time resolution. Moreover the pulse height spectra of each shield in the GRBM energy band are continuously accumulated in 256 channels every 128 s.

THE TRANSPARENCY TO DIFFERENT DIRECTIONS

We show in Figure 1 the relative position of the various detecting units and components of the PDS. The GRBM detectors have been conceived and designed as anticoincidence for the central phoswiches thus, for instance, their crystal frames are flat inside (since they must allow for the collimators rocking) and strengthened outside. Moreover the PDS, which is the heaviest experiment of SAX is located at the center of the satellite. As a consequence the transparency of the entrance window for Gamma Bursts is not even, but is determined from the mass distribution of the PDS experiment, and in general, of the SAX satellite. In Figure 2 we show the PDS relative position with respect to the other SAX experiments from which their shadowing effect on the GRBM can be seen. As a consequence each direction with respect to the satellite frame, and hence every direction in the sky for a certain satellite pointing, is sampled with a different effective area. This condition is far from an ideal, dedicated GRB experiment. Nevertheless the large area, the low background intensity and modulation guarantee a high throughput of information, in particular in the timing domain. Therefore we invested relevant efforts to arrive to a reasonable knowledge of the response of SAX GRBM to photons impinging from different directions.

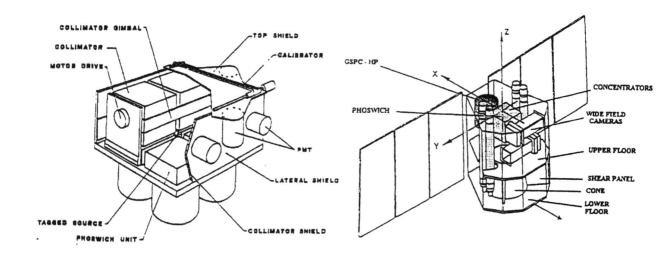

Fig. 1. The structure of the PDS experiment   Fig. 2. A global view of the SAX satellite

On-ground Calibrations

Before the integration in the PDS experiment, the Lateral Shields (LS) spatial responsivities in 30 different points on the crystals surfaces were calibrated with collimated radioactive sources and standard laboratory electronics. After the integration on PDS, one of the shields was calibrated with the flight electronics and with radioactive sources located at a distance of 4 m, also in order to check the impact of thresholds and of event selection logic. After the full integration of SAX (except for the solar panels) the GRBM was calibrated with radioactive sources of previously measured activity, in the 100 μCi range, positioned at a 5 m distance. The sources were mounted in a low-backscattering holder on a graduated vertical bar. SAX was slowly ($\approx$10 arcmin/s) rotated about its pointing axis. Lateral Shields and GRBM ratemeters were recorded, and the energy spectra were collected averaged over 20°, which is suitable to interpret the main variations in the ratemeters in terms of spectral features and to scale the ratemeter calibrations for possible changes of threshold. In this way the azimuth dependence of the GRBM efficiency was explored in detail. By suspending the sources at different heights also a set of polar angles was covered by this calibration. In Figure 3 we show the counting rate of the four GRBM counters as a function of the azimuth angle with a Co^{57} source positioned at the same height of the center of the shields. The count rate from each slab has its maximum when directly exposed to the source but a significant amount of side and back scattering is also present. Some of this is due to the environment of the testing hall where the calibration was performed, and was calibrated with a small detector set in the position of SAX after its removal. In Figure 3 the heavy absorption on LS #2 from the High Presssure Gas Scintillation Proportional Counter and on LS #4 from the two Wide Field Cameras and their masks is apparent. On the other hand LS #1 and #3 see the sky through the carbon fiber tubes of the MECS and LECS which are relatively transparent at high energies. One very appealing conclusion is that the two "good" shields (#1 and #3) are co-aligned with the WFCs (whose field of view is 20° x 20° and energy range 3-30 keV), giving some chance to simultaneously detect a GRB in both instruments.

Simulations

In order to interpret the calibration data, correct them for the finite distance of the source and properly interpolate and extrapolate results to directions not covered from the measurements, we set up a complementary activity of Monte Carlo simulation. The code is based on the release 4.2 of the MCNP Los Alamos code (Forster et al. 1990). A large effort was devoted to include as much detail as possible in the description of SAX. Nowadays the description of SAX includes 168 objects of different materials (excluded collimator cells) and is continuously improving, also with a feed-back from the analysis of calibration data. In Figure 4 we show the simulation data describing the same situation of calibration in Figure 3. Even if the simulation is not yet complete it is already providing the major features of the calibration results. In order to handle the calibration data and have an analytical description of their trends we are also trying to fit detectors efficiencies dependence on energy and direction of incident photons with complex functional forms, partially derived from the underlying Physics, partially semiempirical.

EXPECTED PERFORMANCE

The expected performance of the SAX GRBM was studied at the early stage of its development by Pamini et al. (1990), based on Monte Carlo simulations. The sensitivity was found to be a function of the GRB direction and intensity. Now the on-ground calibration data, the more refined Monte Carlo simulation, and the in-flight calibration data will be used to establish the real sensitivity of the GRBM. At present the cross analysis of the calibration data and of simulations is still in progress, and we can therefore only give a rough estimate of the sensitivity of the GRBM to cosmic events. Using the on-ground calibrations data and the radioactive source activity we have in a first evaluation derived the effective area of LS #3 in their normal direction to be about 185 cm^2 at the Am^{241} energy (60 keV), 610 cm^2 at the Co^{57} energy (122-136 keV), 760 cm^2 at the Ce^{139} energy (166 keV), 680 cm^2 at the Hg^{203} energy (279 keV), 630 cm^2 at the Sn^{113} energy (392 keV) and 520 cm^2 at the Cs^{137} energy (662 keV). The secondary low energy lines of these radioactive sources were subtracted, when not negligible. The above effective areas should be considered as a first order evaluation, since they do not include a correct estimate of the scattering effects in the calibration set-up, that will be studied in detail by means of the Monte Carlo simulation. Similar results are derived for LS #1, while for LS #2 and #4 a more detailed study of the angular dependence is in progress because of the presence of the HPGSPC and WFC, respectively, in their field of view.

From a very preliminary analysis of the in-flight data we derived the background level to be of the order of 0.7 counts cm^{-2} s^{-1} in the 50-600 keV range and an excellent stability ($\pm$7% within one whole orbit). Since a simultaneous trigger from two shields is needed for the GRB detection, we expect that the limiting sensitivity will be driven by the shields which are performing worse.

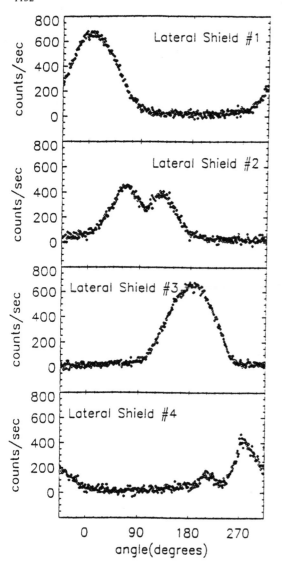

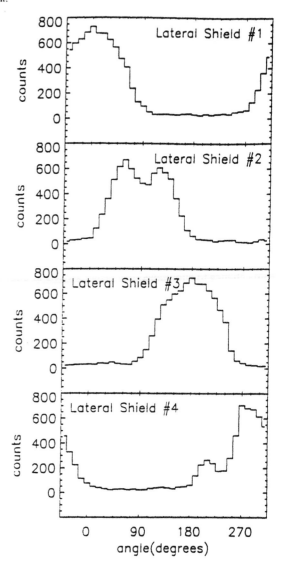

Fig. 3. The four GRBM ratemeters response to a uniform illumination with a Co⁵⁷ source, collected during a satellite 360° rotation

Fig. 4. The same as Figure 3, as simulated by means of the Monte Carlo code with a photon source of 122 keV

REFERENCES

Boella, G., R.C. Butler, G.C. Perola, L. Piro, L. Scarsi, et al., BeppoSAX, the wide band mission for X-Ray Astronomy, *Astron. & Astrophys Suppl. Ser.*, **122**, 299 (1997).

Forster, R.A., R.C. Little, J.F. Briesmeister and J.S. Hendricks, IEEE Trans. on Nucl. Science NS-37, 1378 (1990)

Frontera, F., E. Costa, D. Dal Fiume, M. Feroci, L. Nicastro, et al., The high energy instrument PDS on-board the BeppoSAX X-ray astronomy satellite, *Astron. & Astrophys. Suppl. Ser.*, **122**, 357 (1997).

Pamini, M., L. Natalucci, D. Dal Fiume, F. Frontera, E. Costa, et al., The Gamma-Ray Burst Monitor on Board the SAX Satellite, *Il Nuovo Cimento C*, Vol. **13** N. **2**, 337 (1990).

Adv. Space Res. Vol. 22, No. 7, pp. 1133–1138, 1998
© 1998 COSPAR. Published by Elsevier Science Ltd. All rights reserved
Printed in Great Britain
0273-1177/98 $19.00 + 0.00

Pergamon

PII: S0273–1177(98)00209–9

INFRARED, SUBMILLIMETER, AND MILLIMETER OBSERVATIONS OF THE SOFT GAMMA-RAY REPEATERS

I. A. Smith[1], K. Hurley[2], J. van Paradijs[3,4], L. B. F. M. Waters[3], A. S. B. Schultz[5], P. Durouchoux[6], R. Joyce[7], F. J. Vrba[8], D. Hartmann[9], C. Kouveliotou[10], P. Wallyn[6], and S. Corbel[6]

[1] Rice University, Space Physics and Astronomy, 6100 South Main, Houston, TX 77005, USA
[2] Space Sciences Laboratory, University of California, Berkeley, CA 94720, USA
[3] Astronomical Institute 'Anton Pannekoek', University of Amsterdam and Center for High-Energy Astrophysics, Kruislaan 403, 1098 SJ Amsterdam, The Netherlands
[4] Department of Physics, University of Alabama in Huntsville, Huntsville, AL 35899, USA
[5] NASA Ames Research Center, MS 245-6, Moffett Field, CA 94035-1000, USA
[6] DAPNIA, Service d'Astrophysique, CE Saclay, 91191 Gif sur Yvette Cedex, France
[7] Kitt Peak National Observatory, P. O. Box 26732, Tucson, AZ 85726, USA
[8] U.S. Naval Observatory, Box 1149, Flagstaff Station, Flagstaff, AZ 86002-1149, USA
[9] Clemson University, Department of Physics and Astronomy, Clemson, SC 29634-1911, USA
[10] USRA, NASA Marshall Space Flight Center, ES-62, Huntsville, AL 35812, USA

ABSTRACT

The Soft Gamma-Ray Repeaters appear to be a completely new manifestation of neutron stars. SGR 1806–20 and SGR 1900+14 may have highly unusual stellar counterparts, whose spectra peak in the far infrared. They have not been detected at submillimeter or millimeter wavelengths. Their infrared spectra appear to contain several components: the photospheric emission from star(s) dominates at shorter wavelengths, a bright point source dominates at 25 μm, while an extended source dominates at 60 μm. Their spectra are inconsistent with monoenergetic synchrotron and black body radiation models, but are consistent with simple dust models. We briefly review here our latest millimeter, submillimeter, and infrared observations. We include new upper limits at 235 GHz (1.3 mm) for SGR 1806–20 and SGR 0525–66. The extended IRAS emission detected for SGR 0525–66 is consistent with that expected from heated dust in the supernova remnant N49. Here we show a preliminary analysis of our recent 9.862 μm imaging of the region around SGR 0525–66. We do not detect any point sources, at flux limits lower than would be expected if the other two SGR were placed at the distance of the LMC.   ©1998 COSPAR. Published by Elsevier Science Ltd.

INTRODUCTION

The Soft Gamma-Ray Repeaters (SGR) are sources of brief intense outbursts of low energy gamma rays: it appears that they are new manifestations of neutron stars. The discovery of possible quiescent counterparts has been a significant breakthrough: see Hurley (1996) and Smith (1997) for reviews. SGR 1900+14 and SGR 1806–20 may be associated with highly unusual sources whose spectra peak in the far infrared. However, a counterpart to SGR 0525–66 has only been detected in X-rays. We have embarked on a program of multi-wavelength observations of these quiescent counterparts, in the hope this will lead to an understanding of the source of the bursts of gamma rays. We briefly review here our latest millimeter, submillimeter, and infrared observations, updating our previous reviews (Smith et al. 1996a,b). We include new upper limits at 235 GHz (1.3 mm) for SGR 1806–20 and SGR 0525–66 using the Swedish-ESO Submillimeter Telescope (SEST). We also present a preliminary analysis of 9.862 μm imaging of the region

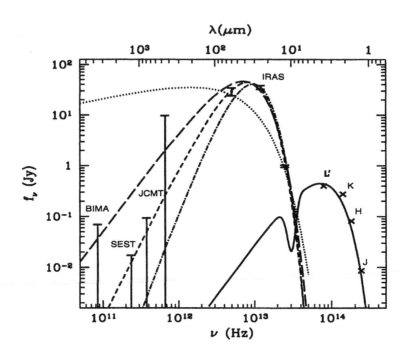

Figure 1. SGR 1806–20 (modified from Smith et al. 1997a). 3σ point source upper limit at 85 GHz (3.5 mm) from the Berkeley-Illinois-Maryland Array (BIMA; Smith, Chernin, & Hurley 1996). 3σ upper limit at 235 GHz (1.3 mm) from the Swedish-ESO Submillimeter Telescope (SEST; this paper): this limit supersedes the one in Wallyn et al. (1995) and Durouchoux et al. (1996). 3σ upper limits at 800 μm and 450 μm from the UKT14 on the James Clerk Maxwell Telescope (JCMT; Smith et al. 1997a). IRAS flux densities are shown with 1σ error bars; the source is point-like at 12 μm and 25 μm, and extended at 60 μm (Van Paradijs et al. 1996). Crosses: L', K, H, and J-band flux densities for star A in Kulkarni et al. (1995). Dotted line is a monoenergetic synchrotron spectrum with $\nu_c = 5 \times 10^{12}$ Hz (Ginzburg & Syrovatskii 1965). Long dashed line is a 120 K black body spectrum. Short dashed line is a Population I dust model ($Q_\nu \propto \nu$) with $T_{gr} = 100$ K (Dwek & Werner 1981). Short dashed-dotted line is a Population II dust model ($Q_\nu \propto \nu^2$) with $T_{gr} = 92$ K (Dwek & Werner 1981). Solid curve is a 30,000 K black body attenuated by the interstellar extinction law of Rieke & Lebofsky (1985) with $A_V = 30$ (Van Kerkwijk et al. 1995).

around SGR 0525–66 using the Thermal Imaging Multi-Mode Instrument (TIMMI) on the 3.6-m telescope at ESO.

REVIEW OF MULTIWAVELENGTH OBSERVATIONS OF SGR 1806–20 AND SGR 1900+14

The current status of the millimeter through near infrared observations of the quiescent counterparts to SGR 1806–20 and SGR 1900+14 are shown in Figures 1 and 2 (modified from Smith et al. 1997a).

The infrared spectra appear to contain several components. The photospheric emission from star(s) dominates at shorter wavelengths. For SGR 1806–20, there is a heavily reddened star, that is consistent with being a rare Luminous Blue Variable (Kulkarni et al. 1995; Van Kerkwijk et al. 1995). For SGR 1900+14, we found a pair of heavily reddened stars that appear to be variable M5 supergiants (Vrba et al. 1996); their spectra are remarkably similar (Vrba et al. 1996; Smith et al. 1997b). An image of SGR 1900+14 at 10 μm showed that the emission is still dominated by the well-resolved stars, with a faint extended component (Van Paradijs et al. 1996). We also performed a detailed study of the IRAS observations of the SGR, and found that the emission consists of multiple components (Van Paradijs et al. 1996): for both

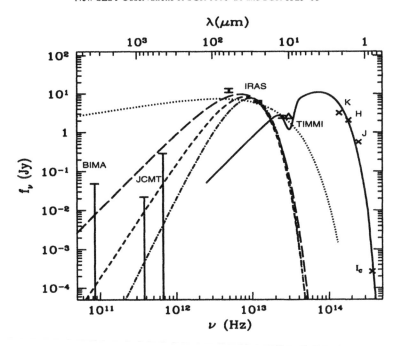

Figure 2. SGR 1900+14 (modified from Smith et al. 1997a). 3σ point source upper limit at 85 GHz (3.5 mm) from BIMA (Smith, Chernin, & Hurley 1996). 3σ upper limits at 800 μm and 450 μm from the UKT14 on the JCMT (Smith et al. 1997a). IRAS flux densities are shown with 1σ error bars; the source is point-like at 12 μm and 25 μm, and extended at 60 μm (Van Paradijs et al. 1996). Triangle: N-band flux density for stars A plus B seen by TIMMI on the 3.6-m at ESO (Van Paradijs et al. 1996). Crosses: K, H, J, and I_C-band flux densities for stars A plus B seen by the USNO (Vrba et al. 1996). Dotted line is a monoenergetic synchrotron spectrum with $\nu_c = 1.3 \times 10^{13}$ Hz (Ginzburg & Syrovatskii 1965). Long dashed line is a 120 K black body spectrum. Short dashed line is a Population I dust model with $T_{gr} = 100$ K (Dwek & Werner 1981). Short dashed-dotted line is a Population II dust model with $T_{gr} = 92$ K (Dwek & Werner 1981). Solid curve is a 2900 K black body attenuated by the interstellar extinction law of Rieke & Lebofsky (1985) with $A_V = 19.2$ (Vrba et al. 1996).

SGR 1806–20 and SGR 1900+14, the emission at 25 μm is point-like and in excess of the photospheric emission, but the emission is extended at 60 μm. Our ISO observations significantly improve on these IRAS observations, and will be presented elsewhere.

Radio emission has been detected for SGR 1806–20 (Kulkarni et al. 1994), with a central 'core' flux density of 4.8 mJy at 3.6 cm (Vasisht, Frail, & Kulkarni 1995), but only upper limits are found for SGR 1900+14 (Hurley et al. 1996). Neither source has been detected in the submillimeter using the James Clerk Maxwell Telescope (Smith et al. 1997a), or in the millimeter using the Berkeley-Illinois-Maryland Array (Smith, Chernin, & Hurley 1996): we discuss new observations using SEST in the next section.

In Smith et al. (1997a) we showed that monoenergetic synchrotron radiation and black body spectra are too broad to be consistent with both the infrared and submillimeter observations. However, simple dust models such as those of Dwek & Werner (1981) can explain the combined observations; see Figures 1 and 2. Our current hypothesis is that the stars may have shed massive dusty shells, and there is also a point source of dust, possibly an opaque dusty accretion disk around a compact object in the system that is responsible for the gamma-ray bursts (Smith et al. 1996a).

NEW SEST OBSERVATIONS OF SGR 1806–20 AND SGR 0525–66

We made new SEST observations of SGR 1806–20 and SGR 0525–66 UT 1996 March 24 and 25.
Using the bolometer at 235 GHz (1.3 mm), we did not detect SGR 1806–20 with a 3σ upper limit of 17.3 mJy. This new limit is included in Figure 1, superseding the one in Wallyn et al. (1995) and Durouchoux et al. (1996). It is the faintest millimeter or submillimeter observation made so far of the SGR.
One suggestion for the radio emission from SGR 1806–20 is that it is due to free-free emission from a stellar wind. Assuming the 4.8 mJy at 3.6 cm is all due to an ionized outflow from the massive star at 14.5 kpc, and a wind expansion velocity of 400 kms^{-1} (Smith et al. 1997b), implies a large mass loss rate of 2×10^{-4} M$_\odot$ per year. If the wind is fully ionized, this would predict a SEST flux density $\gtrsim 35$ mJy (assuming a 0.6 slope for the spectrum, which is about the flattest possible for an ionized outflow that is partially optically thick). This is significantly higher than the new SEST result: unless there is significant variability, this suggests that the combined radio and SEST data rule out the possibility that the spectrum at these frequencies is due to an ionized outflow.
We made the first millimeter observation of SGR 0525–66 using the bolometer at 235 GHz (1.3 mm). No significant signal was detected, with a 3σ upper limit of 37.2 mJy. This limit is included in Figure 3.
Previous millimeter observations have also been made at SEST of the ^{12}CO(1-0) and ^{13}CO(1-0) spectra in the direction of SGR 1806–20 (Wallyn et al. 1995; Durouchoux et al. 1996). Several molecular clouds were found, and a careful analysis indicated that the distance to SGR 1806–20 is 14.5 kpc (Corbel et al. 1997). At this distance, the stellar counterpart could be one of the most luminous stars in our Galaxy.
In our March 1996 SEST run, we looked for CO in the direction of SGR 0525–66, but did not detect anything. This result is not surprising, given the galactic latitude of the source.
CO observations of SGR 1900+14 are currently being analyzed.

NEW TIMMI OBSERVATIONS OF SGR 0525–66

SGR 0525–66 has long been known to have a position consistent with the N49 supernova remnant in the LMC (Cline et al. 1982), and a soft X-ray counterpart has been found (Rothschild et al. 1994). However, the lack of detection at other wavelengths has led to questions about the association with N49 (Dickel et al. 1995). Figure 3 shows the current status of the multiwavelength observations. For comparison, the Crab continuum spectrum and IRAS detections are included.
Because the other two SGR counterparts are brightest in the mid infrared, this is the most promising place to look for one for SGR 0525–66. Our IRAS study found a source that is extended at 12, 25, and 60 μm (Van Paradijs et al. 1996). The IRAS colors are typical of other supernova remnants, and this infrared emission likely originates from heated dust in N49.
To determine if there is a point source in addition to this extended emission, we used TIMMI UT 1995 December 9 on the 3.6-m at ESO. Images were made in the N-band to cover most of N49. Here we give a preliminary analysis of our primary image, which is centered on the X-ray point source: a more detailed study and our full map will be given elsewhere. Figure 4 shows this image; it covers the whole of the gamma-ray error box given in Cline et al. (1982). There do not appear to be any point sources in the TIMMI observations. Our 3σ upper limit is given in Figure 3, though we stress that this is still preliminary. Scaling the 12 μm IRAS flux densities of SGR 1806–20 or SGR 1900+14 to the distance of the LMC would give ~ 100 mJy, which would have been easily detected in our TIMMI run.

ACKNOWLEDGEMENTS

We would like to thank Patrice Bouchet for his excellent job of taking our TIMMI data, and Fred Baas and the JCMT staff for their help with the JCMT observing run. IAS acknowledges financial support from NASA grants NAG 5-1547 and NAG 5-2772 at Rice University, KH from JPL contract 958056 and NASA grant NAG 5-1560 at the University of California, Berkeley, Space Sciences Laboratory, JvP from NASA grant NAG 5-2755 at the University of Alabama, Huntsville, LBFMW from the Royal Netherlands Academy of Arts and Sciences KNAW, and DH from NASA grant NAG 5-1578 at Clemson University. The

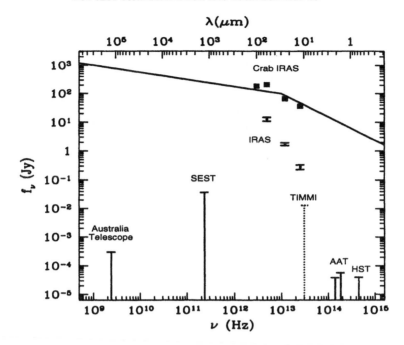

Figure 3. SGR 0525–66. Upper solid curve: Crab continuum spectrum (Marsden et al. 1984). Solid squares: Crab IRAS (Strom & Greidanus 1992). All other points are SGR 0525–66. 3σ point source upper limit at 12.6 cm from the Australia Telescope (Dickel et al. 1995). 3σ upper limit at 235 GHz (1.3 mm) from SEST (this paper). IRAS flux densities are shown with 1σ error bars; the source is extended at all three wavelengths (Van Paradijs et al. 1996). 3σ N-band upper limit from TIMMI on the 3.6-m at ESO (this paper): dotted limit indicates result is preliminary. 3σ upper limit at 2.16 μm and 1.64 μm from IRIS on the 3.9-m Anglo Australian Telescope (Dickel et al. 1995). 3σ upper limit at 656.3 nm from the Hubble Space Telescope (Dickel et al. 1995).

BIMA Hat Creek Interferometer is operated under a joint agreement between the University of California, Berkeley, the University of Illinois and the University of Maryland with support from the NSF grant AST-9320238. The James Clerk Maxwell Telescope is operated by the Royal Observatories on behalf of the Particle Physics and Astronomy Research Council of the United Kingdom, the Netherlands Organisation for Scientific Research, and the National Research Council of Canada. The Swedish-ESO Submillimeter Telescope is operated by the Swedish National Facility for Radio Astronomy, Onsala Space Observatory at Chalmers University of Technology, and by ESO. Kitt Peak National Observatory, National Optical Astronomy Observatories, is operated by the Association of Universities for Research in Astronomy, Inc., under cooperative agreement with the National Science Foundation.

REFERENCES

Cline, T. L., et al. 1982, ApJ, 255, L45
Corbel, S., et al. 1997, ApJ, 478, 624
Dickel, J. R., et al. 1995, ApJ, 448, 623
Durouchoux, P., et al. 1996, in High Velocity Neutron Stars and Gamma Ray Bursts, ed. R. E. Rothschild & R. E. Lingenfelter (New York: AIP), 97
Dwek, E., & Werner, M. W. 1981, ApJ, 248, 138
Ginzburg, V. L., & Syrovatskii, S. I. 1965, ARA&A, 3, 297

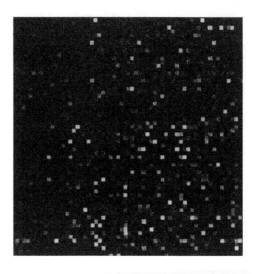

Figure 4. TIMMI N-band image centered on the X-ray counterpart to SGR 0525–66. Recent recalibrations have determined $\lambda_{\text{eff,true}} = 9.862 \ \mu$m (Van der Bliek, Manfroid, & Bouchet 1996). The image is 64×64 pixels, with $0.603''$ per pixel.

Hurley, K., 1996, in Gamma-Ray Bursts: 3rd Huntsville Symposium, ed. C. Kouveliotou, M. S. Briggs, &
    G. J. Fishman (New York: AIP), 889
Hurley, K., et al. 1996, ApJ, 463, L13
Kulkarni, S. R., et al. 1994, Nature, 368, 129
Kulkarni, S. R., et al. 1995, ApJ, 440, L61
Marsden, P. L., et al. 1984, ApJ, 278, L29
Rieke, G. H., & Lebofsky, M. J. 1985, ApJ, 288, 618
Rothschild, R. E., Kulkarni, S. R., & Lingenfelter, R. E. 1994, Nature, 368, 432
Smith, I. A., 1997, in Proceedings of the Fourth Compton Symposium Part One: The Compton Observatory
    in Review, ed. C. D. Dermer, M. S. Strickman, & J. D. Kurfess (New York: AIP), 110
Smith, I. A., Chernin, L. M., & Hurley, K. 1996, A&A, 307, L1
Smith, I. A., et al. 1996a, in High Velocity Neutron Stars and Gamma Ray Bursts, ed. R. E. Rothschild
    & R. E. Lingenfelter (New York: AIP), 79
Smith, I. A., et al. 1996b, in Gamma-Ray Bursts: 3rd Huntsville Symposium, ed. C. Kouveliotou, M. S.
    Briggs, & G. J. Fishman (New York: AIP), 966
Smith, I. A., Schultz, A. S. B., Hurley, K., Van Paradijs, J., & Waters, L. B. F. M. 1997a, A&A, 319, 923
Smith, I. A., et al. 1997b, Proceedings 2nd INTEGRAL Workshop, 'The Transparent Universe', ESA
    SP-382, 191
Strom, R. G., & Greidanus, H. 1992, Nature, 358, 654
Van der Bliek, N. S., Manfroid, J., & Bouchet, P. 1996, A&AS, 119, 547
Van Kerkwijk, M. H., Kulkarni, S. R., Matthews, K., & Neugebauer, G. 1995, ApJ, 444, L33
Van Paradijs, J., et al. 1996, A&A, 314, 146
Vasisht, G., Frail, D. A., & Kulkarni, S. R. 1995, ApJ, 440, L65
Vrba, F. J., et al. 1996, ApJ, 468, 225
Wallyn, P., et al. 1995, Ap&SS, 231, 89

APPENDIX

Pergamon

Adv. Space Res. Vol. 22, No. 7, pp. 1141–1147, 1998
© 1998 COSPAR. Published by Elsevier Science Ltd. All rights reserved
Printed in Great Britain
0273-1177/98 $19.00 + 0.00

PII: S0273–1177(98)00002–7

DEVELOPMENT OF THE SOLAR FLARE PLASMA DENSITY INVESTIGATION METHOD BASED ON THE 2.2 MEV GAMMA-LINE TIME PROFILE ANALYSIS

B. M. Kuzhevskij, S. N. Kuznetsov, and E. V. Troitskaia

D.V. Skobeltsyn Institute of Nuclear Physics, Moscow State University, 119899, Moscow, Russia

ABSTRACT

We propose the method of studying the flare plasma density in different layers of the solar atmosphere, in particular, in the deep layers, which are inaccessible for optical observations. The method is based on the analysis of the time profile of 2.223 MeV gamma-rays, produced by neutron captures by hydrogen. We present the model calculations of the 2.223 MeV gamma-ray production time profiles and depth distribution densities at initial neutron energies 0.1-100 MeV. Both monoenergetic neutrons and power low spectra neutrons are considered at 5 solar plasma density altitude models. The dependence of the resultant curves on the initial neutron energy spectrum and on the altitude profile of the solar atmospheric and subphotospheric density is analysed. Applicability of the method to experimental data is discussed.

©1998 COSPAR. Published by Elsevier Science Ltd.

INTRODUCTION

Flare-generated γ-rays and neutrons carry information about the accelerated particle properties and about the environmental properties of the Sun's atmospheric region wherein nuclear reactions occur (Kuzhevskij, 1982; Ramaty and Murphy, 1987). The works (Kuzhevskij and Troitskaia, 1989; 1995) and (Kuzhevskij et al., 1991) were the first to show that the depth distribution of thermal neutrons immediately after their thermalization and the temporal behaviour of the 2.223 MeV γ-rays produced in flare neutron capture by hydrogen depend on the altitude profile of solar plasma density. We have used the data of the works cited to construct the principles of a method for finding the character of altitude dependence of solar plasma density from γ-ray time profile (Kuzhevskij *et al.*,1995; 1996). It should be noted that neutrons can penetrate deep to the dense photospheric, and even subphotospheric layers. Therefore the neutron-produced γ-rays carry information about the solar plasma density of the deep layers that are yet impossible to observe by the present-day experimental techniques. The density fluctuations at the levels of neutron capture by hydrogen, i.e., in the photosphere and lower chromosphere, may be due to both a proper flare and associated solar activity. This work presents the results of model calculations in terms of five versions of altitude dependence of solar plasma density in the Sun's atmosphere and subphotosphere. The depth distribution evolution of thermal neutrons is analysed. Relevance of the proposed method to the actual experimental results is discussed.

COMPUTATIONAL MODEL; DISCUSSION OF RESULTS

We studied neutrons generated by an instantaneous and, probably, extended source located over the photosphere at altitudes higher than the 10^{15} cm^{-3} layer. The neutrons were assumed to be emitted normally downwards and to belong to a 0.1-100.0 MeV energy range. Figure 1 shows the altitude dependencies of solar plasma density in the lower chromosphere, in the photosphere, and in the convection zone in terms of the basic model (curve 1) and the modified models (curves 2-5) used in our calculations. The only difference between the basic and modified models is in the fragments plotted by the dashed lines. The calculations were made by Monte-Carlo simulation using statistical weights. The calculation techniques and the basic solar atmospheric model are described in detail in (Kuzhevskij *et al.*,1995; 1996). The calculated zero depth corresponds to a 10^{12} cm^{-3} concentration. The calculations allowed for (i) neutron-H nucleus elastic collisions, (ii) np-scattering anisotropy at neutron energies E>10 MeV, (iii) possible ejection of high-energy neutrons from the Sun, (iv) neutron decay, (v) gravitational interaction of neutrons with the Sun at E<2 keV, (vi) thermal motion of thermalized neutrons, and (vii) capture of thermal neutrons by hydrogen with 2.223 MeV γ-emission. We disregarded the nonradiative neutron capture by ^{3}He, for the relevant effect on the temporal behaviour of 2.223 MeV γ-emission has been studied in sufficient detail elsewhere (see review Ramaty and Murphy, 1987). In their calculations, Hua and Lingenfelter (1987) have shown that the amount of ^{3}He-captured neutrons constitute ≈1/3 of the amount of H-captured neutrons at the number density ratio n (^{3}He) / n (^{1}H) = 2×10^{-5} and the capture cross section ratio $\sigma_c(^3He) / \sigma_c(^1H) = 1.61 \times 10^4$. Should a necessity arise, this correction can be introduced in our final results.

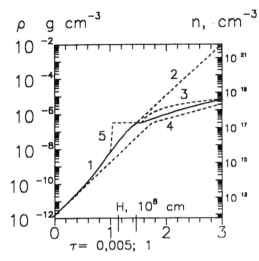

Fig. 1. The main model (curve 1) of solar atmospheric mass density and four of our modified models.

Figures. 2(a,b) present the distribution densities of the 0.1-100.0 MeV primary neutron depths, right after the neutrons decelerate in the solar atmosphere down to the ambient temperature. The same distributions at the moments of γ-emission are also shown. The calculations are in terms of the basic model. The variation in the neutron depth distribution form by the capture moments is defined by thermal neutron motion, by gravity, and by neutron decays. F,om the figure it is seen that the distributions of neutrons of $E_0 \leq 20$ MeV primary energies get shifted to deeper depths, in contrast to the $E_0 > 20$ MeV neutrons whose distributions are not actually shifted. The main effective factor here is the high density of matter at the levels where the $E_0 > 20$ MeV neutrons get thermal. In fact, the mean pre-capture lifetime is $\tau_c = 1 / n\bar{v}\sigma_c$, where $\bar{v}$ is the mean thermal motion velocity; n is number density of H nuclei, σ_c is cross section for radiative neutron capture by hydrogen. At the same time, $\sigma_c = 6.6 \cdot 10^4 / \bar{v}$ b, so $t_c = 1.515 \cdot 10^{19} / n$ s. In the case of $E_0 = 30$ MeV the resultant mean neutron lifetime is 70 s with due allowance for capture by hydrogen and for neutron decays. Within this time, the mean diffusive shift of a neutron is 47 km. From Figure 2(a) it follows that, given low primary energies and having reached thermal velocities, the neutrons keep moving downwards in their drift under gravity. Figures. 2(c,d) show the distribution densities in terms of model (5). In this case, the neutrons of even low energies prove to be in a thermal equilibrium with the ambient medium at high velocities, with null shift to higher densities. Compare between the neutron balances in models (1) and (5). In model (1), the fraction of neutrons ejected from the Sun is (3,8±0,4)%, the fraction of γ-emitting neutrons is (62±1)%, and the decayed neutron fraction is (34±1)%. In model (5), the fractions are (2.9±0,4)%, (73±1)%, and (24±1)%, respectively. The errors correspond to 1σ criterion. Allowing for the absorption by ^{3}He, the fractions of the γ-emitting, ^{3}He-absorbed, and decayed neutrons are, respectively, 51%, 16%, and 29% in Model (1), and 59%, 19%, and 19% in model (5).

Figure 3 shows the distribution densities of 2.223 MeV γ-emission in terms of the five models (see Figure 1) under the power-law primary neutron spectrum $\sim E^{-s}$, where s = 0,1,2,3. Each plot presents the number of generated γ-quanta as calculated per 1 neutron per second. In terms of the dependence of the time profile of the 2.223 MeV γ-ray generation on the function $n(H)$, allowance must be made for the fact that the profile is defined by (1) density of matter at the levels where the radiation is emitted, i.e., by the function *n(H)* in conformity with the relation $\tau_c \propto 1 / n$; (2) depth distribution of the γ-emitting neutrons that, in turn, is defined (i) by the depth distribution of neutrons right after their thermalization, i.e., is again dependent on the function *n(H)*, and (ii) by the variation in the form of the pre-capture time distribution, which was shown above (see Figure 2) to depend also on *n(H)*. It should be noted that the early intervals of the distribution correspond to emission from the layers of the highest density. Thus, the observed time profile of the 2.223 MeV γ-emission makes it possible in principle to restore the character of the altitude dependence of solar plasma density, simultaneously with the spectral index of the primary proton energies.

FEASIBLE APPLICATIONS OF THE METHOD

The time profiles obtained relate to instantaneous neutron injection. Probably, the events with short-term injections (not above 50-60 s) may be used in the analysis. In the case of more persistent injections, the injection time profile may be taken to correspond to the time profile of γ-emission recorded from de-excitation of some nuclei, usually ^{12}C or ^{16}O. If, then, f(t) is the time profile calculated above for the 2.223 MeV γ-emission generated in the instantaneous source under a certain altitude dependence on *n(H)*, the time profile from a persistent source may be calculated as

$$F(t) = \frac{1}{N_0} \int_0^T N(\tau) f(t - \tau) d\tau,$$

where N_0 is the total neutron number, and *T* is the duration of neutron injection.

Another problem is to analyse the potentialities of the present-day instruments in finding the fine details of the 2.223 MeV γ-line. We analysed three instruments based on quantum detection with scintillation crystal, namely, OSSE (CGRO)(Johnson *et al.*, 1993), GRS (SMM) (Chupp *et al.*, 1981), and SONG (KORONAS) (Dmitriev *et al.*, 1993). Figure 3 shows that a 30-s averaging is quite sufficient in the analysis. After that, the relative experimental errors of 1σ are scaled to a 1 cm^{-2}s^{-1} level under a 32-s averaging. The instrumental errors have been calculated from γ-bursts recorded with OSSE (Murphy *et al.*, 1993) and GRS (Johnson *et al.*, 1993) to be 2.5% and 3.2%, respectively. We have estimated the SONG error to be 6% by comparing the computational instrumental error of SONG with that of OSSE. Let the resultant relative errors be compared with the characteristic differences in the 2.223 MeV γ-line time profiles calculated in terms of five models (see Figure 3). The percentage differences Δ are tabulated below. The tabulated data show that the greatest, and some of finer, differences arising in the time profiles from the features of the five models can be detected with the present-day instruments of the above class. This is valid in the case of ~1 cm^{-2}s^{-1} fluxes. From Table 1 it follows, however, that even weaker fluxes

Table 1. Percentage difference between the models

Models	t, s	Δ%
	s=0	
1-5	375	94
1-2	145	38
1-5	145	16
2-5	145	48
	s=3	
1-4	15	15
1-5	15	56
4-5	15	62
1-5	365	69

may still show the greatest differences.

Another method of estimating the potentialities of instruments is to calculate σ_τ, the error in finding the decay constant τ of the calculated time profiles at the levels of fluxes recorded actually. Let the OSSE instrument record an event that realises model 5 at s = 1. If a 1 cm^{-2}s^{-1} flux is recorded in the middle of the interval from t_1=1.75 s to t_2=245 s within a 2.5% error, then $\tau = 102^{+6}_{-8}$ s. The decay constants for s = 1 calculated in the same time interval are 142 s in model (1) and 122 s in model (4) (Kuzhevskij et al., 1996). The differences from model (5) are $6.7\sigma_\tau$ and $3.3\sigma_\tau$, respectively. The instrumental potentialities analysed indicate that the proposed method can still be used, though at the extreme of the potentialities, for sufficiently intensive γ-fluxes under definite conditions with the present-day equipment.

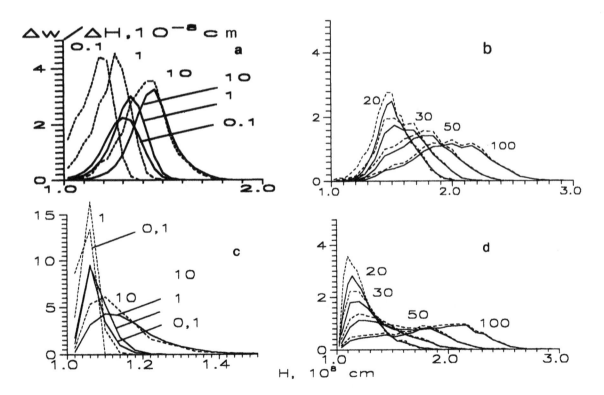

Fig 2. Neutron depth distribution densities just after thermalization (dashed line) and in the moments of γ-ejection in terms of model 1(a,b) and model 5(c,d). Numerals at the curves are initial neutron energy E_{0n} in MeV. ΔH = 40 km, if E_0=0.1-20 MeV and ΔH = 80 km, if E_0=30-100 MeV.

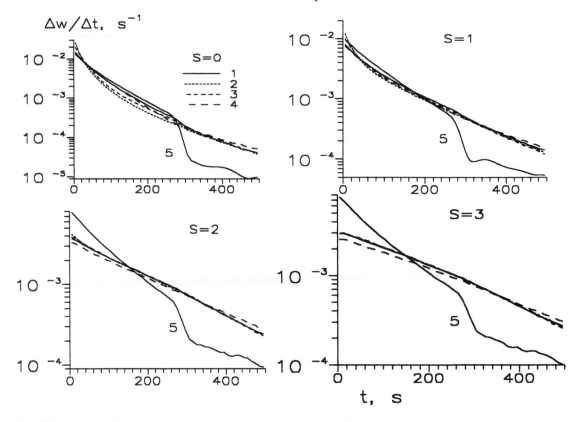

Fig. 3. The time profiles of the 2.223 MeV γ-emission in terms of the five models under the power-low primary neutron spectrum. Δt=10 s.

ACKNOWLEDGEMENT

The authors are indebted to Dr. S.P.Ryumin for kindly providing the SONG characteristics and for informative discussions of the equipment-relevant problems.

REFERENCES

Chupp E. L., D. J. Forrest , J. M. Ryan, M L Cherry, Reppin C, *et al* Observations of the 2.223 MeV Gamma-Ray Line on the SMM Satellite - the Event of 1980 June 7,. //*Astrophys. J.(Lett.)*, **244,** L171 (1981)

Dmitriev A. V., M. A. Kovalevskaya., S. N. Kuznetsov, I. N. Myagkova, S. P. Ryumin, *et al,* Registration of Solar Flare High Energy Emission by SONG-Device in the Project CORONAS, *Proc. 23th ICRC*, Calgary, **3,**. 175.(1993)

Hua, X.-M., and R. E. Lingenfelter, Solar Flare Neutron Production and the Angular Dependence of the Capture Gamma-Ray Emission, *Solar. Phys.*, **107,** 351 (1987)

Johnson W.N., R. L. Kinzer , J. D. Kurfess , M. S. Strickman, W. R. Purcell, *et al.* The Oriented Scintillation Spectrometer Experiment Instrument Description, *Astrophys. J. Suppl.* **86,**. 693.(1993).

Kuzhevskij, B. M., Solar Gamma-Astronomy and Investigations of Solar Cosmic Rays, *Uspekhi Fiz. Nauk,* **137,** 237,.(1982).

Kuzhevskij, B. M., E. I. Kogan-Laskina, and E. V. Troitskaia, *Vestnik MGU ser 3: fiz. astron.*, **32,** 60 (1991).

Kuzhevskij, B.M., S.N. Kuznetsov, and E. V. Troitskaia, The 2.223 MeV Gamma Line Time Dependence as a Probe of Density Inhomogeneity of Solar Atmospheric and Subphotospheric Layers, *Proc. 24th ICRC,* Rome, **4,** 204 (1995).

Kuzhevskij, B.M., S.N.Kuznetsov, and E.V.Troitskaia, *Izv. Ros. Akad. Nauk, ser. fiz.*, **60, 8,** 196 (1996).

Kuzhevskij, B. M., and E. V. Troitskaia, *Solar Flare Neutron Propagation in the Sun's Atmosphere.* Preprint of Institute of Nuclear Physics, MSU, 89-28/105, Moscow (1989).

Kuzhevskij, B. M., and E. V. Troitskaia, *Dependence of Solar Flare 2.223 MeV γ-Line Time Profile on the Solar Plasma Density Inhomogeneity,* Preprint of Institute of Nuclear Physics, MSU, 95-8/372, Moscow (1995).

Murphy R.J, G. H. Share, D. J. Forrest, D. A. Grabelsky, J. E. Jensen, *et al.*, OSSE Observations of the 4 June 1991 Solar Flare, *Proc. 23th ICRC,* Calgary, **3,** 99 (1993).

Ramaty, R., and R. J. Murphy,. Nulear Processes and Accelerated Particles in Solar Flares, *Space Sci. Rev.,* **45,** 213, (1987).

AUTHOR INDEX

Printed and bound by CPI Group (UK) Ltd, Croydon, CR0 4YY

03/10/2024

01040322-0013